J. L. Lehman
Quadratic Ideal Numbers

Also of Interest

J. L. Lehman

Quadratic Ideal Numbers

A Computational Method for Binary Quadratic Forms

DE GRUYTER

Mathematics Subject Classification 2020
11E16, 11E25, 11J70, 11R04, 11R11, 11R29, 11Y40, 11Y65

Author
Prof. J. L. Lehman
Department of Mathematics
University of Mary Washington
1301 College Avenue
Fredericksburg VA 22401
USA

ISBN 978-3-11-131935-3
e-ISBN (PDF) 978-3-11-131936-0
e-ISBN (EPUB) 978-3-11-131979-7

Library of Congress Control Number: 2024947481

Bibliographic information published by the Deutsche Nationalbibliothek
The Deutsche Nationalbibliothek lists this publication in the Deutsche Nationalbibliografie;
detailed bibliographic data are available on the Internet at http://dnb.dnb.de.

© 2025 Walter de Gruyter GmbH, Berlin/Boston
Cover image: oanav / E+ / Getty Images
Typesetting: VTeX UAB, Lithuania

www.degruyter.com
Questions about General Product Safety Regulation:
productsafety@degruyterbrill.com

To Jane

Preface

The concept of *ideal numbers* was introduced in the mid-nineteenth century by Ernst Kummer as a method of rectifying the lack of *uniqueness* in prime factorization in certain domains of algebraic integers, particularly those in cyclotomic fields. Kummer's notion was placed on firmer algebraic ground with Richard Dedekind's definition of *ideals* as a certain type of subset in these domains, and in more general types of rings. The theory of ideals has, of course, become a key concept in the development of abstract algebra since Dedekind's time.

In *Quadratic Number Theory* [1], the author introduced a new notation for ideals of quadratic domains that attempted to recapture the numerical interpretation of ideals. The intent was both to illustrate an application of standard results in elementary number theory and to open up one aspect of the broad area of algebraic number theory (domains of algebraic integers in quadratic fields) for undergraduate students or others with a moderate background in number theory and abstract algebra.

In the current text, we shift the focus by broadening the definition of *quadratic ideal numbers* from a notational device to objects of interest in their own right. We define ideal numbers in Chapter 1, where we also describe connections between ideal numbers and other objects arising from squares of integers: quadratic polynomials, quadratic forms, quadratic numbers, and ideals of quadratic domains. These connections serve as a motivation for the development of operations and relations on sets of ideal numbers in the remainder of the text. In Chapter 3, for example, we define an operation of composition on (congruence classes of) ideal numbers in a way that is consistent with multiplication of ideals in quadratic domains. In Chapters 4 and 5, we define relations of equivalence and genus equivalence that are inspired by relations on quadratic forms of a fixed discriminant. These concepts come together in the definition of the *class group* of ideal numbers of a given discriminant. We develop methods of computing class groups and applying them to numerical questions, for negative discriminants in Chapter 6 and for positive discriminants in Chapter 7.

Quadratic ideal numbers are, at heart, merely solutions of certain quadratic congruences and, as such, they serve as an introduction to, or a review of methods of solving such congruences. To make this text largely self-contained, we include an appendix (Appendix A) that reviews basic concepts in linear and quadratic number theory, including a proof of the quadratic reciprocity theorem. In Chapter 2, methods of solving quadratic congruences with composite moduli are introduced in the context of ideal numbers. These include a variation on the familiar Hensel lifting for solving a quadratic congruence modulo a prime power and a less well-known alternative method (developed by the author) involving *seeding maps* for doing the same calculations when the discriminant of the quadratic polynomial is divisible by the square of the prime in question. We include other appendices to compile results required for developments in the text. These include Appendix B reviewing group theory, with a proof of the fundamental theorem of finite Abelian groups, and Appendix D, which introduces several results concerning

https://doi.org/10.1515/9783111319360-201

continued fractions needed in the development of ideal numbers of positive discriminant. The connection between ideal numbers and ideals of quadratic domains, which formed the focus of [1], is deemphasized in this text in favor of applications to quadratic forms. However, in Appendix C, we review concepts of ideal multiplication needed to prove that ideal number composition is well-defined and possesses the properties of a group operation.

The main goal of this text remains to introduce an aspect of algebraic number theory at a level suitable for an audience of undergraduate mathematics students and nonspecialists. Examples and computational methods are emphasized throughout, with some methods presented in "pseudocode" that can be easily transformed into computer programs by interested readers. Python programs, written by the author, for many of the computational methods introduced in this text are available at a website maintained by De Gruyter.[1]

---

1 https://doi.org/10.1515/9783111319360

Acknowledgments

I deeply appreciate the support that I have received from the editors at De Gruyter, from the inception of this project to its completion. I am particular grateful to: Consulting Editor J. Scott Bentley, for initially contacting me about the possibility of writing a text on number theory and for guiding the process of developing a prospectus for a book on quadratic ideal numbers; Senior Editor Steven Elliot, for his support during the review process and for his willingness to take a chance on this somewhat unconventional project; Content Editor Ute Skambraks, who helped guide me through several technical issues during the preparation of a first draft of this manuscript; and Project Manager Vilma Vaičeliūnienė, who oversaw the editing of the manuscript with great efficiency. The professionalism and the timely communication that I have received at each stage of production has made working with De Gruyter a pleasure.

I am thankful for the opportunities that I have had to work with students, in topics classes, independent studies, and an honors project, while refining aspects of this topic. It is through these interactions that I became convinced that a computational approach to aspects of algebraic number theory could be beneficial to undergraduate mathematics students.

Above all, I am grateful to my wife, Jane, for her support throughout the process of my writing this book. Her patience during the lengthy periods in which this project required much of my time and most of my mental energy made the endeavor feel much easier and more worthwhile than it might have been otherwise. It is to her that I dedicate this book.

Fredericksburg, VA   J. L. Lehman
October 27, 2024

https://doi.org/10.1515/9783111319360-202

Contents

1 Introduction to ideal numbers

In this chapter, we introduce quadratic ideal numbers as our main objects under consideration. Specifically, we associate a collection of ideal numbers to each quadratic discriminant, as defined in Section 1.1. We will see that ideal numbers can be interpreted in several different ways related to squares of integers: as quadratic polynomials (Section 1.3), as binary quadratic forms (Section 1.4), as quadratic numbers (Section 1.5), or as ideals in a domain of quadratic integers (Section 1.6). We also introduce some applications of ideal number notation and terminology in these areas. For example, we will demonstrate that the correspondence between ideal numbers and quadratic polynomials allows us to solve all examples of quadratic congruences in terms of a particular polynomial associated to each discriminant.

1.1 Discriminants and principal polynomials

In this section, we set the groundwork for the definition of ideal numbers, in Section 1.2, by adopting the following notation and terminology.

Definition. If $d \neq 1$ is a square-free integer and g is a positive integer, let

$$\Delta(d, g) = \begin{cases} g^2 d, & \text{if } d \equiv 1 \pmod 4, \\ 4g^2 d, & \text{if } d \equiv 2 \text{ or } 3 \pmod 4. \end{cases} \tag{1.1}$$

We refer to $\Delta = \Delta(d, g)$ as the *(quadratic) discriminant* with *square-free part d* and *index g*. If $g = 1$, we say that Δ is *primitive*.

Every discriminant is congruent to 0 or 1 modulo 4, but is not a square, and every integer with those properties can be written in the form of equation (1.1). The index of a discriminant Δ can be described as the largest integer g so that $\Delta = g^2 \Delta_0$ with Δ_0 also a discriminant. We associate a particular polynomial to each discriminant as follows.

Definition. The *principal polynomial* of discriminant $\Delta = \Delta(d, g)$ is defined to be

$$P(x) = P_\Delta(x) = \begin{cases} x^2 + gx + g^2 \cdot \frac{1-d}{4}, & \text{if } d \equiv 1 \pmod 4, \\ x^2 - g^2 d, & \text{if } d \equiv 2 \text{ or } 3 \pmod 4. \end{cases} \tag{1.2}$$

Example. If $d = -47$, then $\Delta(-47, 1) = -47$ since $-47 \equiv 1 \pmod 4$. The principal polynomial of discriminant -47 is $P(x) = x^2 + 1 \cdot x + 1^2 \cdot \frac{1-(-47)}{4} = x^2 + x + 12$.

Example. If $d = 47$, then $\Delta(47, 1) = 4 \cdot 47 = 188$ since $-47 \equiv 3 \pmod 4$. The principal polynomial of discriminant 188 is $P(x) = x^2 - 1^2 \cdot 47 = x^2 - 47$.

The discriminant of a quadratic polynomial $f(x) = ax^2 + bx + c$ is defined as $\Delta(f) = b^2 - 4ac$. If the coefficients of $f(x)$ are integers, then $f(x)$ factors as a product of linear

https://doi.org/10.1515/9783111319360-001

polynomials with integer coefficients if and only if $\Delta(f)$ is the square of an integer. For a *principal* polynomial $P(x)$, as defined in equation (1.2), one can verify that the coefficients of $P(x)$ are integers, and that $\Delta(P) = \Delta(d, g)$, as the latter is defined in equation (1.1). Thus, a principal polynomial is irreducible over the integers.

We will often use the observations about principal polynomials in the following three propositions. Recall that the *derivative* of a quadratic polynomial $f(x) = ax^2 + bx + c$ is $f'(x) = 2ax + b$.

Proposition 1.1.1. *Let $P(x)$ be the principal polynomial of discriminant Δ, as defined in equation (1.2). If $P'(x)$ is the derivative of $P(x)$, then*

$$P'(x)^2 - 4P(x) = \Delta \tag{1.3}$$

for all x, and

$$P(x) - P(y) = (x - y)(x + y + P'(0)) \tag{1.4}$$

for all x and y.

Proof. Equations (1.3) and (1.4) are special cases of the following more general statements. If $f(x) = ax^2 + bx + c$, then

$$f'(x)^2 - 4af(x) = (2ax + b)^2 - 4a(ax^2 + bx + c)$$
$$= (4a^2x^2 + 4abx + b^2) - (4a^2x^2 + 4abx + 4ac) = b^2 - 4ac$$

and

$$f(x) - f(y) = (ax^2 + bx + c) - (ay^2 + by + c)$$
$$= a(x^2 - y^2) + b(x - y) = (x - y)(a(x + y) + f'(0))$$

since $b = f'(0)$. The leading coefficient of a principal polynomial is always $a = 1$. □

Definition. Let $P(x)$ be the principal polynomial of discriminant $\Delta = \Delta(d, g)$. If x is an integer, define the *conjugate of x with respect to Δ* to be

$$\overline{x} = \overline{x}_\Delta = -x - P'(0) = \begin{cases} -x - g, & \text{if } d \equiv 1 \pmod 4, \\ -x, & \text{if } d \equiv 2 \text{ or } 3 \pmod 4. \end{cases} \tag{1.5}$$

Note that the conjugate of $\overline{x}$ is $-\overline{x} - P'(0) = -(-x - P'(0)) - P'(0) = x$.

Proposition 1.1.2. *Let $P(x)$ be the principal polynomial of discriminant Δ and let $P'(x)$ be its derivative. If $\overline{x}$ is the conjugate of an integer x with respect to Δ, then $P(\overline{x}) = P(x)$ and $P'(\overline{x}) = -P'(x)$.*

Proof. Equation (1.4) and the definition of $\bar{x}$ show that

$$P(x) - P(\bar{x}) = (x - \bar{x})(x + \bar{x} + P'(0)) = (x - \bar{x}) \cdot 0 = 0,$$

so that $P(x) = P(\bar{x})$. For the second claim, note that $P'(x) = 2x + P'(0)$, and so

$$P'(\bar{x}) = 2(-x - P'(0)) + P'(0) = -2x - P'(0) = -P'(x),$$

as we wanted to show. $\square$

Proposition 1.1.3. *Let Δ be a discriminant and let m be a positive integer. Let $P(x)$ and $P_m(x)$ be the principal polynomials of discriminant Δ and $m^2\Delta$ respectively. Then for all x,*

$$P_m(mx) = m^2 P(x) \quad and \quad P'_m(mx) = mP'(x).$$

Proof. From the definition of principal polynomials in equation (1.2), we find that if the principal polynomial of discriminant $\Delta = \Delta(d, g)$ is $P(x) = x^2 + sx + t$ for some integers s and t, then the principal polynomial of discriminant $m^2\Delta = \Delta(d, mg)$ is

$$P_m(x) = x^2 + msx + m^2 t.$$

Thus,

$$P_m(mx) = (mx)^2 + ms(mx) + m^2 t = m^2(x^2 + sx + t) = m^2 P(x)$$

and

$$P'_m(mx) = 2(mx) + ms = m(2x + s) = mP'(x),$$

as we wanted to show. $\square$

1.2 Definition of ideal numbers

We now define our main objects of interest for each quadratic discriminant.

Definition. Let $P(x)$ be the principal polynomial of discriminant Δ. Then a *quadratic ideal number* of discriminant Δ is an expression of the form

$$(a : k) = (a : k)_\Delta$$

where a and k are integers for which a divides $P(k)$. We refer to a and k as the *norm* and *character* of $(a : k)$, respectively. We write the collection of ideal numbers of discriminant Δ as $\mathcal{I}_\Delta$.

Example. It is a trivial matter to find examples of ideal numbers having any given character. For instance, if $P(x) = x^2 + x + 12$, the principal polynomial of discriminant $\Delta = -47$ as noted in Section 1.1, then $P(5) = 42$. Thus, examples of ideal numbers in $\mathcal{I}_{-47}$ having character $k = 5$ include $(42 : 5)$, $(-7 : 5)$, and in general $(a : 5)$ for every positive or negative divisor a of 42.

On the other hand, finding ideal numbers of a given norm may be more difficult. To exhibit an ideal number $(5 : k)$ in $\mathcal{I}_{-47}$, for example, requires finding a value of k so that 5 divides $P(k)$. In this case, trial and error calculations fail to produce a k for which this is true. In fact, no such k exists, as we can confirm by considering the quadratic congruence $P(x) \equiv 0 \pmod{5}$. We will pursue the question of the existence of quadratic ideal numbers with a particular discriminant and norm in Chapter 2.

Proposition 1.2.1. *Let $(a : k)$ be an ideal number of discriminant Δ. If $\bar{k}$ is the conjugate of k with respect to Δ, then $(a : \bar{k})$ is also an element of $\mathcal{I}_\Delta$, as are $(-a : \bar{k})$ and $(-a : k)$.*

Proof. Let $P(x)$ be the principal polynomial of discriminant Δ. If $(a : k)$ is an ideal number of discriminant Δ, then a divides $P(k)$. But $P(\bar{k}) = P(k)$, as we saw in Proposition 1.1.2. Thus, $(a : \bar{k})$ is also an ideal number in $\mathcal{I}_\Delta$. Furthermore, if a divides $P(k) = P(\bar{k})$, then $-a$ divides this common value also. Thus, $(-a : \bar{k})$ and $(-a : k)$ are in $\mathcal{I}_\Delta$. $\qquad\square$

Definition. If $(a : k)$ is an element of $\mathcal{I}_\Delta$, and $\bar{k}$ is the conjugate of k with respect to Δ, we call $(a : \bar{k})$ the *conjugate* of $(a : k)$ in $\mathcal{I}_\Delta$. Based on an observation we will make in Proposition 1.3.2, we call $(-a : \bar{k})$ the *negative* of $(a : k)$. We call $(-a : k)$, which is then the negative of $(a : \bar{k})$, the *negative conjugate* of $(a : k)$.

Example. As noted in a previous example, $(42 : 5)$ is an ideal number in $\mathcal{I}_{-47}$. Since $P(x) = x^2 + x + 12$ is the principal polynomial of discriminant $\Delta = -47$, then the conjugate of an integer x with respect to Δ is $\bar{x} = -x - 1$. The conjugate of $(42 : 5)$ in $\mathcal{I}_{-47}$ is $(42 : -6)$.

Proposition 1.2.2. *Let $(a : k)$ be an ideal number of discriminant Δ. If $\ell = k + aq$ for some integer q, then $(a : \ell)$ is also an element of $\mathcal{I}_\Delta$.*

Proof. Let $P(x)$ be the principal polynomial of discriminant Δ. By equation (1.4),

$$P(\ell) - P(k) = (\ell - k)(\ell + k + P'(0)).$$

If $(a : k)$ is an ideal number in $\mathcal{I}_\Delta$, then a divides $P(k)$, and if $\ell = k + aq$, then a divides $\ell - k$. Thus, a divides $P(\ell)$, and so $(a : \ell)$ is an ideal number of discriminant Δ by definition. $\quad\square$

Example. Since $(42 : 5)$ is an element of $\mathcal{I}_{-47}$, then so are $(42 : 47)$, $(42 : -37)$, and in general $(42 : 5 + 42q)$ for every integer q.

Proposition 1.2.3. *Let $(a : k)$ be an ideal number of discriminant Δ. If m is a positive integer, and b divides $m^2 a$, then $(b : mk)$ is an element of $\mathcal{I}_{m^2\Delta}$.*

Proof. Let $P(x)$ and $P_m(x)$ be the principal polynomials of discriminant Δ and $m^2\Delta$, respectively. If $(a : k)$ is an ideal number of discriminant Δ, then $P(k) = ac$ for some integer c. By Proposition 1.1.3, then $P_m(mk) = m^2 P(k) = m^2 ac$. Thus, if b is a divisor of $m^2 a$, then b divides $P_m(mk)$, and $(b : mk)$ is an ideal number in $\mathcal{I}_{m^2\Delta}$ by definition. □

Example. For $\Delta = \Delta(-47, 2) = -188$, the principal polynomial is $P(x) = x^2 + 2x + 48$, and we find that $P(10) = 168 = P(-12)$. By Proposition 1.2.3, examples of ideal numbers in $\mathcal{I}_{-188}$ include $(42 : 10)$, $(84 : 10)$, and $(168 : 10)$ along with their conjugates $(42 : -12)$, $(84 : -12)$, and $(168 : -12)$, respectively. (We may view these ideal numbers as arising from the ideal number $(42 : 5)$ in $\mathcal{I}_{-47}$, as we have used in previous examples.)

In the remainder of Chapter 1, to motivate the definitions in this section, we introduce several connections between quadratic ideal numbers and other objects related to squares of integers.

1.3 Ideal numbers and quadratic polynomials

As we noted in Section 1.1, a quadratic polynomial $f(x) = ax^2 + bx + c$ with integer coefficients is irreducible over the integers if and only if its discriminant, $\Delta(f) = b^2 - 4ac$, is a *quadratic* discriminant, as that term is defined in equation (1.1). There is a one-to-one correspondence between irreducible quadratic polynomials with integer coefficients and ideal numbers of the same discriminant described by the following proposition.

Proposition 1.3.1. *Let $P(x)$ be the principal polynomial of some quadratic discriminant Δ. Let $(a : k)$ be an ideal number in $\mathcal{I}_\Delta$, so that $P(k) = ac$ for some integer c, and let $P'(k) = b$. Then $f(x) = ax^2 + bx + c$ is a quadratic polynomial of discriminant Δ. Conversely, if $f(x) = ax^2 + bx + c$ is a quadratic polynomial of discriminant Δ, then*

$$k = \frac{f'(0) - P'(0)}{2} = \frac{b - P'(0)}{2}$$

is an integer, and $(a : k)$ is an ideal number in $\mathcal{I}_\Delta$.

Proof. Let $(a : k)$ be an ideal number of discriminant Δ. By definition, then a divides $P(k)$, so that we can write $P(k) = ac$ for some integer c. If we let $P'(k) = b$, then equation (1.3) shows that

$$b^2 - 4ac = P'(k)^2 - 4P(k) - \Delta,$$

and so $f(x) = ax^2 + bx + c$ is a quadratic polynomial with discriminant Δ.

Conversely, let $ax^2 + bx + c$ be a quadratic polynomial with

$$b^2 - 4ac = \Delta = P'(0)^2 - 4P(0),$$

again using equation (1.3). Then b and $P'(0)$ have the same parity, so that $k = \frac{b-P'(0)}{2}$ is an integer. Here, $b = 2k + P'(0) = P'(k)$, so equation (1.3) implies that $ac = \frac{b^2-\Delta}{4} = \frac{P'(k)^2-\Delta}{4} = P(k)$. Thus, a divides $P(k)$ and $(a : k)$ is an ideal number in $\mathcal{I}_\Delta$. $\qquad\square$

Therefore, a quadratic polynomial $f(x) = ax^2 + bx + c$ whose discriminant Δ is not a square can be identified with an ideal number $(a : k)$ in $\mathcal{I}_\Delta$. If Δ is clear from context, we may write $f = (a : k)$ in this case, calling this *ideal number* notation for $f(x)$. Note that $P = (1 : 0)$ is the ideal number expression for the principal polynomial of discriminant Δ.

Example. Let $\Delta = -47$, so that $P(x) = x^2 + x + 12$. In an example in Section 1.2, we saw that $(-7 : 5)$ is an ideal number in $\mathcal{I}_{-47}$. Since $P(5) = 42 = (-7)(-6)$ and $P'(5) = 11$, then $f(x) = -7x^2 + 11x - 6$ is the quadratic polynomial with discriminant -47 having ideal number expression $(-7 : 5)$.

Example. If $f(x) = 3x^2 - 7x + 3$, then $\Delta = (-7)^2 - 4(3)(3) = 13$. The principal polynomial of discriminant $\Delta = 13$ is $P(x) = x^2 + x - 3$, with $P'(x) = 2x + 1$. If $k = \frac{f'(0)-P'(0)}{2} = \frac{-7-1}{2} = -4$, then $(3 : -4)$ is an ideal number in $\mathcal{I}_{13}$, and is the ideal number expression for $f(x)$. To verify this in the reverse direction, we recover the coefficients of $f(x)$ as $b = P'(-4) = -7$ and $c = \frac{1}{a}P(-4) = \frac{9}{3} = 3$.

Proposition 1.3.2. *Let $f(x) = ax^2 + bx + c$ be a quadratic polynomial with discriminant Δ, having $(a : k)$ as its ideal number expression. Let $\bar{k}$ be the conjugate of k with respect to Δ. Then the following statements are true:*
1. *$(a : \bar{k})$ is the ideal number expression for $ax^2 - bx + c$.*
2. *$(-a : k)$ is the ideal number expression for $-ax^2 + bx - c$.*
3. *$(-a : \bar{k})$ is the ideal number expression for $-ax^2 - bx - c$.*

If $f(x) = ax^2 + bx + c$ is a quadratic polynomial, we call $ax^2 - bx + c$ the *conjugate* of $f(x)$ and $-ax^2 + bx - c$ the *negative conjugate* of $f(x)$.

Proof. Let $P(x)$ be the principal polynomial of discriminant Δ. If $(a : k)$ is the ideal number expression for $ax^2 + bx + c$, then $P(k) = ac$ and $P'(k) = b$. From Proposition 1.1.2, then $P(\bar{k}) = P(k) = ac$ and $P'(\bar{k}) = -P'(k) = -b$, from which statement (1) follows. We can also write $P(k) = (-a)(-c) = P(\bar{k})$, implying statements (2) and (3). $\qquad\square$

Are there benefits to introducing ideal number notation for quadratic polynomials, which are, after all, very familiar objects? In the remainder of Section 1.3, we state and prove three propositions that suggest we can answer that question in the affirmative.

Proposition 1.3.3. *Let $P(x)$ be the principal polynomial of discriminant Δ. If $f(x)$ is a quadratic polynomial of the same discriminant, written as $f = (a : k)$ in ideal number notation, then $P(ax + k) = a \cdot f(x)$ for all x.*

Proof. Let $f(x) = ax^2 + bx + c$, where $P(k) = ac$ and $P'(k) = b$. From equation (1.4), we have

$$P(ax + k) - P(k) = (ax + k - k)(ax + k + k + P'(0))$$
$$= ax(ax + 2k + P'(0)) = ax(ax + P'(k)).$$

But then

$$P(ax + k) = ax(ax + P'(k)) + P(k)$$
$$= ax(ax + b) + ac = a(ax^2 + bx + c) = a \cdot f(x),$$

as we wanted to show. □

We can apply Proposition 1.3.3 to the problem of solving quadratic congruences, that is, finding all integers x so that $f(x) \equiv 0$ (mod m), where $f(x) = ax^2 + bx + c$. If $x \equiv y$ (mod m), then $f(x) \equiv f(y)$ (mod m), so it suffices to find all solutions in the set, $\mathbb{Z}_m$, of distinct *congruence classes* modulo m. The following corollary reduces this problem further.

Proposition 1.3.4. *Let $f(x) = ax^2 + bx + c$ be a polynomial with integer coefficients for which $b^2 - 4ac = \Delta$ is not a square, and let $(a : k)$ be the ideal number expression for $f(x)$. Let $P(x)$ be the principal polynomial of discriminant Δ, and let m be a positive integer for which $\gcd(a, m) = 1$. Then the function $T : \mathbb{Z}_m \to \mathbb{Z}_m$ defined by $T(x) = ax + k$ is a one-to-one correspondence between solutions of $f(x) \equiv 0$ (mod m) and solutions of $P(x) \equiv 0$ (mod m). In particular, $f(x) \equiv 0$ (mod m) and $P(x) \equiv 0$ (mod m) have the same number of solutions in $\mathbb{Z}_m$.*

Proof. If $\gcd(a, m) = 1$, then $ax + k \equiv ay + k$ (mod m) if and only if $x \equiv y$ (mod m). (This is an application of the congruence cancellation property in Corollary A.2.5.) Thus, T is a bijection from $\mathbb{Z}_m$ to $\mathbb{Z}_m$ as defined. Proposition 1.3.3 implies that $f(x) \equiv 0$ (mod m) if and only if $P(T(x)) \equiv 0$ (mod m), again using the assumption that $\gcd(a, m) = 1$. Thus, $f(x) \equiv 0$ (mod m) and $P(x) \equiv 0$ (mod m) have the same number of solutions in $\mathbb{Z}_m$. □

Proposition 1.3.4 implies that solving $P(x) \equiv 0$ (mod m), where $P(x)$ is the principal polynomial of discriminant Δ (not a square) allows one to solve each congruence $ax^2 + bx + c \equiv 0$ (mod m) when $\Delta = b^2 - 4ac$ and $\gcd(a, m) = 1$. Thus the general problem of solving quadratic congruences is reduced to a particular polynomial for each discriminant. The following inverse for the function $T : \mathbb{Z}_m \to \mathbb{Z}_m$ defined in Proposition 1.3.4 makes this correspondence more practical. If $\gcd(a, m) = 1$, then $1 = aq + mr$ for some integers q and r, and $T^{-1} : \mathbb{Z}_m \to \mathbb{Z}_m$ is defined by $T^{-1}(x) = q(x - k)$.

Example. Let $\Delta = -47$, so that $P(x) = x^2 + x + 12$. Suppose we know that $P(x) \equiv 0$ (mod 84) has eight distinct solutions in $\mathbb{Z}_{84}$, namely 8, 12, 15, 36, 47, 68, 71, and 75. (We will say more about how we can construct solutions such as these, and be sure the list is complete, in the context of ideal numbers in Chapter 2.) Now note that $f(x) = 17x^2 + 15x + 4$ is a polynomial of discriminant $\Delta = -47$, given by $(17 : 7)$ in ideal number notation. Here,

$\gcd(17, 84) = 1 = 17(5) + 84(-1)$, so the permutation of $\mathbb{Z}_{84}$ sending x to $5(x - 7)$ produces all solutions of $f(x) \equiv 0 \pmod{84}$, as in the following table:

x	8	12	15	36	47	68	71	75
$5(x - 7) \bmod 84$	5	25	40	61	32	53	68	4

Thus, $f(x) \equiv 0 \pmod{84}$ likewise has eight distinct solutions in $\mathbb{Z}_{84}$.

As a second example with $\Delta = -47$, consider $f(x) = 3x^2 + 11x + 14$, which has ideal number expression $(3 : 5)$ in $\mathcal{I}_{-47}$. If $m = 84$ again, we cannot directly apply Proposition 1.3.4, since $\gcd(3, 84) \neq 1$. However, we can still use solutions of $P(x) \equiv 0 \pmod{84}$ to help solve $f(x) \equiv 0 \pmod{84}$ as follows. A solution of $f(x) \equiv 0 \pmod{84}$ satisfies both $f(x) \equiv 0 \pmod 3$ and $f(x) \equiv 0 \pmod{28}$. But $3x^2 + 11x + 14 \equiv 0 \pmod 3$ reduces to $2x + 2 \equiv 0 \pmod 3$, which has $x = 2$ as its only solution modulo 3. On the other hand, we can apply Proposition 1.3.4 to $f(x) \equiv 0 \pmod{28}$. The eight solutions previously calculated for $P(x) \equiv 0 \pmod{84}$ also satisfy $P(x) \equiv 0 \pmod{28}$, although only four of them are now distinct modulo 28, namely 8, 12, 15, and 19. (Both 47 and 71 are congruent to 19 modulo 28, for example.) Now with $\gcd(3, 28) = 1 = 3(-9) + 28(1)$, we find that the permutation of $\mathbb{Z}_{28}$ sending x to $-9(x - 5)$ produces all solutions of $f(x) \equiv 0 \pmod{28}$:

x	8	12	15	19
$-9(x - 5) \bmod 28$	1	21	22	14

Finally, all solutions of $f(x) \equiv 0 \pmod{84}$ are obtained by solving $x \equiv s \pmod 3$ and $x \equiv t \pmod{28}$ where $s = 2$ and $t = 1, 14, 21$, or 22. Each such system has a unique solution modulo 84, which one can verify are 29, 14, 77, and 50, respectively.

We conclude Section 1.3 with a second consequence of Proposition 1.3.3, which we will use in Section 1.5.

Proposition 1.3.5. *Let $f(x)$ be a quadratic polynomial whose discriminant Δ is not a square and let $P(x)$ be the principal polynomial of discriminant Δ. If w is a root of $P(x)$ and $f = (a : k)$ is the ideal number expression for $f(x)$, then $v = \frac{w-k}{a}$ is a root of $f(x)$.*

Proof. If $v = \frac{w-k}{a}$, then $w = av + k$, so that $P(w) = P(av + k) = a \cdot f(v)$ by Proposition 1.3.3. Since Δ is not a square, then $P(k) \neq 0$ and so $a \neq 0$. Thus, if $P(w) = 0$, then $f(v) = 0$, as we wanted to show. $\square$

1.4 Ideal numbers and quadratic forms

If $f(x) = ax^2 + bx + c$ is a quadratic polynomial, then

$$f(x,y) = y^2 \cdot f\left(\frac{x}{y}\right) = ax^2 + bxy + cy^2$$

is a degree two homogeneous function of two variables, which we call a *(binary) quadratic form*. Conversely, if $f(x,y) = ax^2 + bxy + cy^2$ is a quadratic form, then $f(x) = f(x,1) = ax^2 + bx + c$ is a quadratic polynomial. We again assume that a, b, and c, the *coefficients* of $f(x,y)$, are integers and define the *discriminant* of $f(x,y)$ to be $\Delta = \Delta(f) = b^2 - 4ac$. When Δ is a square, $f(x,y)$ factors as a product of two linear homogeneous functions in x and y. As with polynomials, we will restrict our attention to the opposite case, so that Δ is a quadratic discriminant. Thus, there is a one-to-one correspondence between quadratic polynomials and quadratic forms of a fixed discriminant. We will typically use the same letter for a quadratic polynomial and its associated quadratic form, with the number of variables determining its interpretation. For instance, if $P(x)$ is the principal polynomial of discriminant Δ, as defined in equation (1.2), then there is a corresponding *principal form*, $P(x,y)$, of the same discriminant.

We can associate each quadratic form with an ideal number in the same way that we did for quadratic polynomials.

Proposition 1.4.1. *Let $P(x)$ be the principal polynomial of some quadratic discriminant Δ. If $(a : k)$ is an ideal number in $\mathcal{I}_\Delta$, so that $P(k) = ac$ for some integer c, and we let $P'(k) = b$, then $f(x,y) = ax^2 + bxy + cy^2$ is a quadratic form of discriminant Δ. Conversely, if $f(x,y) = ax^2 + bxy + cy^2$ is a quadratic form of discriminant Δ, then $k = \frac{b - P'(0)}{2}$ is an integer and $(a : k)$ is an ideal number in $\mathcal{I}_\Delta$.*

Proof. See the proof of Proposition 1.3.1. □

We write $f = (a : k)$ as *ideal number* notation for $f(x,y) = ax^2 + bxy + cy^2$. Context will determine whether we regard $f = (u : k)$ as a quadratic form or a quadratic polynomial.

Definition. Let $f(x,y)$ be a quadratic form. If $f(q, r) = m$ for some integers q and r, we say that $f(x,y)$ *represents* m. We say that $f(x,y)$ *properly represents* m if $f(q, r) = m$ and $\gcd(q, r) = 1$. We also say that (q, r) is a (proper) representation of m by $f(x,y)$ if these equations hold.

If $f(q, r) = m$, then $f(dq, dr) = d^2 m$. Conversely, if $f(q, r) = m$ and $\gcd(q, r) = d$, then d^2 divides m, and f properly represents m/d^2. We will use the algebraic observations of the following proposition on occasion.

Proposition 1.4.2. *Let $f(x,y) = ax^2 + bxy + cy^2$ be a quadratic form with $\Delta = b^2 - 4ac$. If $f(q, r) = m$ for some integers q and r, then*

$$4am = (2aq + br)^2 - \Delta r^2 \quad and \quad 4cm = (2cr + bq)^2 - \Delta q^2. \tag{1.6}$$

Proof. The formulas in equations (1.6) are obtained by completing the square and can be verified by direct calculation. We omit the details. □

The question of which integers are (properly) represented by a particular quadratic form is one that we will revisit often throughout the text, in terms of ideal numbers. In the remainder of Section 1.4, we prove some statements that again suggest the benefits of having ideal number notation available as an alternative to standard notation for quadratic forms.

Proposition 1.4.3. *Let $P(x, y)$ be the principal form of discriminant Δ. Let $f(x, y)$ be a quadratic form of the same discriminant, given by $f = (a : k)$ in ideal number notation. Then for all x and y,*

$$P(ax + ky, y) = a \cdot f(x, y) \quad and \quad f(x - ky, ay) = a \cdot P(x, y).$$

Proof. Let $P(x)$ be the principal polynomial of discriminant Δ. Then we can write

$$P(x, y) = x^2 + P'(0)xy + P(0)y^2 \quad and \quad f(x, y) = ax^2 + bxy + cy^2,$$

where $P(k) = ac$ and $P'(k) = b$. Thus,

$$\begin{aligned}
P(ax + ky, y) &= (ax + ky)^2 + P'(0)(ax + ky)y + P(0)y^2 \\
&= a^2x^2 + 2akxy + k^2y^2 + aP'(0)xy + P'(0)ky^2 + P(0)y^2 \\
&= a^2x^2 + a(2k + P'(0))xy + (k^2 + P'(0)k + P(0))y^2 \\
&= a^2x^2 + aP'(k)xy + P(k)y^2.
\end{aligned}$$

Therefore,

$$P(ax + ky, y) = a^2x^2 + abxy + acy^2 = a \cdot f(x, y),$$

implying the first claim.

Likewise,

$$\begin{aligned}
f(x - ky, ay) &= a(x - ky)^2 + b(x - ky)ay + c(ay)^2 \\
&= a((x^2 - 2kxy + k^2y^2) + (bxy - bky^2) + (acy^2)) \\
&= a(x^2 + (b - 2k)xy + (k^2 - bk + ac)y^2).
\end{aligned}$$

But $b - 2k = P'(k) - 2k = P'(0)$. Then from equation (1.4),

$$P(0) - P(k) = (0 - k)(0 + k + P'(0)) = -k(k + (b - 2k)) = k^2 - bk.$$

Hence, $P(0) = k^2 - bk + P(k) = k^2 - bk + ac$, the coefficient of y^2 in the final equation above. Thus,

$$f(x - ky, ay) = a(x^2 + P'(0)xy + P(0)y^2) = a \cdot P(x,y),$$

establishing the second claim. □

Proposition 1.4.4. *Let $P(x,y)$ be the principal form of discriminant Δ and let $f(x,y)$ be a quadratic form of the same discriminant, given by $f = (a : k)$ in ideal number notation. Let m be an integer. Then the following statements are true:*

1. *If (q,r) is a proper representation of m by $f(x,y)$, then $d = \gcd(aq + kr, r)$ is a common divisor of a and m.*
2. *If (q,r) is a proper representation of m by $P(x,y)$, then $d = \gcd(q - kr, ar)$ is a common divisor of a and m.*

Proof. For statement (1), let $f(x,y) = ax^2 + bxy + cy^2$, as in the preceding proof. If $\gcd(q,r) = 1$, then there are integers s and t so that $qs + rt = 1$. Now

$$(aq + kr)s + r(-ks + at) = aqs + krs - krs + art = a(qs + rt) = a.$$

Thus, a is a combination of $aq + kr$ and r, implying that $d = \gcd(aq + k, r)$ divides a. But then d also divides $m = f(q,r) = aq^2 + bqr + cr^2$, since d divides both a and r. This establishes statement (1).

For statement (2), let $P(x,y) = x^2 + P'(0)xy + P(0)y^2$ as above. Assume again that $qs + rt = 1$ for some integers s and t. Here,

$$(q - kr)as + (ar)(ks + t) = aqs - akrs + akrs + art = a(qs + rt) = a,$$

implying that $d = \gcd(q - kr, ar)$ divides a. Since d divides $q - kr$, we can write $q = kr + dv$ for some integer v. Now

$$
\begin{aligned}
m = P(q,r) &= (kr + dv)^2 + P'(0)(kr + dv)r + P(0)r^2 \\
&= (k^2r^2 + 2dkrv + d^2v^2) + P'(0)kr^2 + P'(0)drv + P(0)r^2 \\
&= (k^2 + P'(0)k + P(0))r^2 + d(2krv + dv^2 + P'(0)rv) \\
&= P(k)r^2 + d(2krv + dv^2 + P'(0)rv)
\end{aligned}
$$

Since d divides a, and so divides $P(k) = ac$, then d divides m, completing the proof of statement (2). □

Corollary 1.4.5. *Let $P(x,y)$ be the principal form of discriminant Δ. Let $f(x,y)$ be a quadratic form of the same discriminant, given by $f = (a : k)$ in ideal number notation. Then the following statements are true:*

1. *If $P(x,y)$ properly represents m, and $\gcd(a,m) = 1$, then $f(x,y)$ properly represents am.*
2. *If $f(x,y)$ properly represents m, and $\gcd(a,m) = 1$, then $P(x,y)$ properly represents am.*

3. If (q, r) is a proper representation of m by $P(x, y)$, and $d = \gcd(q - kr, ar)$, then $(\frac{q-kr}{d}, \frac{ar}{d})$ is a proper representation of $\frac{a}{d} \cdot \frac{m}{d}$ by $f(x, y)$.

4. If (q, r) is a proper representation of m by $f(x, y)$, and $d = \gcd(aq + kr, r)$, then $(\frac{aq+kr}{d}, \frac{r}{d})$ is a proper representation of $\frac{a}{d} \cdot \frac{m}{d}$ by $P(x, y)$.

Proof. Proposition 1.4.3 shows that if $P(q, r) = m$, then $f(q - kr, ar) = am$. If $\gcd(q, r) = 1$, then Proposition 1.4.4 states that $d = \gcd(q - kr, ar)$ divides both a and m. Thus, if $\gcd(a, m) = 1$, then $d = 1$, implying that $f(x, y)$ properly represents am. This establishes statement (1). More generally, if $d = \gcd(q - kr, ar)$, then $\gcd(\frac{q-kr}{d}, \frac{ar}{d}) = 1$. Since $f(x, y)$ is a homogeneous function of degree two, then

$$f\left(\frac{q - kr}{d}, \frac{ar}{d}\right) = \frac{1}{d^2} \cdot f(q - kr, ar) = \frac{a}{d} \cdot \frac{m}{d}.$$

This establishes statement (3). The proofs of statements (2) and (4) are entirely similar and are omitted. □

Example. The quadratic form $P(x, y) = x^2 + xy + 12y^2$ is the principal form of discriminant $\Delta = -47$, and we find that $P(2, -3) = 106$. Now $(7 : 5)$ is an ideal number in $\mathcal{I}_{-47}$, with $f(x, y) = 7x^2 + 11xy + 6y^2$ its corresponding quadratic form. (We can check that $P(5) = 42 = 7 \cdot 6$ and $P'(5) = 11$.) Since $\gcd(7, 106) = 1$, then $f(x, y)$ properly represents $7 \cdot 106 = 742$. From statement (3) of Corollary 1.4.5, which also applies when $d = 1$, we have that $(q - kr, ar) = (2 - 5(-3), 7(-3)) = (17, -21)$ is a proper representation of 742 by $f(x, y)$ (as one can verify).

In the reverse direction, note that $f(2, 1) = 56$. Since $\gcd(a, m) = \gcd(7, 56) = 7$, we cannot immediately conclude that $P(x, y)$ properly represents $7 \cdot 56 = 392$. But here we find that $d = \gcd(aq + kr, r) = \gcd(7(2) + 5(1), 1) = \gcd(19, 1) = 1$, and so $(19, 1)$ is a proper representation of 392 by $P(x, y)$.

1.5 Ideal numbers and quadratic numbers

A complex number v is called a *quadratic number* if v is a root of a nonzero quadratic polynomial $f(x)$ having integer coefficients. Every rational number $v = \frac{q}{n}$ is a quadratic number, being a root of $f(x) = nx^2 - qx$ for instance. As an application of the quadratic formula, every irrational quadratic number can be written uniquely as $v = \frac{q + r\sqrt{d}}{n}$, where $d \neq 1$ is a squarefree integer, q, r, and n are integers with $r \neq 0$ and n positive, and $\gcd(q, r, n) = 1$. In this section, we establish a correspondence between ideal numbers and *irrational* quadratic numbers. We begin as follows.

Definition. Let $P(x)$ be the principal polynomial of discriminant Δ and let $w = \frac{-P'(0) + \sqrt{\Delta}}{2}$. If $(a : k)$ is an ideal number of discriminant Δ, define the *major root* of $(a : k)$ to be

$$v = \text{root}(a : k) = \frac{w - k}{a} = \frac{-P'(k) + \sqrt{\Delta}}{2a}.$$

The alternative expressions for $\text{root}(a : k)$ in this definition are equal since

$$\frac{w - k}{a} = \frac{1}{a}\left(\frac{-P'(0) + \sqrt{\Delta}}{2} - k\right) = \frac{-(2k + P'(0)) + \sqrt{\Delta}}{2a} = \frac{-P'(k) + \sqrt{\Delta}}{2a}.$$

The following proposition ensures that $\text{root}(a : k)$ is an irrational quadratic number.

Proposition 1.5.1. *Let* $v = \text{root}(a : k)$ *be defined as above for an ideal number* $(a : k)$. *Then* v *is a root of* $f(x)$, *the quadratic polynomial given by* $(a : k)$ *in ideal number notation.*

Proof. Note that $w = \frac{-P'(0) + \sqrt{\Delta}}{2}$ is a root of $P(x)$, since we can write $P(x) = x^2 + P'(0)x + P(0)$ for all x, and $P'(0)^2 - 4P(0) = \Delta$. Thus, $v = \frac{w-k}{a}$ is a root of the polynomial $f(x)$ given as $(a : k)$ in ideal number notation by Proposition 1.3.5. Here, v is irrational since the discriminant, Δ, of $f(x)$ is not a square. $\qquad\square$

Note that $z = \frac{-P'(0) - \sqrt{\Delta}}{2}$ is the second root of $P(x)$, so that $\frac{z-k}{a}$ is also a root of $f(x)$ by Proposition 1.3.5. Since $w + z = -P'(0)$, we find that

$$\frac{z - k}{a} = \frac{-w - P'(0) - k}{a} = \frac{-w + \overline{k}}{a} = \frac{w - \overline{k}}{-a},$$

where $\overline{k} = -k - P'(0)$ is the conjugate of k with respect to Δ. Thus, the second or "minor" root of $f(x)$ can also be described as the major root of $(-a : \overline{k})$. Recall from Proposition 1.3.2 that if $f(x) = ax^2 + bx + c$ is the polynomial given by $(a : k)$, then $(-a : \overline{k})$ is the ideal number expression for $-f(x) = -ax^2 - bx - c$. The polynomials $f(x)$ and $-f(x)$ have the same two roots but, by our definition, we associate one root to the ideal number $(a : k)$ and the other to $(-a : \overline{k})$, arbitrarily but consistently.

Example. Let $\Delta = -47$, so that $P(x) = x^2 + x + 12$ and $P'(x) = 2x + 1$. Since $P(5) = 42 = P(-6)$, then $(42 : 5)$ and $(42 : -6)$ are ideal numbers in $\mathcal{I}_{-47}$, having major roots

$$\text{root}(42 : 5) = \frac{-P'(k) + \sqrt{\Delta}}{2a} = \frac{-11 + \sqrt{-47}}{84}$$

and

$$\text{root}(42 : -6) = \frac{11 + \sqrt{-47}}{84},$$

respectively. For $(-42 : -6)$ and $(-42 : 5)$, the respective major roots are

$$\text{root}(-42 : 5) = \frac{-(-11) + \sqrt{-47}}{-84} = \frac{-11 - \sqrt{-47}}{84}$$

and

$$\text{root}(-42:-6) = \frac{-11 + \sqrt{-47}}{-84} = \frac{11 - \sqrt{-47}}{84}.$$

Note that root$(42:5)$ and root$(-42:-6)$ are the two roots of $f(x) = 42x^2 + 11x + 1$, whereas root$(42:-6)$ and root$(-42:5)$ are the two roots of $g(x) = 42x^2 - 11x + 1$.

We can also associate an ideal number with each quadratic number, but here some caution is required. Consider first the following calculation.

Example. Let $\Delta = \Delta(-47, 2) = -188$, with $P(x) = x^2 + 2x + 48$ as its principal polynomial. Since $P(10) = 168$, we find that $(84:10)$ is an ideal number in $\mathcal{I}_{-188}$. The major root of $(84:10)$ is

$$\text{root}(84:10) = \frac{-P'(k) + \sqrt{\Delta}}{2a} = \frac{-22 + \sqrt{-188}}{168} = \frac{-11 + \sqrt{-47}}{84}.$$

Note that this is the same as the major root of $(42:5)$ in $\mathcal{I}_{-47}$, as seen in the preceding example. Thus, this quadratic number can be associated with at least two ideal numbers.

We introduce some terminology for ideal numbers needed to make this correspondence precise. If $f(x)$ is a polynomial with integer coefficients, then the greatest common divisor of the coefficients of $f(x)$ is called the *content* of the polynomial. Since an ideal number can be interpreted as a quadratic polynomial, as we saw in Section 1.3, we can also apply this terminology to ideal numbers as follows.

Definition. Let $P(x)$ be the principal polynomial of discriminant Δ and let $(a:k)$ be an ideal number in $\mathcal{I}_\Delta$. If $P(k) = ac$ and $P'(k) = b$, then we define the *content* of $(a:k)$ to be cont$(a:k) = \gcd(a, b, c)$. If cont$(a:k) = 1$, we say that $(a:k)$ is *primitive*.

We can describe all ideal numbers for which cont$(a:k) > 1$ in terms of primitive ideal numbers. We begin with a lemma.

Lemma 1.5.2. *Let $(a:k)$ be an ideal number of discriminant $\Delta = \Delta(d, g)$. If* cont$(a:k) = m$*, then m divides g, the index of Δ.*

Proof. Let $P(x)$ be the principal polynomial of discriminant Δ, with $P(k) = ac$ and $P'(k) = b$. If cont$(a:k) = m$, then we can write $a = ma_1$, $b = mb_1$, and $c = mc_1$ for some integers a_1, b_1, and c_1. Here, $\Delta = b^2 - 4ac = m^2(b_1^2 - 4a_1c_1)$, and $\Delta_1 = b_1^2 - 4a_1c_1$ is a discriminant with the same square-free part as Δ. So, then $\Delta = \Delta(d, g) = m^2\Delta(d, h)$ for some integer h, which implies that $g = mh$. $\square$

Lemma 1.5.2 implies that if Δ is a primitive discriminant, then all ideal numbers in $\mathcal{I}_\Delta$ are primitive.

Proposition 1.5.3. *Let $(a:k)$ be a primitive ideal number of discriminant Δ. If m is a positive integer, then $(ma:mk)$ is an ideal number with discriminant $m^2\Delta$ and content m. Every ideal number of content m is obtained from a primitive ideal number in this way.*

Proof. Let $P(x)$ and $P_m(x)$ be the principal polynomials of discriminant Δ and $m^2\Delta$, respectively. Let $(a : k)$ be a primitive ideal number of discriminant Δ, so that $P(k) = ac$ and $P'(k) = b$ with $\gcd(a, b, c) = 1$. Proposition 1.1.3 implies that $P_m(mk) = m^2P(k) = (ma)(mc)$ and $P'_m(mk) = mP'(k) = mb$. Thus, $(ma : mk)$ is an ideal number of discriminant $m^2\Delta$ and $\mathrm{cont}(ma : mk) = \gcd(ma, mb, mc) = m$.

Conversely, let Δ be a discriminant and suppose that $(a_m : k_m)$ is an ideal number of discriminant $m^2\Delta$ having content m. (We know that the discriminant of $(a_m : k_m)$ has this form by Lemma 1.5.2.) Let $P(x)$ and $P_m(x)$ be the principal polynomials of discriminant Δ and $m^2\Delta$, respectively, as above. By the definition of content, we have $P_m(k_m) = (ma)(mc)$ and $P'_m(k_m) = mb$ for some integers a, b, and c with $\gcd(a, b, c) = 1$. Since

$$P'_m(k_m)^2 - 4P_m(k_m) = m^2\Delta$$

by equation (1.3), we find that $b^2 - 4ac = \Delta$. We also have $P'(0)^2 - 4P(0) = \Delta$, again by equation (1.3), so that b and $P'(0)$ have the same parity. If we let k be the integer $\frac{b-P'(0)}{2}$, then

$$mk = \frac{mb - mP'(0)}{2} = \frac{P'_m(k_m) - P'_m(0)}{2} = \frac{2k_m + P'_m(0) - P'_m(0)}{2} = k_m.$$

Thus, $(a_m : k_m) = (ma : mk)$ where $f = (a : k)$ is a primitive ideal number of discriminant Δ. □

Proposition 1.5.4. *If $(a : k)$ is a primitive ideal number of discriminant Δ, then*

$$\mathrm{root}(a : k) = \mathrm{root}(ma : mk)$$

for every positive integer m.

Proof. Let $P(x)$ and $P_m(x)$ be the principal polynomials of discriminant Δ and $m^2\Delta$, respectively. If $w = \frac{-P'(0) + \sqrt{\Delta}}{2}$, then

$$mw = \frac{-mP'(0) + m\sqrt{\Delta}}{2} = \frac{-P'_m(0) + \sqrt{m^2\Delta}}{2},$$

using the fact that $P'_m(mx) = mP'(x)$ from Lemma 1.1.3. But then, $\mathrm{root}(a : k) = \frac{w-k}{a}$ equals $\mathrm{root}(ma : mk) = \frac{mw-mk}{ma}$. □

Thus, we can restrict our attention to primitive ideal numbers when considering corresponding quadratic numbers. We conclude Section 1.5 by showing that every irrational quadratic number v can be identified with a unique primitive ideal number.

Theorem 1.5.5. *Let $v = \frac{q + r\sqrt{d}}{n}$ where $q, r \neq 0$, and $n > 0$ are integers, $d \neq 1$ is a square-free integer, and $\gcd(q, r, n) = 1$. Let $m = \pm\gcd(n^2, -2nq, q^2 - r^2d)$, where the sign of m is selected to be the same as the sign of r, and let $\Delta = (\frac{2nr}{m})^2 \cdot d$. If*

$$a = \frac{n^2}{m} \quad \text{and} \quad k = \begin{cases} -n(q+r)/m, & \text{if } d \equiv 1 \pmod 4, \\ -nq/m, & \text{if } d \equiv 2 \text{ or } 3 \pmod 4, \end{cases} \tag{1.7}$$

then $(a : k)$ is a primitive ideal number of discriminant Δ for which $v = \text{root}(a : k)$.

Proof. If $v = \frac{q + r\sqrt{d}}{n}$ as above and we let $\bar{v} = \frac{q - r\sqrt{d}}{n}$, then

$$(x - v)(x - \bar{v}) = x^2 - (v + \bar{v})x + v \cdot \bar{v} = x^2 - \frac{2q}{n}x + \frac{q^2 - r^2 d}{n^2}.$$

Thus, if $m = \pm \gcd(n^2, -2nq, q^2 - r^2 d)$, then v and $\bar{v}$ are roots of

$$f(x) = \frac{n^2}{m}x^2 - \frac{2nq}{m}x + \frac{q^2 - r^2 d}{m}, \tag{1.8}$$

which has integer coefficients, content 1, and discriminant

$$\Delta = \frac{(-2nq)^2 - 4n^2(q^2 - r^2 d)}{m^2} = \left(\frac{2nr}{m}\right)^2 \cdot d.$$

We can write $\Delta = \Delta(d, g)$ where $g = \frac{2nr}{m}$ if $d \equiv 1 \pmod 4$ and $g = \frac{nr}{m}$ if $d \equiv 2$ or 3 (mod 4). (The sign choice of m ensures that g is positive, as must be true of the index of a discriminant.) If $(a : k)$ is the ideal number expression for $f(x)$, as given in (1.8), then $a = \frac{n^2}{m}$ in every case. To determine the character k and to verify that $\text{root}(a : k) = v$, we consider two cases.

If $d \equiv 1 \pmod 4$, then the linear coefficient of the principal polynomial of discriminant Δ is $P'(0) = \frac{2nr}{m}$. Thus,

$$k = \frac{f'(0) - P'(0)}{2} = \frac{1}{2}\left(-\frac{2nq}{m} - \frac{2nr}{m}\right) = -\frac{n(q+r)}{m}.$$

Here,

$$w = \frac{-P'(0) + \sqrt{\Delta}}{2} = \frac{1}{2}\left(-\frac{2nr}{m} + \frac{2nr}{m}\sqrt{d}\right) = \frac{-nr + nr\sqrt{d}}{m}.$$

(Again, it is important that m is selected so that $\frac{2nr}{m}$ is positive.) We then find that

$$\frac{w - k}{a} = \frac{(-nr + nr\sqrt{d} + n(q+r))/m}{n^2/m} = \frac{nq + nr\sqrt{d}}{n^2} = \frac{q + r\sqrt{d}}{n} = v,$$

as claimed.

If $d \equiv 2$ or 3 (mod 4), then $P'(0) = 0$ and

$$k = \frac{f'(0) - P'(0)}{2} = \frac{1}{2}\left(-\frac{2nq}{m} - 0\right) = -\frac{nq}{m}.$$

Now

$$w = \frac{-P'(0) + \sqrt{\Delta}}{2} = \frac{\sqrt{\Delta}}{2} = \frac{nr\sqrt{d}}{m}$$

and so

$$\frac{w - k}{a} = \frac{(nr\sqrt{d} + nq)/m}{n^2/m} = \frac{q + r\sqrt{d}}{n} = v$$

in this case as well. □

Definition. If v is an irrational quadratic number, then the polynomial $f(x)$ defined in equation (1.8) is called the *minimal polynomial* of v. The discriminant of $f(x)$ is also called the *discriminant* of v, written as $\Delta(v)$. If v is a rational number, it is convenient to define its (quadratic) determinant to be 0.

We illustrate Theorem 1.5.5 with two examples that confirm previous calculations from this section in the reverse direction.

Example. If $v = \frac{-11 + \sqrt{-47}}{84}$, then $q = -11$, $r = 1$, $n = 84$, and $d = -47$. With r positive, then $m = \gcd(n^2, -2nq, q^2 - r^2 d) = \gcd(7056, 1848, 168) = 168$. Thus, $\Delta = \left(\frac{2nr}{m}\right)^2 d = \left(\frac{2 \cdot 84 \cdot 1}{168}\right)^2 \cdot -47 = -47$, and since $-47 \equiv 1 \pmod 4$, we find that

$$a = \frac{n^2}{m} = \frac{7056}{168} = 42 \quad \text{and} \quad k = -\frac{n(q + r)}{m} = -\frac{84(-10)}{168} = 5.$$

Theorem 1.5.5 implies that $(42 : 5)$ is an ideal number in $\mathcal{I}_{-47}$ for which

$$\text{root}(42 : 5) = \frac{-11 + \sqrt{-47}}{84}.$$

Example. If $v = \frac{-11 - \sqrt{-47}}{84}$, then q, n, and d are the same as in the preceding example, while $r = -1$. Since r is negative, then

$$m = -\gcd(n^2, -2nq, q^2 - r^2 d) = -\gcd(7056, 1848, 168) = -168.$$

We find again that $\Delta = -47$, but now

$$a = \frac{n^2}{m} = \frac{7056}{-168} = -42 \quad \text{and} \quad k = -\frac{n(q + r)}{m} = -\frac{84(-12)}{-168} = -6.$$

Thus, $(-42 : -6)$ is a primitive ideal number in $\mathcal{I}_{-47}$ for which

$$\text{root}(-42 : -6) = \frac{-11 - \sqrt{-47}}{84}.$$

1.6 Ideal numbers and ideals in quadratic domains

In this section, we briefly introduce one more interpretation of quadratic ideal numbers. We provide only a broad outline of this development in Section 1.6, gathering together all details and verifications of claims in Appendix C. Our starting point is a type of quadratic number defined below.

A quadratic number v is called a *quadratic integer* if v is the root of a *monic* quadratic polynomial, $x^2 + bx + c$, with integer coefficients. A rational number v is a quadratic integer if and only if v is an integer in the usual sense. (Elements of $\mathbb{Z}$ are often called *rational integers* in this context.) If v is an irrational quadratic number, then v is a quadratic integer if and only if its minimal polynomial, defined as in equation (1.8), has leading coefficient 1 or -1. The discriminant of v, written as $\Delta(v)$, is defined to be the discriminant of its minimal polynomial, with $\Delta(v) = 0$ if v is rational.

If $P(x)$ is the principal polynomial of a fixed quadratic discriminant Δ and

$$z = z_\Delta = \frac{P'(0) + \sqrt{\Delta}}{2}, \tag{1.9}$$

then z is a quadratic integer having discriminant Δ, and the *quadratic domain* of discriminant Δ is defined as

$$D = D(\Delta) = \{q + rz \mid q, r \in \mathbb{Z}\}. \tag{1.10}$$

The set D is an integral domain and can also be defined as the set of all quadratic integers v for which $\Delta(v) = r^2\Delta$ for some rational integer r. A nonempty subset A of D is called an *ideal* of D if A is closed under addition and subtraction and has the following "strong closure" property for multiplication: For every v in A and x in D, the product vx is an element of A. The set $A = \{0\}$ satisfies these conditions—all other ideals of D are called *nontrivial*.

If $P(x)$ is the principal polynomial of discriminant Δ, and a and k are rational integers, then the set

$$A = \{m(a) + n(k + z) \mid m, n \in \mathbb{Z}\} = \{(ma + nk) + nz \mid m, n \in \mathbb{Z}\} \tag{1.11}$$

is an ideal of $D = D(\Delta)$ if and only if a divides $P(k)$, that is, if and only if $(a : k)$ is an ideal number of discriminant Δ. (This is proved as Theorem C.3.3 in Appendix C.) Furthermore, every nontrivial ideal of D can be written as gA, where g is a positive rational integer and A is defined as in (1.11). Thus, all nontrivial ideals of a quadratic domain can be classified in terms of ideal numbers.

Example. If $\Delta = \Delta(-1, 1) = -4$, so that $P(x) = x^2 + 1$, then $z = \frac{0 + \sqrt{-4}}{2} = i$ and

$$D = D(-4) = \{q + ri \mid q, r \in \mathbb{Z}\}.$$

This collection of complex numbers is called the domain of *Gaussian integers*, also written as $\mathbb{Z}[i]$. Since $P(5) = 26 = 13 \cdot 2$, so that $(13 : 5)$ is an ideal number of discriminant -4, then

$$A = \{m(13) + n(5 + i) \mid m, n \in \mathbb{Z}\} = \{(13m + 5n) + ni \mid m, n \in \mathbb{Z}\}$$

is an ideal of $\mathbb{Z}[i]$. (We can also describe the typical element of A as any $q + ri$ for which $q \equiv 5r \pmod{13}$.) On the other hand, the set

$$B = \{m(3) + n(2 + i) \mid m, n \in \mathbb{Z}\} = \{q + ri \in \mathbb{Z}[i] \mid q \equiv 2r \pmod{3}\}$$

is not an ideal of $\mathbb{Z}[i]$ since 3 does not divide $P(2) = 5$. As a specific counterexample to the strong closure property, note that $2+i$ is in B and $2-i$ is in $\mathbb{Z}[i]$, but that $(2+i)(2-i) = 5$ is not in B.

Every ideal number of discriminant Δ produces an ideal of $D(\Delta)$ as in equation (1.11). However, the same ideal is obtained from more than one ideal number. We will use the following proposition to make the correspondence between ideals and ideal numbers more precise.

Proposition 1.6.1. *Let* $D = \{q + rz \mid q, r \in \mathbb{Z}\}$ *be the quadratic domain of discriminant* Δ. *Let* a, b, k, *and* ℓ *be rational integers and let* A *and* B *be the following subsets of* D:

$$A = \{m(a) + n(k + z) \mid m, n \in \mathbb{Z}\} \quad and \quad B = \{m(b) + n(\ell + z) \mid m, n \in \mathbb{Z}\}.$$

Then the following are true:
1. *B is a subset of A if and only if a divides both b and $\ell - k$.*
2. *A is equal to B if and only if $|a| = |b|$ and a divides $\ell - k$.*

We do not assume that $(a : k)$ and $(b : \ell)$ are necessarily ideal numbers in this statement.

Proof. Suppose first that B is a subset of A. Then b, an element of B, can be written as

$$b = b + 0z = m(a) + n(k + z) = (ma + nk) + nz$$

for rational integers m and n. Comparing coefficients, then $n = 0$, and so $b = ma$. Similarly,

$$\ell + z = (ma + nk) + nz$$

for some rational integers m and n, implying that $n = 1$ and $\ell = ma + k$. Thus, a divides both b and $\ell - k$.

Conversely, suppose that $b = aq$ and $\ell - k = ar$ for some rational integers q and r. Then the typical element of B,

$$m(b) + n(\ell + z) = m(aq) + n((k + ar) + z) = (mq + nr)(a) + n(k + z),$$

is an element of A. Therefore, B is a subset of A.

For statement (2), $A = B$ if and only if $A \subseteq B$ and $B \subseteq A$. By statement (1), $B \subseteq A$ if and only if a divides b and $\ell - k$, and likewise $A \subseteq B$ if and only if b divides a and $k - \ell$. But b divides a and a divides b if and only if $|a| = |b|$, and then the other conditions are satisfied if and only if a divides $\ell - k$. $\qquad\square$

In Chapter 2, we define ideal numbers $(a : k)$ and $(b : \ell)$ in $\mathcal{I}_\Delta$ to be *congruent* if $|a| = |b|$ and a divides $\ell - k$, and we show that congruence is an equivalence relation on $\mathcal{I}_\Delta$. The resulting set of *congruence classes* can be identified with ideals of quadratic domains. In Appendix C, we develop properties of an operation of multiplication on ideals. We will see, beginning in Chapter 3, that we can carry this operation over to ideal numbers, providing an important algebraic structure to these objects.

---

2 Congruence classes of ideal numbers

Inspired by the classification of ideals of quadratic domains as ideal numbers, as introduced in Section 1.6, we now define an equivalence relation called *congruence* on the set of ideal numbers of a fixed discriminant. In effect, ideal numbers of discriminant Δ are congruent exactly when they are equal as ideals of the quadratic domain D_Δ. We define congruence of ideal numbers in Section 2.1, along with a partial order relation of divisibility on the resulting sets of equivalence classes under congruence. We develop methods of listing congruence classes of a fixed discriminant in the remainder of Chapter 2. In particular, we will see how to determine a pattern for all congruence classes whose norm is a power of a given prime, via the concepts of the *level* of a prime number in a discriminant and *seeding maps* to build from one level to the next. Determining all congruence classes of a given norm a and discriminant Δ is the same as solving the quadratic congruence $P(x) \equiv 0 \pmod{a}$, where $P(x)$ is principal polynomial of discriminant Δ. We will assume the standard results on quadratic congruences modulo primes, reviewed in Appendix A, as needed.

2.1 Congruence and divisibility of ideal numbers

The following general observation allows us to define an equivalence relation on ideal numbers and a partial order relation on the resulting equivalence classes.

Proposition 2.1.1. *Let $\mathcal{R}$ be a relation on a set A that has both the reflexive and transitive properties. Then there is an equivalence relation $\sim$ defined on A by saying that $a \sim b$ if and only if $a\mathcal{R}b$ and $b\mathcal{R}a$ are both true. If $A_\sim = \{[a] \mid a \in A\}$ denotes the set of equivalence classes of elements of A under $\sim$, and we say that $[a] \preceq [b]$ if and only if $a\mathcal{R}b$, then $\preceq$ is a well-defined partial order relation on $A_\sim$.*

Proof. We first show that $\sim$ is an equivalence relation as defined. Let a, b, and c be elements of A.

1. $a \sim a$ is always true, since $\mathcal{R}$ is reflexive, so that $a\mathcal{R}a$ and $a\mathcal{R}a$ are both true.
2. If $a \sim b$, so that $a\mathcal{R}b$ and $b\mathcal{R}a$, then it is also true that $b\mathcal{R}a$ and $a\mathcal{R}b$, which implies that $b \sim a$.
3. Suppose that $a \sim b$ and $b \sim c$, so that $a\mathcal{R}b$, $b\mathcal{R}a$, $b\mathcal{R}c$, and $c\mathcal{R}b$ are all true. Since $\mathcal{R}$ is transitive, $a\mathcal{R}b$ and $b\mathcal{R}c$ imply that $a\mathcal{R}c$, and $c\mathcal{R}b$ and $b\mathcal{R}a$ imply that $c\mathcal{R}a$. Thus, $a \sim c$.

Now consider the relation $\preceq$ defined on the set of equivalence classes of elements of A under $\sim$ as above. To show that $\preceq$ is well-defined, suppose that $[a] = [c]$ and $[b] = [d]$ for elements a, b, c, and d in A. This occurs if and only if $a \sim c$ and $b \sim d$, so that $a\mathcal{R}c$, $c\mathcal{R}a$, $b\mathcal{R}d$, and $d\mathcal{R}b$ are all true. If $[a] \preceq [b]$, so that $a\mathcal{R}b$, then $c\mathcal{R}d$ is also true, and thus

https://doi.org/10.1515/9783111319360-002

$[c] \preceq [d]$. (Here, we use the statements $c\mathcal{R}a$, $a\mathcal{R}b$, and $b\mathcal{R}d$, together with the transitivity of $\mathcal{R}$.)

Finally, let $[a]$, $[b]$, and $[c]$ be elements of $A_\sim$.

1. $[a] \preceq [a]$ is true since $\mathcal{R}$ is reflexive, so that $a\mathcal{R}a$.
2. If $[a] \preceq [b]$ and $[b] \preceq [a]$, then $a\mathcal{R}b$ and $b\mathcal{R}a$. But this means that $a \sim b$ is true, so that $[a] = [b]$.
3. If $[a] \preceq [b]$ and $[b] \preceq [c]$, then $a\mathcal{R}b$ and $b\mathcal{R}c$. Since $\mathcal{R}$ is transitive, then $a\mathcal{R}c$, so that $[a] \preceq [c]$.

Thus $\preceq$ is reflexive, antisymmetric, and transitive. By definition, then $\preceq$ is a partial order relation on $A_\sim$. $\qquad\square$

In this case, we say that $\sim$ is the equivalence relation and that $\preceq$ is the partial order relation *induced* from the reflexive and transitive relation $\mathcal{R}$ on A.

Proposition 2.1.2. *Let $\mathcal{R}$ be defined on the set $\mathcal{I}_\Delta$ of ideal numbers of discriminant Δ by saying that $(a : k)\mathcal{R}(b : \ell)$ if a divides both b and $\ell - k$. Then $\mathcal{R}$ is a reflexive and transitive relation on $\mathcal{I}_\Delta$.*

Proof. We have that $(a : k)\mathcal{R}(a : k)$ since a divides both a and $k - k = 0$. If $(a : k)\mathcal{R}(b : \ell)$ and $(b : \ell)\mathcal{R}(c : m)$, then a divides both b and $\ell - k$, while b divides both c and $m - \ell$. But then a divides both c and $m - k = (m - \ell) + (\ell - k)$ by properties of divisibility. $\qquad\square$

Thus, $\mathcal{R}$ induces an equivalence relation on the set of all ideal numbers of fixed discriminant as in Proposition 2.1.1. We present this relation, which we call *congruence*, more directly as follows.

Definition. Let $(a : k)$ and $(b : \ell)$ be ideal numbers of discriminant Δ. We say that $(a : k)$ is *congruent* to $(b : \ell)$, and write $(a : k) \equiv (b : \ell)$, if $|a| = |b|$ and a divides $\ell - k$.

We denote the equivalence class of an ideal number $(a : k)$ in $\mathcal{I}_\Delta$ under the congruence relation as $\langle a : k \rangle$ (i.e., with diagonal brackets rather than parentheses), and call this the *congruence class* of $(a : k)$. We will always take a to be positive when using this notation, and we refer to that value as the *norm* of $\langle a : k \rangle$. The *character* of $\langle a : k \rangle$ will mean the congruence class of k modulo a, which we can regard as an element of $\mathbb{Z}_a$. This value can be written in more than one way, and there are advantages to allowing these different possibilities. (See, e. g., the statement of Proposition 2.1.4.) However, when $P(x)$ is the principal polynomial of discriminant Δ, we will usually select k to satisfy

$$\frac{-a - P'(0)}{2} < k \leq \frac{a - P'(0)}{2}. \tag{2.1}$$

These are a incongruent values of k for which $P(k)$ is as small as possible. We will say that the congruence class $\langle a : k \rangle$ is in *minimal form* if the inequalities of (2.1) are satisfied. We write the set of all distinct congruence classes of ideal numbers of discriminant Δ as $\mathcal{Q}_\Delta$. That is,

$$\mathcal{Q}_\Delta = \{\langle a : k \rangle \mid a > 0 \text{ and } a \text{ divides } P(k)\},$$

with the understanding that $\langle a : k \rangle = \langle a : \ell \rangle$ if and only if $\ell \equiv k \pmod{a}$. The following proposition shows that we can refer to the *content* of a congruence class, or say that $\langle a : k \rangle$ is or is not *primitive*, without ambiguity.

Proposition 2.1.3. *Congruent ideal numbers have the same content.*

Proof. First, note that $\gcd(a, b, c) = \gcd(\pm a, \pm b, \pm c)$, so that an ideal number has the same content as its negative conjugate (as well as its conjugate and its negative). Thus, we can assume that the norms of the given congruent ideal numbers are equal, say $\langle a : k \rangle$ and $\langle a : k_1 \rangle$, where $k_1 = k + an$ for some integer n. Let $P(x)$ be the principal polynomial of discriminant Δ, with $P(k) = ac$ and $P'(k) = b$. Equation (1.4) and direct calculation show that $P(k_1) = ac_1$ and $P'(k_1) = b_1$ where $b_1 = 2an + b$ and $c_1 = an^2 + bn + c$. It follows that $\gcd(a, b, c)$ divides b_1 and c_1, and so divides $\gcd(a, b_1, c_1)$. But we also find that $b = b_1 - 2an$ and $c = c_1 - b_1 n + an^2$, and similarly conclude that $\gcd(a, b_1, c_1)$ divides $\gcd(a, b, c)$. Thus, $\gcd(a, b, c)$ equals $\gcd(a_1, b_1, c_1)$, that is, $\langle a : k \rangle$ and $\langle a : k_1 \rangle$ have the same content. □

Propositions 2.1.1 and 2.1.2 imply that the following is a well-defined partial order relation on $\mathcal{Q}_\Delta$.

Definition. If $\langle a : k \rangle$ and $\langle b : \ell \rangle$ are congruence classes of ideal numbers with a particular discriminant, we say that $\langle a : k \rangle$ *divides* $\langle b : \ell \rangle$, and write $\langle a : k \rangle \leq \langle b : \ell \rangle$, if a divides both b and $\ell - k$.

Proposition 2.1.4. *Let $\langle b : \ell \rangle$ be an element of $\mathcal{Q}_\Delta$ for some discriminant Δ. Then*

$$\{\langle a : \ell \rangle \mid a \text{ is a positive divisor of } b\}$$

is the collection of all ideal numbers that divide $\langle b : \ell \rangle$.

Proof. Let $P(x)$ be the principal polynomial of discriminant Δ. If $\langle b : \ell \rangle$ is an element of $\mathcal{Q}_\Delta$, then b divides $P(\ell)$. If a is a positive divisor of b, then a likewise divides $P(\ell)$, so that $\langle a : \ell \rangle$ is an element of $\mathcal{Q}_\Delta$. Here, $\langle a : \ell \rangle \leq \langle b : \ell \rangle$ since a divides both b and $\ell - \ell = 0$. Conversely, suppose that $\langle a : k \rangle$ divides $\langle b : \ell \rangle$, so that a divides both b and $\ell - k$. Since $\ell \equiv k \pmod{a}$, we have that $\langle a : k \rangle = \langle a : \ell \rangle$. Thus, $\{\langle a : \ell \rangle \mid a \text{ is a positive divisor of } b\}$ is the entire set of ideal numbers in $\mathcal{Q}_\Delta$ that divide $\langle b : \ell \rangle$. □

To conclude Section 2.1, we illustrate a "brute force" method for listing representatives of all congruence classes of ideal numbers $\langle a : k \rangle$ for a fixed a, or for all a less than or equal to some fixed upper bound u. We begin with an observation that reduces our calculations considerably.

Proposition 2.1.5. *Let $P(x)$ be the principal polynomial of discriminant Δ. For all integers x, let $\overline{x} = -x - P'(0)$ be the conjugate of x with respect to Δ.*

1. If Δ is even, then $m = -\frac{P'(0)}{2}$ is an integer for which $P(m) = -\frac{\Delta}{4}$. If $x = m + n$ for some nonnegative integer n, then $P(x + 1) = P(x) + (2n + 1)$ and $\bar{x} = m - n$.
2. If Δ is odd, then $m = \frac{1-P'(0)}{2}$ is an integer for which $P(m) = \frac{1-\Delta}{4}$. If $x = m + n$ for some nonnegative integer n, then $P(x + 1) = P(x) + (2n + 2)$ and $\bar{x} = m - n - 1$.

Proof. Equation (1.3) shows that $P'(0)$ has the same parity as Δ, so that m is an integer in both cases. If $x = m + n$, then equation (1.4) shows that

$$P(x + 1) - P(x) = 1 \cdot (x + 1 + x + P'(0)) = (2n + 1) + (2m + P'(0)). \qquad (2.2)$$

Furthermore,

$$\bar{x} = -(m + n) - P'(0) = (m - n) - (2m + P'(0)). \qquad (2.3)$$

If Δ is even, then $m = -\frac{P'(0)}{2}$ is an integer for which $2m + P'(0) = P'(m) = 0$. Thus, $P(m) = -\frac{\Delta}{4}$ by equation (1.3), and equations (2.2) and (2.3) show that $P(x + 1) = P(x) + (2n + 1)$ and $\bar{x} = m - n$. On the other hand, if Δ is odd, then $m = \frac{1-P'(0)}{2}$ is an integer for which $2m + P'(0) = P'(m) = 1$. The same equations show that $P(m) = \frac{1-\Delta}{4}$, $P(x+1) = P(x)+(2n+2)$, and $\bar{x} = m - n - 1$. □

Proposition 2.1.5 makes it is easy to calculate as many values of a principal polynomial $P(x)$ as needed, which we can then use to find distinct congruence classes of ideal numbers with norm not exceeding some upper bound. The following steps can be easily transformed into a computer program, which we will write as **IdNumList**(Δ, u), for this process.

Given a discriminant Δ with principal polynomial $P(x)$ and a positive integer u:
1. Create a list of ordered pairs (x, y), beginning with an empty list, as follows:
    - If Δ is even, let $x = -P'(0)/2$, let $y = -\Delta/4$, and append (x, y) to the list
    - If Δ is odd, let $x = (1 - P'(0))/2$, let $y = (1 - \Delta)/4$, and append both (x, y) and $(-x - P'(0), y)$ to the list
    - While the number of ordered pairs in the list is smaller than u:
        - Replace y by $y + (2x + 1 + P'(0))$
        - Replace x by $x + 1$
        - Append both (x, y) and $(-x - P'(0), y)$ to the list
2. Create a list of $\langle a : k \rangle$, beginning with an empty list, as follows:
    - For $1 \le a \le u$:
        - For $1 \le i \le a$:
            - Let (k, y) be item i in the list of ordered pairs
            - If a divides y, append $\langle a : k \rangle$ to the new list
3. Return the list of $\langle a : k \rangle$

Example. We illustrate this algorithm with $\Delta = -35$, so that $P(x) = x^2 + x + 9$ is the principal polynomial, and $u = 9$. In Step 1, we first have $x = 0$ and $y = 9$, and append

$(0, 9)$ and $(-1, 9)$ to an empty list of ordered pairs. Since the list has fewer than $u = 9$ items, we replace y by $y + (2x + 1 + P'(0)) = 9 + 2 = 11$, replace x by $x + 1 = 1$, and append $(1, 11)$ and $(-2, 11)$ to the list. After Step 1, we have the following list of ten ordered pairs:

$$(0, 9), (-1, 9), (1, 11), (-2, 11), (2, 15), (-3, 15), (3, 21), (-4, 21), (4, 29), (-5, 29).$$

One can verify that each ordered pair has the form $(k, P(k))$.

In Step 2, we let a vary from 1 to 9. For each one, we look at the first a pairs in the preceding list and test whether a divides the y-component, say y_i. If so, and if $k = x_i$ for the ordered pair (x_i, y_i), we append $\langle a : k \rangle$ to an initially empty list. (We omit even values of a below for convenience, since all $P(k)$ values calculated are odd. The algorithm would, of course, test each such a.)

- $a = 1$ divides $y_1 = 9$, so we append $\langle 1 : 0 \rangle$ to the list.
- $a = 3$ divides $y_1 = 9$ and $y_2 = 9$, but not $y_3 = 11$. We append $\langle 3 : 0 \rangle$ and $\langle 3 : -1 \rangle$.
- $a = 5$ divides only $y_5 = 15$, among the first five ordered pairs. We append $\langle 5 : 2 \rangle$.
- $a = 7$ divides only $y_7 = 21$. We append $\langle 7 : 3 \rangle$.
- $a = 9$ divides $y_1 = 9 = y_2$, but no other y_i values among the first nine ordered pairs. We append both $\langle 9 : 0 \rangle$ and $\langle 9 : -1 \rangle$.

At the conclusion, we have the following list of seven distinct congruence classes:

$$\langle 1 : 0 \rangle \quad \langle 3 : 0 \rangle \quad \langle 3 : -1 \rangle \quad \langle 5 : 2 \rangle \quad \langle 7 : 3 \rangle \quad \langle 9 : 0 \rangle \quad \langle 9 : -1 \rangle.$$

(Note that $\langle 5 : -3 \rangle = \langle 5 : 2 \rangle$ and that $\langle 7 : -4 \rangle = \langle 7 : 3 \rangle$.) For later reference, we extend this example to the upper bound $u = 40$ in the following list:

$\langle 1 : 0 \rangle$	$\langle 3 : 0 \rangle$	$\langle 3 : -1 \rangle$	$\langle 5 : 2 \rangle$	$\langle 7 : 3 \rangle$	$\langle 9 : 0 \rangle$
$\langle 9 : -1 \rangle$	$\langle 11 : 1 \rangle$	$\langle 11 : -2 \rangle$	$\langle 13 : 5 \rangle$	$\langle 13 : -6 \rangle$	$\langle 15 : 2 \rangle$
$\langle 15 : -3 \rangle$	$\langle 17 : 6 \rangle$	$\langle 17 : -7 \rangle$	$\langle 21 : 3 \rangle$	$\langle 21 : -4 \rangle$	$\langle 27 : 8 \rangle$
$\langle 27 : -9 \rangle$	$\langle 29 : 4 \rangle$	$\langle 29 : -5 \rangle$	$\langle 33 : 9 \rangle$	$\langle 33 : -10 \rangle$	$\langle 33 : 12 \rangle$
$\langle 33 : -13 \rangle$	$\langle 35 : 17 \rangle$	$\langle 39 : 5 \rangle$	$\langle 39 : -6 \rangle$	$\langle 39 : 18 \rangle$	$\langle 39 : -19 \rangle.$

2.2 Ideal numbers of prime power norm and level zero

If Δ is a discriminant and a is a positive integer, we will write $\mathcal{Q}_\Delta(a)$ for the set of congruence classes of ideal numbers having norm a. We write the number of elements in $\mathcal{Q}_\Delta(a)$ as $n_\Delta(a)$, which we can also describe as the number of solutions of the quadratic congruence $P(x) \equiv 0 \pmod{a}$, where $P(x)$ is the principal polynomial of discriminant Δ. We may refer to solutions of $P(x) \equiv 0 \pmod{a}$ as *roots* of $P(x)$ modulo a. In the remainder of Chapter 2, we develop a formula for $n_\Delta(a)$ that depends on the prime factorizations of Δ and a. We begin with the following observation.

Proposition 2.2.1. *If Δ is a discriminant and a and b are integers with $\gcd(a, b) = 1$, then*

$$n_\Delta(ab) = n_\Delta(a) \cdot n_\Delta(b).$$

Proof. Let $P(x)$ be the principal polynomial of discriminant Δ. If $\mathcal{Q}_\Delta(ab)$ is nonempty, we show that the function

$$f : \mathcal{Q}_\Delta(ab) \to \mathcal{Q}_\Delta(a) \times \mathcal{Q}_\Delta(b) \quad \text{defined by} \quad f(\langle ab : m \rangle) = (\langle a : m \rangle, \langle b : m \rangle)$$

is a well-defined bijection if $\gcd(a, b) = 1$. If ab divides $P(m)$, then a and b both divide $P(m)$, so that $\langle a : m \rangle$ and $\langle b : m \rangle$ are elements of $\mathcal{Q}_\Delta$. Likewise, if $m \equiv n \pmod{ab}$, then $m \equiv n \pmod{a}$ and $m \equiv n \pmod{b}$. When a and b are relatively prime, then the converses of these statements are true as well. Thus, f is a well-defined and injective. In the reverse direction, if $\langle a : k \rangle$ and $\langle b : \ell \rangle$ are elements of $\mathcal{Q}_\Delta$, and $\gcd(a, b) = 1$, then there is an m for which $m \equiv k \pmod{a}$ and $m \equiv \ell \pmod{b}$. Note that $\langle a : k \rangle = \langle a : m \rangle$ and $\langle b : \ell \rangle = \langle b : m \rangle$ in this case. Here, $\langle ab : m \rangle$ is an element of $\mathcal{Q}_\Delta(ab)$ sent to $(\langle a : k \rangle, \langle b : \ell \rangle)$ by f. Therefore, f is surjective as well. (This also shows that $\mathcal{Q}_\Delta(ab)$ cannot be empty if $\mathcal{Q}_\Delta(a) \times \mathcal{Q}_\Delta(b)$ is nonempty.) □

Example. If $\Delta = -35$, then using data from an example in Section 2.1, we have that

$$\mathcal{Q}_\Delta(33) = \{\langle 33 : 9 \rangle, \langle 33 : -10 \rangle, \langle 33 : 12 \rangle, \langle 33 : -13 \rangle\}$$

while

$$\mathcal{Q}_\Delta(3) = \{\langle 3 : 0 \rangle, \langle 3 : -1 \rangle\} \quad \text{and} \quad \mathcal{Q}_\Delta(11) = \{\langle 11 : 1 \rangle, \langle 11 : -2 \rangle\}.$$

As one example, the correspondence described in the proof of Proposition 2.2.1 sends $\langle 33 : 9 \rangle$ to $(\langle 3 : 0 \rangle, \langle 11 : -2 \rangle)$, since $9 \equiv 0 \pmod 3$ and $9 \equiv -2 \pmod{11}$.

Proposition 2.2.1 implies that, when establishing a formula for $n_\Delta(a)$, it suffices to compute $n_\Delta(p^e)$ for every prime number p and positive integer e. We begin with $e = 1$ as follows.

Proposition 2.2.2. *If Δ is a discriminant and p is prime, then $n_\Delta(p) \le 2$, with $n_\Delta(p) = 1$ if and only if p divides Δ.*

Proof. Let $P(x)$ be the principal polynomial of discriminant Δ and suppose that p divides $P(k)$ for some prime number p and integer k. Recall that if $\bar{k} = -k - P'(0)$ is the conjugate of k with respect to Δ, then $P(k) = P(\bar{k})$. Equation (1.4) shows that

$$P(x) - P(k) = (x - k)(x - \bar{k})$$

for every integer x. Since p is prime, it follows that p divides $P(x)$ if and only if $x \equiv k \pmod p$ or $x \equiv \bar{k} \pmod p$. Thus, there are no more than two solutions of $P(x) \equiv 0 \pmod p$.

Suppose that $P(x) \equiv 0 \pmod{p}$ has precisely one solution, say k. Since $P(k) = P(\overline{k})$, this is possible only when $k \equiv \overline{k} \pmod{p}$. In that case, p divides $k - \overline{k} = k - (-k - P'(0)) = 2k + P'(0) = P'(k)$. But then p divides $P'(k)^2 - 4P(k) = \Delta$. Conversely, suppose that p divides Δ. If p is odd, then p divides $P(x)$ if and only if p divides $P'(x)$, and the linear congruence $P'(x) \equiv 0 \pmod{p}$ has a unique solution. If $p = 2$ divides Δ, then $P'(x)$ is even for both $x = 0$ and $x = 1$. But then equation (1.4) shows that $P(1) - P(0) = P'(0) + 1$ is odd. Since $P(0)$ and $P(1)$ have opposite parity, then $P(x) \equiv 0 \pmod{p}$ has precisely one solution in this case as well. □

When p does not divide Δ, there is a systematic method to determine whether $n_\Delta(p) = 2$ or $n_\Delta(p) = 0$, outlined in Appendix A. Assuming that we know $n_\Delta(p)$, using this method or trial-and-error, we consider $n_\Delta(p^e)$ for $e > 1$. Here, it is convenient to introduce the following notation and terminology.

Notation. If Δ is a discriminant and p is a prime number, then we write $\mathcal{Q}_\Delta(p^\infty)$ for the set of all elements of $\mathcal{Q}_\Delta$ whose norm is a power of p. That is,

$$\mathcal{Q}_\Delta(p^\infty) = \bigcup_{e=0}^{\infty} \mathcal{Q}_\Delta(p^e).$$

Definition. If p is a prime number and $\Delta = \Delta(d, g)$ is a discriminant, then define the *level* of p in Δ, or the *p-level* of Δ, to be the largest nonnegative integer ℓ, so that p^ℓ divides g. We denote ℓ as $\mathrm{lev}_p(\Delta)$.

In the remainder of this section, we restrict our attention to discriminants Δ and primes p for which $\mathrm{lev}_p(\Delta) = 0$. For reasons we will see later, it will be important to describe the *primitive* elements of each set $\mathcal{Q}_\Delta(p^\infty)$. In this case, the following proposition shows that this collection is all of $\mathcal{Q}_\Delta(p^\infty)$.

Proposition 2.2.3. *If $\Delta = \Delta(d, g)$ is a discriminant and p is a prime number for which $\mathrm{lev}_p(\Delta) = 0$, then every $\langle p^e : k \rangle$ in $\mathcal{Q}_\Delta$ is primitive.*

Proof. The content of $\langle p^e : k \rangle$ is the greatest common divisor of a collection of integers that includes p^e. But by Lemma 1.5.2, this content also divides g, the index of Δ. Since p does not divide g when $\mathrm{lev}_p(\Delta) = 0$, the content of $\langle p^e : k \rangle$ equals 1. □

When the level of p in Δ is zero, we can classify the sets $\mathcal{Q}_\Delta(p^\infty)$ in terms of $n_\Delta(p)$, which equals 0, 1, or 2 by Proposition 2.2.2.

Proposition 2.2.4. *Let Δ be a discriminant and let p be a prime number for which $\mathrm{lev}_p(\Delta) = 0$ and $n_\Delta(p) = 0$. Then $\mathcal{Q}_\Delta(p^\infty) = \{\langle 1 : 0 \rangle\}$.*

Proof. Let $P(x)$ be the principal polynomial of discriminant Δ. If $\langle p^e : k \rangle$ is in $\mathcal{Q}_\Delta(p^\infty)$ with $e \geq 1$, then p^e divides $P(k)$. But then p divides $P(k)$, so that $\langle p : k \rangle$ is also in $\mathcal{Q}_\Delta(p^\infty)$. This is impossible if $n_\Delta(p) = 0$. Thus, $\langle 1 : 0 \rangle$ is the only element in $\mathcal{Q}_\Delta(p^\infty)$. □

Proposition 2.2.5. *Let Δ be a discriminant, with principal polynomial $P(x)$. Let p be a prime number for which* $\mathrm{lev}_p(\Delta) = 0$ *and* $n_\Delta(p) = 1$. *Then*

$$\mathcal{Q}_\Delta(p^\infty) = \{\langle 1 : 0 \rangle, \langle p : k \rangle\},$$

where k is the unique root of $P(x)$ modulo p.

Proof. If $n_\Delta(p) = 1$, then p divides Δ by Proposition 2.2.2. Suppose that p^e divides $P(x)$ for some integer x and some $e > 1$. From the equation $P'(x)^2 - 4P(x) = \Delta$, it follows that p divides $P'(x)$, and then that p^2 divides Δ. But if $\mathrm{lev}_p(\Delta) = 0$, then $\Delta = \Delta(d, g)$ with p not dividing g. This is impossible if p is odd, since d is square-free. If $p = 2$, so that g is odd, this implies that d is congruent to 2 or 3 modulo 4. But 4 cannot divide $P(x) = x^2 - g^2 d$ since a square is congruent to 0 or 1 modulo 4, so this is again impossible. Thus, in all cases where $\mathrm{lev}_p(\Delta) = 0$ and $n_\Delta(p) = 1$, we conclude that $\mathcal{Q}_\Delta(p^\infty)$ contains only $\langle 1 : 0 \rangle$ and $\langle p : k \rangle$. □

It remains to consider primes p for which $n_\Delta(p) = 2$. Here, we can describe the elements of $\mathcal{Q}_\Delta(p^\infty)$ in the form of an algorithm, as follows. The notation introduced here will be useful for later calculations with these elements.

Theorem 2.2.6. *Let Δ be a discriminant with principal polynomial $P(x)$. Let p be a prime number for which* $\mathrm{lev}_p(\Delta) = 0$ *and* $n_\Delta(p) = 2$, *with k one of the roots of $P(x)$ modulo p. Then there is an integer s such that $s \cdot P'(k) \equiv 1 \pmod{p}$, and we can define k_n for all integers n as follows. Let $k_0 = 0$ and $k_1 = k$. For $n \geq 1$, let $k_{-n} = \overline{k_n}$, the conjugate of k_n with respect to Δ, and let $k_{n+1} = k_n - sP(k_n)$. Then $\mathcal{Q}_\Delta(p^\infty)$ consists precisely of all $\langle p^{|n|} : k_n \rangle$ with n an integer.*

We illustrate this algorithm with an example before proving Theorem 2.2.6.

Example. Let $\Delta = -35$, with principal polynomial $P(x) = x^2 + x + 9$, and let $p = 3$. There are two solutions of $P(x) \equiv 0 \pmod 3$, namely 0 and -1. If we let $k = 0$, then $P'(k) = 1$, and $s = 1$ satisfies $s \cdot P'(k) \equiv 1 \pmod p$. Then $k_{n+1} = k_n - sP(k_n) = -k_n^2 - 9$ for $n \geq 1$, with $k_1 = k = 0$. The following table keeps track of these calculations for $1 \leq n \leq 6$. The entry for k_{-n} is the conjugate of k_n with respect to $\Delta = -35$, in this case, $k_{-n} = -1 - k_n$. The entry for k_{n+1} is $-k_n^2 - 9$ reduced to its minimal absolute value modulo 3^{i+1}. This value becomes the entry in the k_n column for the next n:

n	k_n	k_{-n}	$-k_n^2 - 9$	k_{n+1}
1	0	−1	−9	0
2	0	−1	−9	−9
3	−9	8	−90	−9
4	−9	8	−90	−90
5	−90	89	−8109	−90
6	−90	89	−8109	639

The algorithm of Theorem 2.2.6 is a variation on the following standard method, called a *Hensel lifting*, of constructing solutions of $f(x) \equiv 0 \pmod{p^{e+1}}$ from solutions of $f(x) \equiv 0 \pmod{p^e}$ when $f(x)$ is a polynomial with integer coefficients. If $f(r) \equiv 0 \pmod{p^e}$, then $s = r + p^e t$ satisfies $f(x) \equiv 0 \pmod{p^{e+1}}$ if and only if t satisfies the linear congruence $f'(r)t \equiv -\frac{f(r)}{p^e} \pmod{p}$. In our algorithm, we give a formula for t, and then for $r + p^e t$, based on the assumption that p does not divide $f'(r)$.

We can illustrate (Figure 2.1) the set $\mathcal{Q}_{-35}(3^\infty)$ with the following Hasse diagram for the partial order relation of divisibility, as defined in Section 2.1, drawn horizontally to save space. Each vertex corresponds to an ideal number $\langle 3^e : k \rangle$, but for readability, k is the only label on each vertex, with 3^e noted at the top of the diagram. The pattern of two conjugate ideal numbers for each norm 3^e continues as e increases beyond 6.

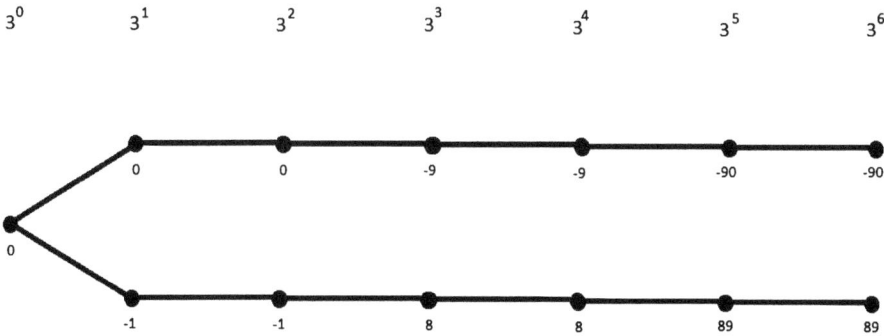

Figure 2.1: Partial Hasse diagram for $\mathcal{Q}_{-35}(3^\infty)$.

Proof of Theorem 2.2.6. Let $P(x)$ be the principal polynomial of discriminant Δ and let p be a prime number such that $n_\Delta(p) = 2$. We first observe inductively that $n_\Delta(p^n)$ is no larger than 2 for any positive integer n. Suppose that $y \equiv x \pmod{p^n}$ and that p^{n+1} divides both $P(x)$ and $P(y)$. We can write $y = x + p^n t$ for some integer t, and then equation (1.4) shows that

$$P(y) - P(x) = (y - x)(y + x + P'(0))$$
$$= (y - x)(p^n t + 2x + P'(0)) = (y - x)(p^n t + P'(x)).$$

Since p does not divide $\Delta = P'(x)^2 - 4P(x)$ when $n_\Delta(p) = 2$, then p does not divide $p^n t + P'(x)$ if $n > 0$. Thus, if p^{n+1} divides both $P(x)$ and $P(y)$, then p^{n+1} divides $y - x$, and each solution of $P(x) \equiv 0 \pmod{p^n}$ produces no more than one solution of $P(x) = 0 \pmod{p^{n+1}}$.

Now let k and $\bar{k}$ be the solutions of $P(x) \equiv 0 \pmod{p}$. Since p does not also divide $P'(k)$, the linear congruence $P'(k)x \equiv 1 \pmod{p}$ has a solution, which we label as s. With $P'(\bar{k}) = -P'(k)$, we can use $-s$ as the solution of $P'(\bar{k})x \equiv 1 \pmod{p}$.

For an integer $n \geq 1$, suppose that k_n is an integer for which p^n divides $P(k_n)$, with $k_n \equiv k \pmod{p}$, and let $k_{n+1} = k_n - s \cdot P(k_n)$. Then

$$P(k_{n+1}) - P(k_n) = (k_{n+1} - k_n)(k_{n+1} + k_n + P'(0))$$
$$= (-s \cdot P(k_n))(-s \cdot P(k_n) + P'(k_n)),$$

which we can rearrange as

$$P(k_{n+1}) = P(k_n)(1 - s \cdot P'(k_n)) + s^2 \cdot P(k_n)^2. \tag{2.4}$$

Now p^{2n} divides $P(k_n)^2$, and with $n + 1 \leq 2n$, it follows that

$$P(k_{n+1}) \equiv P(k_n)(1 - s \cdot P'(k_n)) \pmod{p^{n+1}}.$$

But p divides $1 - s \cdot P'(k_n)$, since $P'(k_n) \equiv P'(k) \pmod{p}$, so we conclude that p^{n+1} divides $P(k_{n+1})$ and $\langle p^{n+1} : k_{n+1} \rangle$ is an ideal number of discriminant Δ. Furthermore,

$$\overline{k_n} - (-s)P(\overline{k_n}) = -P'(0) - k_n + sP(k_n) = -P'(0) - k_{n+1} = \overline{k_{n+1}}.$$

Note that k_{n+1} and $\overline{k_{n+1}}$ are congruent modulo p^n to k_n and $\overline{k_n}$, respectively.

Therefore, we conclude that $\mathcal{Q}_\Delta(p^\infty)$ contains exactly two ideal numbers of norm p^e for each integer $e > 0$. These ideal numbers have the form $\langle p^{|n|} : k_n \rangle$, where k_n is defined for each integer n as in Theorem 2.2.6. $\qquad\square$

2.3 Seeding maps

In Section 2.2, we described the set $\mathcal{Q}_\Delta(p^\infty)$ when the level of p in Δ is zero. We will extend these results to arbitrary discriminants in the remainder of Chapter 2, using a collection of functions defined in Theorem 2.3.1. We will adopt the following notation for simplicity in the statement of this theorem and throughout Section 2.3.

Notation. Let p be a prime number, let $\Delta = \Delta(d, g)$ be a discriminant for which p does not divide g, and let $P(x)$ be the principal polynomial of discriminant Δ. Then for every nonnegative integer ℓ, let $\Delta_\ell = \Delta(d, p^\ell g)$, a discriminant for which the level of p is ℓ, and let $P_\ell(x)$ be the principal polynomial of discriminant Δ_ℓ. (Note that $\Delta_0 = \Delta$ and $P_0(x) = P(x)$.) Finally, we write the set $\mathcal{Q}_{\Delta_\ell}(p^\infty)$ simply as Q_ℓ.

Theorem 2.3.1. *Let p be a prime number and let Δ_ℓ and Q_ℓ be the discriminants and the sets of congruence classes of ideal numbers defined for $\ell \geq 0$ as above. Then*

$$S : Q_\ell \to Q_{\ell+1} \quad \text{defined by} \quad S(\langle p^e : k \rangle) = \langle p^{e+1} : pk \rangle \tag{2.5}$$

is a well-defined injective function, and the following statements are true:

1. *An element of $Q_{\ell+1}$ is primitive if and only if it is not the image of an element of Q_ℓ under S.*
2. *If $\langle p^e : k \rangle$ is an element of Q_ℓ, then there are p distinct ideal numbers in $Q_{\ell+1}$ of the form $\langle p^{e+2} : pk + p^{e+1}t \rangle$, where $0 \leq t < p$. Aside from $\langle 1 : 0 \rangle$ and $\langle p : 0 \rangle$, all ideal numbers in $Q_{\ell+1}$ are obtained from elements of Q_ℓ in this way.*

We refer to S defined in (2.5) as a *seeding map*, explaining that terminology with examples before proving Theorem 2.3.1.

Example. Let $p = 3$ and $\Delta_0 = \Delta(-35, 1) = -35$, so that $\Delta_1 = \Delta, (-35, 3) = -315$. In the notation of Theorem 2.3.1, let $Q_0 = \mathcal{Q}_{-35}(3^\infty)$ and $Q_1 = \mathcal{Q}_{-315}(3^\infty)$. In Section 2.2, we presented a Hasse diagram for Q_0 under the partial order relation of divisibility. Figure 2.2 illustrates the seeding map $S : Q_0 \rightarrow Q_1$ as thin lines drawn from each vertex in the Hasse diagram for Q_0 to certain vertices in the Hasse diagram for Q_1. For instance, $S(\langle 27 : 8 \rangle) = \langle 81 : 24 \rangle$ is indicated by the diagonal line from the vertex labeled as 8 under 3^3 to the vertex labeled as 24 under 3^4. An implication of statement (2) in Theorem 2.3.1 is that each vertex "seeded" by S in this way produces $p = 3$ new vertices; in this case, $\langle 243 : 24 \pm 81t \rangle$. On the other hand, aside from $\langle 1 : 0 \rangle$, the "unseeded" vertices in Q_1, such as $\langle 81 : -3 \rangle$, produce no new ideal numbers at the next level, and so die out. These vertices, however, represent the primitive ideal numbers in Q_1. (The labeling of vertices in the Hasse diagram for Q_1 is carried out after the fact, with the placement of the seeded nodes chosen simply to make the diagram look better.) The overall implication of Theorem 2.3.1 is that the Hasse diagram for Q_1 (aside from the initial vertex for $\langle 1 : 0 \rangle$) can be constructed in full from the Hasse diagram of Q_0 via the seeding map.

For discriminants with larger p-level, we can repeat application of the seeding maps.

Example. If we wish to describe the set $\mathcal{Q}_{192}(2^\infty)$, we can build up from $\Delta_0 = \Delta(3, 1) = 12$ to $\Delta_1 = \Delta(3, 2^1 \cdot 1) = 48$, and then to $\Delta_2 = \Delta(3, 2^2 \cdot 1) = 192$, using seeding maps. Again following the notation of Theorem 2.3.1, let Q_0, Q_1, and Q_2 be the collections of congruence classes of ideal numbers whose norm is a power of 2, having discriminant 12, 48, and 192, respectively. Since $p = 2$ divides $\Delta = 12$, Proposition 2.2.5 implies that $Q_0 = \{\langle 1 : 0 \rangle, \langle 2 : 1 \rangle\}$. The seeding map $S : Q_0 \rightarrow Q_1$ takes $\langle 1 : 0 \rangle$ to $\langle 2 : 0 \rangle$, which then produces $p = 2$ ideal numbers of norm 4, namely $\langle 4 : 0 \rangle$ and $\langle 4 : 2 \rangle$. We also have $S(\langle 2 : 1 \rangle) = \langle 4 : 2 \rangle$, which likewise produces $p = 2$ ideal numbers of norm 8, $\langle 8 : 2 \rangle$ and $\langle 8 : -2 \rangle$. However, $\langle 4 : 0 \rangle$ is not the image of an element of Q_0 under S, and produces no new ideal numbers. The top of the diagram in Figure 2.3 illustrates this map and its effect.

Now a second seeding map can be applied to each of the six elements in Q_1, as illustrated at the bottom of the diagram. Each image branches off into $p = 2$ new nodes while those vertices that are not the image of an element of Q_1 under S die out. We find fourteen elements in total in Q_2. The eight ideal numbers that are not an image of an element of Q_1 under S are all primitive. The two ideal numbers that have the form $S(S(\langle a : k \rangle))$

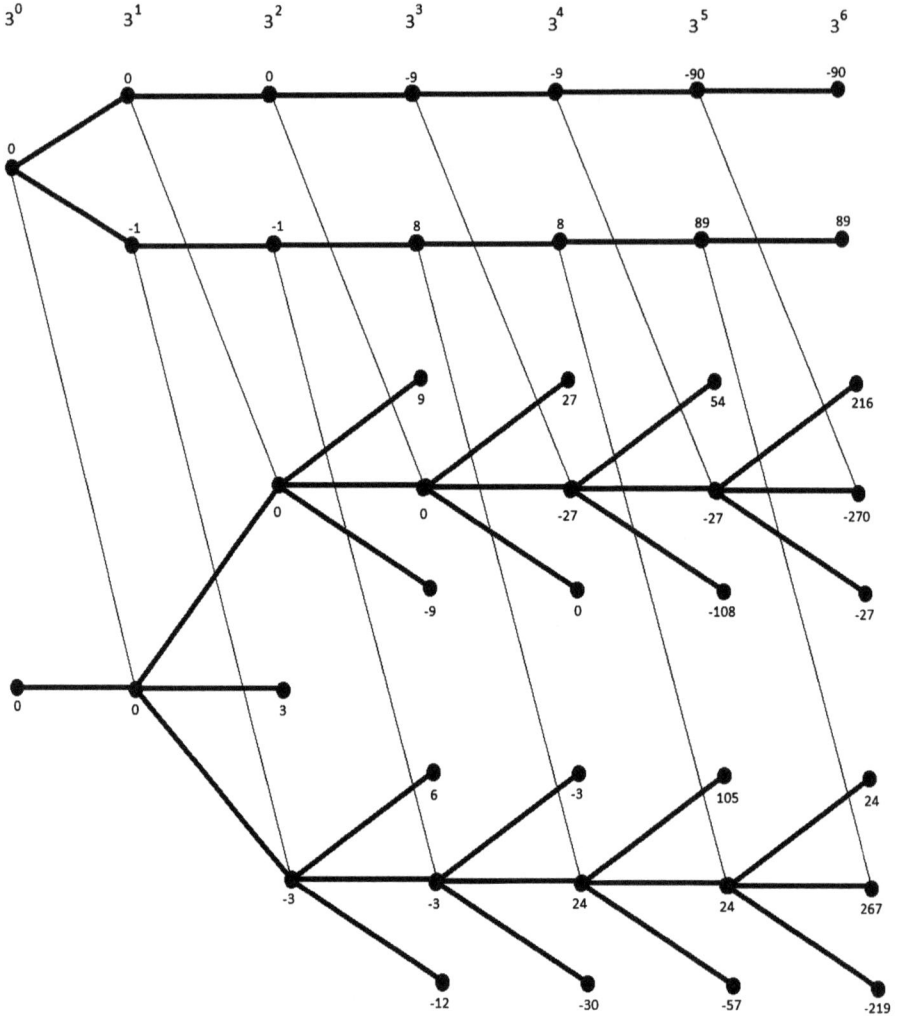

Figure 2.2: Seeding map from $\mathcal{Q}_{-35}(3^{\infty})$ to $\mathcal{Q}_{-315}(3^{\infty})$.

for an ideal number $\langle a : k \rangle$ in Q_0 have content 4, while the remaining ideal numbers have content 2. These results are as predicted in Theorem 2.3.1.

Proof of Theorem 2.3.1. For each $\ell \geq 0$, let $P_{\ell}(x)$ be the principal polynomial of discriminant Δ_{ℓ}. Proposition 1.1.3 implies that $P_{\ell+1}(px) = p^2 P_{\ell}(x)$ and $P'_{\ell+1}(px) = p P'_{\ell}(x)$ for all x. Now for a fixed nonnegative integer ℓ, let $\langle p^e : k \rangle$ be an ideal number of discriminant Δ_{ℓ}, so that $P_{\ell}(k) = p^e c$ for some integer c. Then $P_{\ell+1}(pk) = p^2 P_{\ell}(k) = p^{e+1}(pc)$, confirming that $S(\langle p^e : k \rangle) = \langle p^{e+1} : pk \rangle$ is an ideal number of discriminant $\Delta_{\ell+1}$. Since $k \equiv k' \pmod{p^e}$ if and only if $pk \equiv pk' \pmod{p^{e+1}}$, as an application of the congruence cancellation property, S is well-defined and injective.

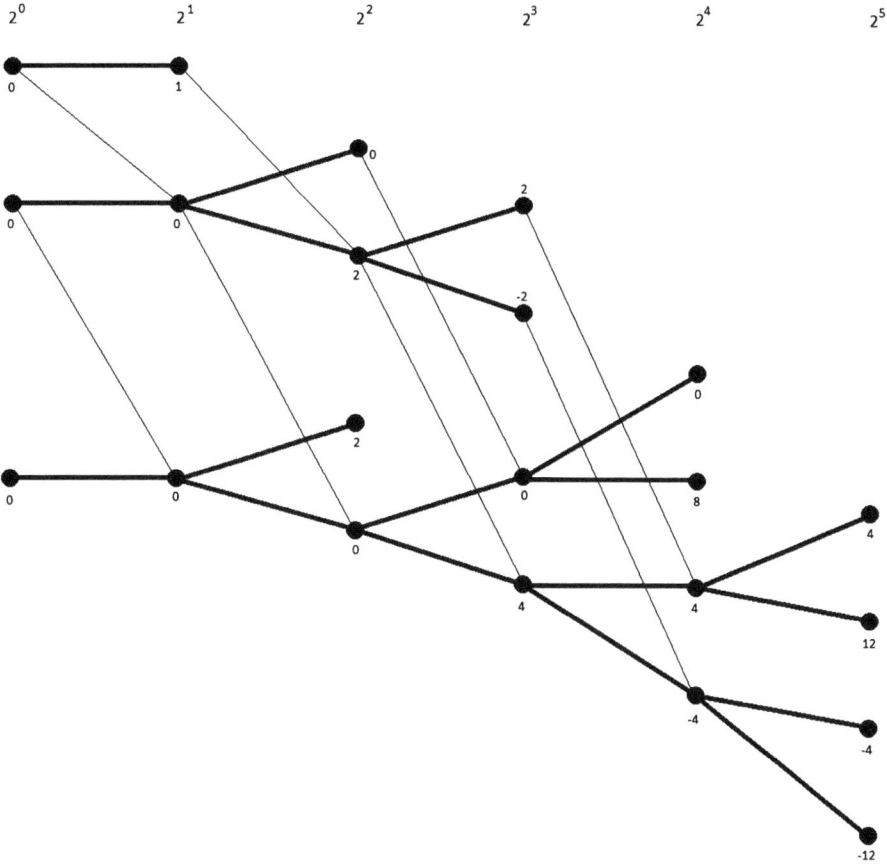

Figure 2.3: Seeding maps from $\mathcal{Q}_{12}(2^\infty)$ to $\mathcal{Q}_{48}(2^\infty)$ to $\mathcal{Q}_{192}(2^\infty)$.

The ideal number $\langle 1 : 0 \rangle$ in $Q_{\ell+1}$ is primitive and is not the image of an element of Q_ℓ under S. Since p divides the index of $\Delta_{\ell+1}$, which in turn divides the linear and constant coefficients of $P_{\ell+1}(x)$, then every element of $Q_{\ell+1}$ other than $\langle 1 : 0 \rangle$ can be written as $\langle p^{e+1} : pk \rangle$ for some $e \geq 0$ and integer k. In this case, $p^2 P_\ell(k) = P_{\ell+1}(pk) = p^{e+1}q$ for some integer q. Here, p divides q if and only if p^e divides $P_\ell(k)$, in which case $\langle p^e : k \rangle$ is an element of Q_ℓ sent to $\langle p^{e+1} : pk \rangle$ by S. If $q = pc$, then the content of $\langle p^e : k \rangle$ is $\gcd(p^e, b, c)$ where $b = P'_\ell(k)$, while that of $\langle p^{e+1} : pk \rangle$ is $\gcd(p^{e+1}, pb, pc) = p \cdot \gcd(p^e, b, c)$, not equal to 1. On the other hand, if p does not divide q, then the content of $\langle p^{e+1} : pk \rangle$ is the greatest common divisor of a collection of integers including p^{e+1} and q, so must equal 1. This completes the proof of statement (1).

Let $\langle p^e : k \rangle$ be an element of Q_ℓ, so that p^e divides $P_\ell(k)$. If t is an integer, then

$$P_{\ell+1}(pk + p^{e+1}t) = P_{\ell+1}(p(k + p^e t)) = p^2 P_\ell(k + p^e t).$$

But $P_\ell(k + p^e t) \equiv P_\ell(k)$ (mod p^e) since $k + p^e t \equiv k$ (mod p^e). It follows that p^{e+2} divides $P_{\ell+1}(pk + p^{e+1}t)$ for all t, so that $\langle p^{e+2} : pk + p^{e+1}t \rangle$ is an element of $Q_{\ell+1}$. Since $pk + p^{e+1}t \equiv pk + p^{e+1}s$ (mod p^{e+2}) if and only if $t \equiv s$ (mod p), there are precisely p distinct ideal numbers of this form in $Q_{\ell+1}$.

Finally, consider an element of $Q_{\ell+1}$ whose norm is at least p^2. Such an element can be expressed as $\langle p^{e+2} : pk \rangle$ for some integers $e \geq 0$ and k. But now p^{e+2} divides $P_{\ell+1}(pk) = p^2 P_\ell(k)$, so that $\langle p^e : k \rangle$ is an element of Q_ℓ. This shows that every ideal number of $Q_{\ell+1}$ other than $\langle 1 : 0 \rangle$ or $\langle p : 0 \rangle$ (the unique elements of norm 1 or p, respectively) is obtained from an ideal number of Q_ℓ as described above, and completes the proof of statement (2). $\qquad\square$

2.4 Counting congruence classes of ideal numbers

The seeding maps defined in Section 2.3 give us information about the number of elements of a set Q_Δ, or its subset of primitive elements, having a particular prime power norm. For example, Figure 2.2 implies that Q_{-315} has precisely six elements of norm 3^e for every $e \geq 3$, with four of them primitive. In this section, we provide formulas for similar counts with arbitrary discriminants and prime powers.

Recalling our notation, let p be a prime number and let $\Delta = \Delta(d, g)$ be a discriminant for which the p-level is zero (i. e., with p not dividing g). Then for every nonnegative integer ℓ, let $\Delta_\ell = \Delta(d, p^\ell g)$, a discriminant with p-level ℓ. We let $n_{\Delta_\ell}(p^e)$ and $m_{\Delta_\ell}(p^e)$ denote respectively the number of congruence classes and the number of primitive congruence classes of ideal numbers of discriminant Δ_ℓ and norm p^e. If p and Δ are clear from context, we abbreviate these numbers as $n(\ell, e)$ and $m(\ell, e)$, respectively. Note that $n(0, 1)$ equals $n_\Delta(p)$, the number of roots of $P(x)$, the principal polynomial of discriminant Δ, modulo p.

Propositions 2.2.4 and 2.2.5 and Theorem 2.2.6 compute the numbers $n(0, e)$ for all e, depending on whether $n_\Delta(p) = 0, 1$, or 2. Statement (2) in Theorem 2.3.1 provides the following general fact for nonnegative integers ℓ and e:

$$n(\ell + 1, 0) = 1, \quad n(\ell + 1, 1) = 1, \quad \text{and} \quad n(\ell + 1, e + 2) = p \cdot n(\ell, e). \tag{2.6}$$

For every prime number p and discriminant Δ_0 for which the p-level is zero, we can now calculate $n(\ell, e)$ for all $\ell \geq 0$ and $e \geq 0$. We will present our results first as tables depending on the three possible values of $n_\Delta(p)$. In all tables, the row and column headings are values of ℓ and e, respectively. Equations (2.6) show that all rows after the first begin with 1 and 1, and each subsequent entry is p times the entry two places to the left in the preceding row. (The tables are presented for $0 \leq \ell \leq 3$ and $0 \leq e \leq 8$, but could be extended indefinitely in both directions by following these patterns.)

1. If $n_\Delta(p) = 0$, then Proposition 2.2.4 implies that $n(0, e) = 0$ for all $e \geq 1$:

	0	1	2	3	4	5	6	7	8
0	1	0	0	0	0	0	0	0	0
1	1	1	p	0	0	0	0	0	0
2	1	1	p	p	p^2	0	0	0	0
3	1	1	p	p	p^2	p^2	p^3	0	0

2. If $n_\Delta(p) = 1$, then Proposition 2.2.5 implies that $n(0, e) = 0$ for all $e \geq 2$:

	0	1	2	3	4	5	6	7	8
0	1	1	0	0	0	0	0	0	0
1	1	1	p	p	0	0	0	0	0
2	1	1	p	p	p^2	p^2	0	0	0
3	1	1	p	p	p^2	p^2	p^3	p^3	0

3. If $n_\Delta(p) = 2$, then Theorem 2.2.6 implies that $n(0, e) = 2$ for all $e \geq 1$:

	0	1	2	3	4	5	6	7	8
0	1	2	2	2	2	2	2	2	2
1	1	1	p	$2p$	$2p$	$2p$	$2p$	$2p$	$2p$
2	1	1	p	p	p^2	$2p^2$	$2p^2$	$2p^2$	$2p^2$
3	1	1	p	p	p^2	p^2	p^3	$2p^3$	$2p^3$

We summarize these results in the following theorem.

Theorem 2.4.1. *Let p be a prime number and let $\Delta = \Delta(d, g)$ be a discriminant for which p does not divide g. Let $n_\Delta(p)$ be the number of elements in $\mathcal{Q}_\Delta$ having norm p. For each nonnegative integer ℓ, let $\Delta_\ell = \Delta(d, p^\ell g)$. If e is a nonnegative integer, let $n(\ell, e)$ denote the number of congruence classes of ideal numbers having discriminant Δ_ℓ and norm p^e. Then the following statements are true:*
1. *If $0 \leq e \leq 2\ell$ and $q = \lfloor \frac{e}{2} \rfloor$, then $n(\ell, e) = p^q$.*
2. *If $n_\Delta(p) = 2$ and $e > 2\ell$, then $n(\ell, e) = 2p^\ell$.*
3. *If $n_\Delta(p) = 1$, then $n(\ell, 2\ell + 1) = p^\ell$, and $n(\ell, e) = 0$ if $e > 2\ell + 1$.*
4. *If $n_\Delta(p) = 0$ and $e > 2\ell$, then $n(\ell, e) = 0$.*

Proof. The proof is by induction on ℓ. If $\ell = 0$, then the claims of Theorem 2.4.1 are consequences of Propositions 2.2.4 and 2.2.5 and Theorem 2.2.6. Suppose then that statements (1)–(4) have been established for some nonnegative integer ℓ and all nonnegative integers e. We show that the same statements are then true of $\ell + 1$. First, note that $n(\ell + 1, 0) = 1 = n(\ell + 1, 1)$ by equations (2.6). This is as claimed in statement (1), since $q = 0$ if $e = 0$ or $e = 1$, and both of these values are smaller than $2(\ell + 1)$. Now $n(\ell + 1, e + 2) = p \cdot n(\ell, e)$ for $e \geq 0$, again by equations (2.6). If $e \leq 2\ell$ and $q = \lfloor \frac{e}{2} \rfloor$, then $e + 2 \leq 2(\ell + 1)$ and $q + 1 = \lfloor \frac{e+2}{2} \rfloor$. Then $n(\ell + 1, e + 2) = p \cdot n(\ell, e) = p \cdot p^q = p^{q+1}$ is as claimed in statement (1). On the other hand, if $e > 2\ell$ so that $e + 2 > 2(\ell + 1)$, then

$$n(\ell + 1, e + 2) = p \cdot n(\ell, e)$$
$$= \begin{cases} p \cdot 2p^\ell = 2p^{\ell+1}, & \text{if } n_\Delta(p) = 2, \\ p \cdot p^\ell = p^{\ell+1}, & \text{if } n_\Delta(p) = 1 \text{ and } e = 2\ell + 1, \text{ so that } e + 2 = 2(\ell + 1) + 1 \\ p \cdot 0 = 0, & \text{if } n_\Delta(p) = 1 \text{ and } e + 2 > 2(\ell + 1) + 1, \text{ or if } n_\Delta(p) = 0. \end{cases}$$

Thus, Theorem 2.4.1 is true for all ℓ by induction. □

Recall from Proposition 2.2.3 that if the level of a prime p in a discriminant Δ is zero, then all ideal numbers of norm p^e in $\mathcal{Q}_\Delta$ are primitive. In the current notation, this means that $m(0, e) = n(0, e)$ for all $e \geq 0$. The ideal number $\langle 1 : 0 \rangle$ is always primitive, so that $m(\ell, 0) = n(\ell, 0) = 1$ for all $\ell \geq 0$. Finally, statement (1) of Theorem 2.3.1 implies that for all prime numbers p and discriminants Δ_i with $i > 0$, the subset of primitive elements in $\mathcal{Q}_{\Delta_i}(p^\infty)$ is the complement of the image of $\mathcal{Q}_{\Delta_{i-1}}(p^\infty)$ under the seeding map S. Since S is injective and takes elements of norm p^{e-1} to elements of norm p^e, we obtain the following general statement for positive integers ℓ and e:

$$m(\ell, e) = n(\ell, e) - n(\ell - 1, e - 1). \tag{2.7}$$

We will again present our results for primitive ideal numbers first as tables for $m(\ell, e)$, this time with $0 \leq \ell \leq 2$ and $0 \leq e \leq 6$. Each of the following m-tables is obtained from the corresponding n-table by leaving the row for $\ell = 0$ and the column for $e = 0$ unchanged, and replacing all other entries by the difference between that entry and the one in the preceding row and column.

1. If $n_\Delta(p) = 0$:

	0	1	2	3	4	5	6
0	1	0	0	0	0	0	0
1	1	0	p	0	0	0	0
2	1	0	$p-1$	0	p^2	0	0

2. If $n_\Delta(p) = 1$:

	0	1	2	3	4	5	6
0	1	1	0	0	0	0	0
1	1	0	$p-1$	p	0	0	0
2	1	0	$p-1$	0	$p^2 - p$	p^2	0

3. If $n_\Delta(p) = 2$:

	0	1	2	3	4	5	6
0	1	2	2	2	2	2	2
1	1	0	$p-2$	$2p-2$	$2p-2$	$2p-2$	$2p-2$
2	1	0	$p-1$	0	$p^2 - 2p$	$2p^p$	$2p^p$

The following theorem summarizes these results.

Theorem 2.4.2. *Let p be a prime number and let $\Delta = \Delta(d, g)$ be a discriminant for which p does not divide g. Let $n_\Delta(p)$ be the number of elements having norm p in $\mathcal{Q}_\Delta$. For each nonnegative integer ℓ, let $\Delta_\ell = \Delta(d, p^\ell g)$. If e is a nonnegative integer, let $n(\ell, e)$ and $m(\ell, e)$ denote the total number of congruence classes and the number of primitive congruence classes, respectively, of ideal numbers having discriminant Δ_ℓ and norm p^e. Then $m(0, e) = n(0, e)$ for all $e \geq 0$ and $m(\ell, 0) = 1$ for all $\ell \geq 0$. If ℓ and e are both positive, then the following statements are true:*

1. *If $0 < e < 2\ell$, then $m(\ell, e) = 0$ if e is odd and $m(\ell, e) = p^q - p^{q-1}$ if $e = 2q$.*
2. *If $e = 2\ell$, then $m(\ell, e) = p^{\ell-1}(p - n_\Delta(p))$.*
3. *If $n_\Delta(p) = 2$ and $e > 2\ell$, then $m(\ell, e) = 2p^\ell - 2p^{\ell-1}$.*
4. *If $n_\Delta(p) = 1$, then $m(\ell, 2\ell + 1) = p^\ell$, and $m(\ell, e) = 0$ if $e > 2\ell + 1$.*
5. *If $n_\Delta(p) = 0$ and $e > 2\ell$, then $m(\ell, e) = 0$.*

Proof. The claims about $m(0, e)$ and $m(\ell, 0)$ have already been noted, so we can assume that ℓ and e are both positive. In this case, we use equation (2.7) together with the formulas for $n(\ell, e)$ in Theorem 2.4.1.

(1) If $0 < e < 2\ell$, then $0 \leq e - 1 \leq 2(\ell - 1)$, so that statement (1) of Theorem 2.4.1 applies to both $n(\ell, e)$ and $n(\ell - 1, e - 1)$. If $e = 2q + 1$, then $\lfloor \frac{e}{2} \rfloor = q = \lfloor \frac{e-1}{2} \rfloor$, while if $e = 2q$, then $\lfloor \frac{e}{2} \rfloor = q$ and $\lfloor \frac{e-1}{2} \rfloor = q - 1$. Thus,

$$m(\ell, e) = n(\ell, e) - n(\ell - 1, e - 1) = \begin{cases} p^q - p^q = 0, & \text{if } e \text{ is odd,} \\ p^q - p^{q-1}, & \text{if } e = 2q. \end{cases}$$

(2) If $e = 2\ell$, then $e - 1 > 2(\ell - 1)$. Then statement (1) of Theorem 2.4.1 tells us that $n(\ell, e) = p^\ell$, while statement (2), (3), or (4) applies to $n(\ell - 1, e - 1)$. In each of these cases, we can write $n(\ell - 1, e - 1)$ as $n_\Delta(p) \cdot p^{\ell-1}$. Thus,

$$m(\ell, e) = p^\ell - np^{\ell-1} = p^{\ell-1}(p - n).$$

If $e \geq 2\ell+1$, then $e-1 \geq 2\ell > 2(\ell-1)$. Thus, statements (2), (3), and (4) of Theorem 2.4.1 apply to the remaining cases of this theorem.

(3) If $n_\Delta(p) = 2$ and $e > 2\ell$, then

$$m(\ell, e) = n(\ell, e) - n(\ell - 1, e - 1) = 2p^\ell - 2p^{\ell-1}.$$

(4) If $n_\Delta(p) = 1$ and $e = 2\ell + 1$, then $e - 1 = 2\ell > 2(\ell - 1) + 1$. In that case,

$$m(\ell, 2\ell + 1) = n(\ell, 2\ell + 1) - n(\ell - 1, 2\ell) = p^\ell - 0 = p^\ell.$$

On the other hand, if $e > 2\ell + 1$, then $e - 1 > 2(\ell - 1) + 1$ and so

$$m(\ell, e) = n(\ell, e) - n(\ell - 1, e - 1) = 0 - 0 = 0.$$

(5) If $n_\Delta(p) = 0$ and $e > 2\ell$, then

$$m(\ell, e) = n(\ell, e) - n(\ell - 1, e - 1) = 0 - 0 = 0.$$

This establishes all claims of Theorem 2.4.2. $\square$

3 Composition of ideal numbers

In Chapter 2, we defined an equivalence relation of congruence on ideal numbers of a fixed discriminant so that congruence classes can be identified with ideals in a particular quadratic domain. Ideals can be multiplied in a natural way that is defined and described in Appendix C. In this chapter, we apply that definition to a certain subset of congruence classes of each discriminant. In this context, we refer to this operation as *composition*, defined purely in terms of ideal numbers in Section 3.1. We will demonstrate that, in many cases, this composition operation can be performed by the same techniques used to solve quadratic congruences modulo composite integers.

The subset of *primitive* congruences classes of ideal numbers forms a group under composition, which we describe in Chapter 3. We will see that it suffices to calculate subgroups of these congruence classes whose norm is a power of a prime number. The seeding maps introduced in Chapter 2 give us useful information about these subgroups, as we will demonstrate in the concluding sections of Chapter 3.

3.1 Composition of congruence classes

Recall that we denote the set of all congruence classes of ideal numbers of discriminant Δ as $\mathcal{Q}_\Delta$. We write the typical element of $\mathcal{Q}_\Delta$ as $\langle a : k \rangle$ with a positive, where it is understood that $\langle a : k \rangle = \langle a : \ell \rangle$ if and only if $k \equiv \ell \pmod{a}$. In Proposition 2.1.3, we saw that congruent ideal numbers have the same content. Thus, the property of a congruence class of ideal numbers being or not being *primitive* is well-defined. We denote the subset of primitive elements of $\mathcal{Q}_\Delta$ as $\mathcal{G}_\Delta$ and define the following operation on $\mathcal{G}_\Delta$.

Definition. Let $P(x)$ be the principal polynomial of discriminant Δ and let $\langle a : k \rangle$ and $\langle b : \ell \rangle$ be elements of $\mathcal{G}_\Delta$. Let $u = k + \ell + P'(0)$ and $v = k\ell - P(0)$, and let $g = \gcd(a, b, u) = ar + bs + ut$ for some integers r, s, and t. Then define $\langle a : k \rangle \circ \langle b : \ell \rangle$ to equal $\langle c : m \rangle$, where

$$c = \frac{a}{g} \cdot \frac{b}{g} \quad \text{and} \quad m = \frac{a\ell r + bks + vt}{g}. \tag{3.1}$$

We refer to $\langle c : m \rangle$ as the *composite* of $\langle a : k \rangle$ and $\langle b : \ell \rangle$ and to this operation on $\mathcal{G}_\Delta$ as *composition*.

Example. Let $\Delta = \Delta(-35, 1) = -35$, with $P(x) = x^2 + x + 9$ as its principal polynomial. Since Δ is primitive, all elements of $\mathcal{Q}_\Delta$ are primitive. We see, for instance, that $\langle 9 : -1 \rangle$ and $\langle 15 : -3 \rangle$ are elements of $\mathcal{G}_{-35}$, for which

$$u = k + \ell + P'(0) = -1 - 3 + 1 = -3 \quad \text{and} \quad v = k\ell - P(0) = (-1)(-3) - 9 = -6.$$

Here, $g = \gcd(9, 15, -3) = 3 = 9r + 15s - 3t$ with $r = 0$, $s = 0$, and $t = -1$, among other possibilities. Thus,

https://doi.org/10.1515/9783111319360-003

$$c = \frac{9}{3} \cdot \frac{15}{3} = 15 \quad \text{and} \quad m = \frac{(9 \cdot -3 \cdot 0) + (15 \cdot -1 \cdot 0) + (-6 \cdot -1)}{3} = 2,$$

so that $\langle 9 : -1 \rangle \circ \langle 15 : -3 \rangle = \langle 15 : 2 \rangle$.

On the other hand, if we want the composite of $\langle 9 : -1 \rangle$ and $\langle 15 : 2 \rangle$, we begin with $u = -1 + 2 + 1 = 2$ and $v = -1 \cdot 2 - 0 = -11$. Now $g = \gcd(9, 15, 2) = 1 = 9r + 15s + 2t$ for $r = 2$, $s = -1$, and $t = -1$, again among many other possibilities. Then

$$c = 9 \cdot 15 = 135 \quad \text{and} \quad m = (9 \cdot 2 \cdot 2) + (15 \cdot -1 \cdot -1) + (-11 \cdot -1) = 62,$$

and so $\langle 9 : -1 \rangle \circ \langle 15 : 2 \rangle = \langle 135 : 62 \rangle$. (Note that $P(62) = 135 \cdot 29$, confirming that $\langle 135 : 62 \rangle$ is an element of $\mathcal{G}_{-35}$.)

The formula for composition is derived from an operation of multiplication of ideals in quadratic domains. The details of this definition are provided in Appendix C. In particular, it is shown in Corollary C.7.2 that composition does not depend on the various choices that are made in its definition, and in Theorem C.7.1 that $\mathcal{G}_\Delta$ is closed under composition and has the properties of an Abelian group. We will call $\mathcal{G}_\Delta$ under composition the *primitive ideal group* of discriminant Δ. In this section, we develop some general properties of ideal number composition in this group. We first verify the existence of an identity element and of the inverse of a given element.

Proposition 3.1.1. *Let $\mathcal{G}_\Delta$ be the set of congruence classes of primitive ideal numbers of discriminant Δ. Then $\langle 1 : 0 \rangle$ is an identity element for composition in $\mathcal{G}_\Delta$ and every $\langle a : k \rangle$ in $\mathcal{G}_\Delta$ has an inverse under composition, namely its conjugate $\langle a : \overline{k} \rangle$.*

Proof. We will assume the commutative property of composition in both parts. Note also that the content of $\langle 1 : 0 \rangle$ is always 1, and that $\text{cont}\langle a : k \rangle = \text{cont}\langle a : \overline{k} \rangle$. For the composite $\langle a : k \rangle \circ \langle 1 : 0 \rangle$, we have $g = \gcd(a, 1, u) = 1 = a(0) + 1(1) + u(0)$, no matter what u is. Thus, we find that $\langle a : k \rangle \circ \langle 1 : 0 \rangle = \langle a : k \rangle$, directly from the formula in (3.1).

Recalling that $\overline{k} = -k - P'(0)$, then for the composite $\langle a : k \rangle \circ \langle a : \overline{k} \rangle$ we have $u = k + \overline{k} + P'(0) = 0$. Thus, $g = \gcd(a, a, 0) = a$, and so the norm of $\langle a : k \rangle \circ \langle a : \overline{k} \rangle$ is $\frac{a}{a} \cdot \frac{a}{a} = 1$. All integers are congruent to 0 modulo 1, hence $\langle a : k \rangle \circ \langle a : \overline{k} \rangle = \langle 1 : 0 \rangle$. $\square$

While the formula for composition in equation (3.1) seems rather unwieldy and not intuitive, there are many situations in which this operation can be carried out using more familiar calculations. We illustrate this with an example after the following proposition.

Proposition 3.1.2. *Let $P(x)$ be the principal polynomial of discriminant Δ. Let $\langle a : k \rangle$ and $\langle b : \ell \rangle$ be elements of $\mathcal{G}_\Delta$, with $u = k + \ell + P'(0)$, and suppose that $\gcd(a, b, u) = 1$. Then*

$$\langle a : k \rangle \circ \langle b : \ell \rangle = \langle ab : m \rangle$$

where m satisfies $m \equiv k \pmod{a}$ and $m \equiv \ell \pmod{b}$.

Proof. Since $P(x) = x^2 + P'(0)x + P(0)$ for all x, we have

$$uk - P(k) = k^2 + k\ell + P'(0)k - k^2 - P'(0)k - P(0) = k\ell - P(0).$$

A similar calculation shows that $u\ell - P(\ell) = k\ell - P(0)$, so we have the following alternative expressions for v, as it is given in the definition of ideal number composition:

$$v = k\ell - P(0), \quad v = uk - P(k), \quad v = u\ell - P(\ell). \tag{3.2}$$

If $\gcd(a, b, u) = 1 = ar + bs + ut$, then from the formula for ideal number composition in (3.1), we have $c = ab$ and $m = a\ell r + bks + vt$. Substituting the second expression in (3.2) for v, we have

$$m = a\ell r + bks + (uk - P(k))t = a\ell r - P(k)t + (bs + ut)k \equiv k \pmod{a},$$

since a divides $P(k)$ and $1 = ar + bs + ut \equiv bs + ut \pmod{a}$. Similarly, substituting the third expression in (3.2) for v,

$$m = a\ell r + bks + (u\ell - P(\ell))t = (ar + ut)\ell + bks - P(\ell)t \equiv \ell \pmod{b}.$$

Thus, m satisfies $m \equiv k \pmod{a}$ and $m \equiv \ell \pmod{b}$. $\qquad\square$

In particular, Proposition 3.1.2 applies when $\gcd(a, b) = 1$, in which case the solution of $m \equiv k \pmod{a}$ and $m \equiv \ell \pmod{b}$ is uniquely determined modulo ab. If $ar + bs = 1$, we can write $\langle a : k \rangle \circ \langle b : \ell \rangle = \langle ab : m \rangle$ where $m = bsk + ar\ell$. Thus, we can compose ideal numbers of relatively prime norm as an application of the Euclidean algorithm, as in the following example.

Example. To calculate $\langle 11 : 1 \rangle \circ \langle 13 : 5 \rangle$ in $\mathcal{G}_{-35}$, we apply the Euclidean algorithm to write

$$1 = \gcd(11, 13) = 11(6) + 13(-5).$$

The ideal number composite equals $\langle c : m \rangle$ where $c = 11 \cdot 13 = 143$ and $m = 13 \cdot -5 \cdot 1 + 11 \cdot 6 \cdot 5 = 265$. That is, $\langle 11 : 1 \rangle \circ \langle 13 : 5 \rangle = \langle 143 : 265 \rangle = \langle 143 : -21 \rangle$.

We can also reverse this composition as follows.

Proposition 3.1.3. *Let Δ be a discriminant. If $\langle c : m \rangle$ is an element of $\mathcal{G}_\Delta$ and $c = ab$ with $\gcd(a, b) = 1$, then $\langle a : m \rangle$ and $\langle b : m \rangle$ are also elements of $\mathcal{G}_\Delta$, and we can write $\langle c : m \rangle = \langle a : m \rangle \circ \langle b : m \rangle$.*

Proof. Let $P(x)$ be the principal polynomial of discriminant Δ. If $\langle c : m \rangle$ is a primitive element of $\mathcal{Q}_\Delta$, then we can write $P(m) = cx$ and $P'(m) = y$, with $\gcd(c, y, x) = 1$. Now $P(m) = a(bx) = b(ax)$, so that $\langle a : m \rangle$ and $\langle b : m \rangle$ are also elements of $\mathcal{Q}_\Delta$. The content of $\langle a : m \rangle$ is $\gcd(a, y, bx)$. Since $\gcd(a, b) = 1$, we can see that any prime common divisor of a, y, and bx would also divide c, y, and x. Thus, $\langle a : m \rangle$ is primitive, as is $\langle b : m \rangle$ by a similar observation. But then

$$\langle a : m \rangle \circ \langle b : m \rangle = \langle c : m \rangle,$$

immediately from Proposition 3.1.2. $\qquad\qquad\qquad\qquad\qquad\qquad\qquad\qquad\qquad\qquad\Box$

Thus, if $a = p_1^{e_1} \cdot p_2^{e_2} \cdots p_t^{e_t}$ is the prime factorization of a positive integer a into distinct prime powers, then we can *decompose* a congruence class $\langle a : k \rangle$ as follows:

$$\langle a : k \rangle = \langle p_1^{e_1} : k \rangle \circ \langle p_2^{e_2} : k \rangle \circ \cdots \circ \langle p_t^{e_t} : k \rangle.$$

However, the following example shows that the condition that $\gcd(a, b) = 1$ is necessary, in general, for decomposition of $\langle ab : m \rangle$.

Example. Let $\Delta = \Delta(-35, 3) = -315$, with $P(x) = x^2 + 3x + 81$ as its principal polynomial, and consider $\langle 9 : 3 \rangle$. Here, $P(3) = 99 = 9 \cdot 11$ and $P'(3) = 9$, so that the content of $\langle 9 : 3 \rangle$ is $\gcd(9, 9, 11) = 1$. However, we cannot write $\langle 9 : 3 \rangle$ as $\langle 3 : 3 \rangle \circ \langle 3 : 3 \rangle$. Although $\langle 3 : 3 \rangle$ is an element of $\mathcal{Q}_{-315}$, it is not primitive, as its content is $\gcd(3, 9, 33) = 3$.

In an example at the conclusion of Section C.6, it is illustrated why we should not regard $\langle 3 : 3 \rangle \circ \langle 3 : 3 \rangle$ as equal to $\langle 9 : 3 \rangle$, even if the definition of composition were extended to all ideal numbers.

Proposition 3.1.3 implies the following decomposition in each group $\mathcal{G}_\Delta$.

Proposition 3.1.4. *Let $\mathcal{G}_\Delta$ be the primitive ideal group of discriminant Δ. If p is a prime number, let*

$$H = \{\langle a : k \rangle \mid a = p^e \text{ for some } e \geq 0\} \quad and \quad K = \{\langle a : k \rangle \mid \gcd(a, p) = 1\}.$$

Then $\mathcal{G}_\Delta$ is the internal direct product of the subgroups H and K.

Proof. Both H and K contain the identity element $\langle 1 : 0 \rangle$ of $\mathcal{G}_\Delta$. Since a primitive ideal number and its inverse (i. e., conjugate) have the same norm, we see that H and K both contain the inverse of each of their elements. Finally, if $\langle a : k \rangle \circ \langle b : \ell \rangle = \langle c : m \rangle$, then c is a divisor of ab, as seen in equations (3.1). Thus, if a and b are powers of p, then c is also a power of p. Likewise, if a and b are relatively prime to p, then c is relatively prime to p. Thus, H and K are both subgroups of $\mathcal{G}_\Delta$.

Since $\mathcal{G}_\Delta$ is Abelian, it now suffices to show that $H \cap K$ contains only the identity element and that $HK = \mathcal{G}_\Delta$. The first claim is true since 1 is the only power of p that is relatively prime to p. For the second claim, if $\langle b : k \rangle$ is an element of $\mathcal{G}_\Delta$, and we write $b = p^e a$ with $\gcd(a, p) = 1$, then $\langle b : k \rangle = \langle p^e : k \rangle \circ \langle a : k \rangle$, an element of HK, by Proposition 3.1.3. Thus, $\mathcal{G}_\Delta$ is the internal direct product of H and K by definition. $\qquad\Box$

If Δ is a discriminant and p is a prime number, recall that $\mathcal{Q}_\Delta(p^\infty)$ denotes the set of all congruence classes of ideal numbers of discriminant Δ whose norm is a power of p. We will similarly write $\mathcal{G}_\Delta(p^\infty)$ for the subset of primitive elements in $\mathcal{Q}_\Delta(p^\infty)$, that is, for each of the subgroups H as defined in Proposition 3.1.4. We refer to $\mathcal{G}_\Delta(p^\infty)$ as a *prime subgroup*, or as the *p-prime subgroup*, of discriminant Δ.

3.2 Prime subgroups of level zero

Proposition 3.1.4 shows that to describe a primitive ideal group $\mathcal{G}_\Delta$, it suffices to determine each of its prime subgroups. We do so in the remainder of Chapter 3, following the approach that we took to describe the elements of these sets in Chapter 2. In this section, we determine the group structure of $\mathcal{G}_\Delta(p^\infty)$ when the level of p in Δ is zero, that is, when $\Delta = \Delta(d, g)$ with g not divisible by p. We use seeding polynomials to build from this case to arbitrary discriminants at the conclusion of this chapter.

Recall that when $\mathrm{lev}_p(\Delta) = 0$, then every ideal number of discriminant Δ and norm p^e is primitive, so that $\mathcal{Q}_\Delta(p^\infty) = \mathcal{G}_\Delta(p^\infty)$. In Section 2.2, we determined the elements of $\mathcal{Q}_\Delta(p^\infty)$ under the assumption that $\mathrm{lev}_p(\Delta) = 0$. We describe the resulting groups in the following theorem.

Theorem 3.2.1. *Let $\Delta = \Delta(d, g)$ be a discriminant and let p be a prime number that does not divide g. Let $n_\Delta(p)$ equal the number of elements of norm p in $\mathcal{Q}_\Delta$. Then the following statements are true:*
1. *If $n_\Delta(p) = 0$, then $\mathcal{G}_\Delta(p^\infty)$ is a trivial group.*
2. *If $n_\Delta(p) = 1$, then $\mathcal{G}_\Delta(p^\infty)$ is a cyclic group of order two.*
3. *If $n_\Delta(p) = 2$, then $\mathcal{G}_\Delta(p^\infty)$ is an infinite cyclic group.*

Proof. Case (1) is immediate from Proposition 2.2.4, which shows that

$$\mathcal{Q}_\Delta(p^\infty) = \mathcal{G}_\Delta(p^\infty) = \{\langle 1 : 0 \rangle\}$$

when $n_\Delta(p) = 0$. Case (2) is likewise a consequence of Proposition 2.2.5, together with the observation that $\langle p : k \rangle$ is its own conjugate, and hence its own inverse under composition, when $n_\Delta(p) = 1$.

For case (3), let $P(x)$ be the principal polynomial of discriminant Δ. Assuming that $n_\Delta(p) = 2$, let k be one of the two roots of $P(x)$ modulo p. Let $k_1 = k$ and define k_n for all integers n as in Theorem 2.2.6, so that

$$\mathcal{Q}_\Delta(p^\infty) = \mathcal{G}_\Delta(p^\infty) = \{\langle p^{|n|} : k_n \rangle \mid n \in \mathbb{Z}\}.$$

We will show that $\langle p^n : k_n \rangle = \langle p : k \rangle^n$, the composite of n copies of $\langle p : k \rangle$, for every *positive* integer n. This is sufficient to show that $\mathcal{G}_\Delta(p^\infty)$ equals the cyclic subgroup generated by $\langle p : k \rangle$, as k_{-n} is the conjugate of k_n with respect to Δ, meaning that

$\langle p^n : k_{-n} \rangle$ is the inverse of $\langle p^n : k_n \rangle$ under composition for all $n \geq 1$. (It is also true that $\langle p^0 : k_0 \rangle = \langle 1 : 0 \rangle = \langle p : k \rangle^0$, since the zero exponent of a group element is defined to be the identity element of the group.)

It is clear that $\langle p^1 : k_1 \rangle = \langle p : k \rangle^1$. Suppose that $\langle p^n : k_n \rangle = \langle p : k \rangle^n$ for some positive integer n. To simplify notation, let $\ell = k_n$. We can then rewrite $\langle p : k \rangle$ as $\langle p : \ell \rangle$, since $k_n \equiv k \pmod{p}$ for every positive integer n. Below we apply the composition formula to $\langle p^n : \ell \rangle \circ \langle p : \ell \rangle$.

Let $u = \ell + \ell + P'(0) = P'(\ell)$ and write $v = u\ell - P(\ell)$, as in equation (3.2). Here, p^n divides $P(\ell)$ by assumption. We cannot also have $P'(\ell)$ a multiple of p, since then p would divide $\Delta = P'(\ell)^2 - 4P(\ell)$, which is not the case when $n_\Delta(p) = 2$. Thus, we can write

$$g = \gcd(p^n, p, u) = 1 = p^n(0) + ps + ut$$

for some integers s and t. Therefore, $\langle p^n : \ell \rangle \circ \langle p : \ell \rangle = \langle p^{n+1} : m \rangle$, where

$$m = p\ell s + vt = p\ell s + (u\ell - P(\ell))t = (ps + ut)\ell - P(\ell)t = \ell - P(\ell)t.$$

Finally, observe that t is an integer for which $P'(k)t \equiv P'(\ell)t \equiv ut \equiv 1 \pmod{p}$. Thus, $m = k_{n+1}$, as that value is defined in Theorem 2.2.6. We conclude that

$$\langle p : k \rangle^n = \langle p^n : k_n \rangle$$

for all positive integers n by induction, and so $\mathcal{G}_\Delta(p^\infty)$ is an infinite cyclic group, generated by $\langle p : k \rangle$. □

As we saw in Section 2.2, Theorem 2.2.6 provides a step-by-step algorithm for constructing $\langle p : k \rangle^n = \langle p^n : k_n \rangle$ for every positive integer n, which can then be extended to negative integers via conjugation. The composition formulas of (3.1) may provide a quicker alternative in some cases, as illustrated in the following example.

Example. Let $\Delta = -35$, with principal polynomial $P(x) = x^2 + x + 9$. Suppose we are interested in calculating the ideal number $\langle 3 : 0 \rangle^7$. We can do so in three steps as follows:
1. For $\langle 3 : 0 \rangle \circ \langle 3 : 0 \rangle$, we have $u = 0 + 0 + 1 = 1$, $v = 0 \cdot 0 - 9 = -9$, and

$$g = \gcd(3, 3, 1) = 1 = 3(0) + 3(0) + 1(1).$$

So then, $c = 3 \cdot 3 = 9$ and $m = v(1) = -9$. We conclude that

$$\langle 3 : 0 \rangle^2 = \langle 9 : -9 \rangle = \langle 9 : 0 \rangle.$$

2. Now $\langle 3 : 0 \rangle^4 = \langle 3 : 0 \rangle^2 \circ \langle 3 : 0 \rangle^2 = \langle 9 : 0 \rangle \cdot \langle 9 : 0 \rangle$. Again, $u = 1$, $v = -9$, and

$$g = \gcd(9, 9, 1) = 1 = 9(0) + 9(0) + 1(1).$$

Thus, $c = 9 \cdot 9 = 81$ and $m = v(1) = -9$, and we conclude that

$$\langle 3 : 0 \rangle^4 = \langle 81 : -9 \rangle.$$

3. Finally, we can calculate $\langle 3 : 0 \rangle^8$ as $\langle 81 : -9 \rangle \circ \langle 81 : -9 \rangle$. Here, $u = -17$, $v = 72$, and

$$g = \gcd(81, 81, -17) = 1 = 81(4) + 81(0) + (-17)(19),$$

using the Euclidean algorithm. Therefore, $c = 81 \cdot 81 = 6561$ and $m = (81 \cdot 4 \cdot -9) + (72 \cdot 19) = -1548$, so that

$$\langle 3 : 0 \rangle^8 = \langle 6561 : -1548 \rangle.$$

Now we can write $\langle 3 : 0 \rangle^i$ as $\langle 3^i : -1548 \rangle$ for $1 \le i \le 8$. In particular,

$$\langle 3 : 0 \rangle^7 = \langle 2187 : -1548 \rangle = \langle 2187 : 639 \rangle.$$

These computations confirm results from a previous example illustrating the algorithm of Theorem 2.2.6.

3.3 Prime kernels of a discriminant

In the remainder of Chapter 3, we turn our attention to prime subgroups, $\mathcal{G}_\Delta(p^\infty)$, when the p-level of Δ is positive. We will build to our main results by introducing certain subgroups of $\mathcal{G}_\Delta(p^\infty)$ that help to describe the full group.

The details of the construction of prime subgroups of positive level will not be required in later chapters of the text. Thus the remainder of Chapter 3 can be regarded as optional.

We begin the description of prime subgroups of positive level by defining a group of matrices that we can associate to every discriminant and every prime number. If p is a prime number, then there is a field $\mathbb{F}_p$, unique up to isomorphism, containing precisely p elements. In this setting, if p is odd, we will often write

$$\mathbb{F}_p = \left\{ 0, 1, -1, 2, -2, \ldots, \frac{p-1}{2}, -\frac{p-1}{2} \right\},$$

with addition and multiplication defined modulo p. The set of *invertible* 2×2 matrices with entries in $\mathbb{F}_p$, written as $GL_2(\mathbb{F}_p)$, is a group under matrix multiplication. It is straightforward to check that the set $N = \{ aI_2 \mid a \in \mathbb{F}_p^* \}$ of nonzero scalar matrices is a normal subgroup of $GL_2(\mathbb{F}_p)$. We write G for the quotient group $GL_2(\mathbb{F}_p)/N$, that is,

$$G = \left\{ \begin{bmatrix} a & b \\ c & d \end{bmatrix} \Big| \ a, b, c, d \in \mathbb{F}_p \text{ and } ad - bc \ne 0 \text{ in } \mathbb{F}_p \right\}, \tag{3.3}$$

under matrix multiplication, with the understanding that two matrices in G are considered equal if one is a nonzero scalar multiple of the other.

Definition. Let p be a prime number and let G be the group defined above. Let Δ be a discriminant with principal polynomial $P(x)$. For each k in $\mathbb{F}_p$, let

$$A_k = \begin{bmatrix} k & -P(0) \\ 1 & k + P'(0) \end{bmatrix}. \tag{3.4}$$

Then let

$$\mathcal{K}_\Delta(p) = \{I_2\} \cup \{A_k \mid k \in \mathbb{F}_p \text{ and } P(k) \neq 0 \text{ in } \mathbb{F}_p\}. \tag{3.5}$$

We call $\mathcal{K}_\Delta(p)$ a *prime kernel* of the discriminant Δ, or say that it is the *p-kernel* of Δ.

Note that the determinant of A_k is $k^2 + P'(0)k + P(0) = P(k)$, so the requirement that $P(k) \neq 0$ in $\mathbb{F}_p$ ensures that A_k is invertible. Thus, $\mathcal{K}_\Delta(p)$, as defined in equation (3.5), is a subset of the group G given in equation (3.3). We take this further in the following proposition.

Proposition 3.3.1. *Let Δ be a discriminant with principal polynomial $P(x)$ and let p be a prime number. Then $\mathcal{K}_\Delta(p)$, the p-kernel of Δ defined in equation (3.5), is an Abelian group under matrix multiplication (modulo the subgroup N of nonzero scalar matrices). The order of this group is $|\mathcal{K}_\Delta(p)| = (p + 1) - n_\Delta(p)$, where $n_\Delta(p)$ is the number of solutions of $P(x) \equiv 0 \pmod{p}$.*

Proof. There are p choices for k in $\mathbb{F}_p$, together with the identity matrix, so a maximum of $p + 1$ elements in $\mathcal{K}_\Delta(p)$. However, as noted above, we must eliminate all choices of k for which $P(k) = 0$ in $\mathbb{F}_p$. By definition, there are $n_\Delta(p)$ such elements. We obtain the formula for $|\mathcal{K}_\Delta(p)|$ given in this proposition.

To show that $\mathcal{K}_\Delta(p)$ is a group, it suffices to show that $\mathcal{K}_\Delta(p)$ is closed under the operation of matrix multiplication modulo N. Let k and ℓ be elements of $\mathbb{F}_p$ for which $P(k) \neq 0$ and $P(\ell) \neq 0$ in $\mathbb{F}_p$. Let $u = k + \ell + P'(0)$ and let $v = k\ell - P(0)$ in $\mathbb{F}_p$. Then we find that

$$\begin{aligned}
A_k \cdot A_\ell &= \begin{bmatrix} k & -P(0) \\ 1 & k + P'(0) \end{bmatrix} \cdot \begin{bmatrix} \ell & -P(0) \\ 1 & \ell + P'(0) \end{bmatrix} \\
&= \begin{bmatrix} k\ell - P(0) & -P(0)(k + \ell + P'(0)) \\ k + \ell + P'(0) & (k\ell - P(0)) + P'(0)(k + \ell + P'(0)) \end{bmatrix} \\
&= \begin{bmatrix} v & -P(0)u \\ u & v + P'(0)u \end{bmatrix}.
\end{aligned}$$

If $u = 0$, then $A_k \cdot A_\ell$ is a nonzero scalar multiple of the identity matrix. (We can be sure that v is not also 0 since G is closed under matrix multiplication or from the observation that $v = uk - P(k)$, as seen in equation (3.2).) If $u \neq 0$ in $\mathbb{F}_p$, then u has an inverse in $\mathbb{F}_p$, and we find that $A_k \cdot A_\ell = A_{u^{-1}v}$. (The closure of G again ensures that $P(u^{-1}v) \neq 0$, so that this product is in $\mathcal{K}_\Delta(p)$.) Finally, since the calculations of u and v do not depend on the order of in which k and ℓ appear, then $A_k \cdot A_\ell = A_\ell \cdot A_k$ in every case. □

Notation. In a prime kernel $\mathcal{K}_\Delta(p)$, we will write the identity matrix I_2 as A_∞. We may shorten notation by simply writing the subscripts of the matrices A_k as the elements of $\mathcal{K}_\Delta(p)$. If so, we will write the operation of $\mathcal{K}_\Delta(p)$ as $*$ to avoid confusion with the multiplication operation of $\mathbb{F}_p$. Hence, the identity element of each group $\mathcal{K}_\Delta(p)$ is ∞. The inverse of a nonidentity element k is $\overline{k}$, the conjugate of k with respect to Δ. (If $\ell = -k - P'(0)$, then $u = k + \ell + P'(0) = 0$.) If ℓ is neither ∞ nor $\overline{k}$, then $k * \ell = u^{-1}v$, where $u = k+\ell+P'(0)$ and $v = k\ell-P(0)$. We can also write $k*\ell = k-u^{-1}P(k) = \ell-u^{-1}P(\ell)$, using the alternative expressions for v in equation (3.2).

Example. Let $\Delta = -15$, with principal polynomial $P(x) = x^2 + x + 4$, and let $p = 5$. We find that $P(x) \equiv 0 \pmod 5$ has one solution, $x = 2$, and so

$$\mathcal{K}_{-15}(5) = \{\infty, 0, -1, 1, -2\}.$$

A group with five elements is necessarily cyclic, with any nonidentity element as a generator. If $k = 0$, for instance, then for every ℓ, we have $u = \ell+1$ and $v = -4 = 1$ in $\mathbb{F}_5$. So, $0 * \ell = (\ell + 1)^{-1}$ if $\ell \neq -1$. Thus, we verify that

- $0^2 = 0 * 0 = 1^{-1} = 1$
- $0^3 = 0 * 1 = 2^{-1} = -2$
- $0^4 = 0 * (-2) = (-1)^{-1} = -1$
- $0^5 = 0 * (-1) = \infty$

Example. If $\Delta = -15$ and $p = 7$, we find that $(\frac{\Delta}{p}) = (\frac{-15}{7}) = -1$, so that $P(x) \equiv 0 \pmod 7$ has no solutions. Thus,

$$\mathcal{K}_{-15}(7) = \{\infty, 0, -1, 1, -2, 2, -3, 3\},$$

with eight elements. Note that 3 is the only element of order two in this group, that is, the only number that equals its conjugate, -4, in $\mathbb{F}_7$. Thus, $\mathcal{K}_{-15}(7)$ is cyclic. Trial-and-error shows that $k = 1$ generates this group, with

$$1, \quad -1, \quad 2, \quad 3, \quad -3, \quad 0, \quad -2, \quad \infty$$

the sequence of powers of k.

3.4 The kernel of a prime subgroup

In this section, we apply the prime kernels of a discriminant defined in Section 3.3 to groups of ideal numbers, particularly a subgroup that can be formed in a p-prime subgroup $\mathcal{G}_\Delta(p^\infty)$ when the level of p in Δ is positive. We will fix the notation used in the following proposition throughout this section.

Proposition 3.4.1. *Let p be a prime number, let Δ be a discriminant, and let $\Delta_1 = p^2\Delta$. Let $G = \mathcal{G}_\Delta(p^\infty)$ and $G_1 = \mathcal{G}_{\Delta_1}(p^\infty)$ be the p-prime subgroups of discriminant Δ and Δ_1, respectively. Then, aside from the identity element $\langle 1 : 0 \rangle$, all elements of G_1 can be written as $\langle p^{e+2} : pk \rangle$ for some integer $e \geq 0$ and integer k. In this case, $\langle p^e : k \rangle$ is an element of G.*

Proof. We can write the principal polynomials of discriminant Δ and Δ_1, respectively as

$$P(x) = x^2 + sx + t \quad \text{and} \quad P_1(x) = x^2 + psx + p^2t, \tag{3.6}$$

for some integers s and t. Then p divides $P_1(\ell)$ if and only if p divides ℓ. If $\ell = pk$, then $P_1(\ell) = p^2c$ for some integer c, and $P_1'(\ell) = pb$ for $b = 2k + s$. Thus, a *primitive* ideal number of discriminant Δ_1 whose norm is a positive power of p has the form $\langle p^{e+2} : pk \rangle$. (That is, its norm must be divisible by p^2.)

In this case, p^{e+2} divides $P_1(pk) = p^2P(k)$, so that p^e divides $P(k)$. But then $\langle p^e : k \rangle$ is an ideal number of discriminant Δ. If $P_1(pk) = p^{e+2}c$ and $P_1'(pk) = pb$, then the content of $\langle p^{e+2} : pk \rangle$ is $\gcd(p^{e+2}, pb, c)$, while the content of $\langle p^e : k \rangle$ is $\gcd(p^e, b, c)$. If $\langle p^{e+2} : pk \rangle$ is primitive, then p does not divide c, and so $\langle p^e : k \rangle$ is also primitive. ☐

Proposition 3.4.2. *Let Δ be a discriminant and let $\Delta_1 = p^2\Delta$ for some prime number p, with G_1 the p-prime subgroup of discriminant Δ_1. Then the set*

$$K = \{\langle p^e : \ell \rangle \in G_1 \mid e \leq 2\} \tag{3.7}$$

is a subgroup of G_1, and is isomorphic to $\mathcal{K}_\Delta(p)$, the p-kernel of Δ as defined in equation (3.5).

Proof. By Proposition 3.4.1, every element of K, aside from the identity element $\langle 1 : 0 \rangle$, can be written as $\langle p^2 : pk \rangle$. Note that $\langle p^2 : pk \rangle = \langle p^2 : p\ell \rangle$ if and only if $k \equiv \ell \pmod{p}$. Thus, counting the identity element, there are no more than $p + 1$ elements in K. Let $P(x)$ and $P_1(x)$ be the principal polynomials of discriminant Δ and Δ_1, respectively, as expressed in equation (3.6). Here, $P_1(pk) = p^2P(k)$ and $P_1'(pk) = pP'(k)$, and the content of $\langle p^2 : pk \rangle$ is $\gcd(p^2, pP'(k), P(k))$. Thus, the ideal number $\langle p^2 : pk \rangle$ is primitive if and only if p does not divide $P(k)$, and so the number of elements in K is $(p + 1) - n_\Delta(p)$, the same as for $\mathcal{K}_\Delta(p)$.

Write $\mathcal{K}_\Delta(p)$ as $\{\infty\} \cup \{k \in \mathbb{F}_p \mid P(k) \neq 0\}$, with the operation written as $*$, as in Section 3.3. Consider the function $T : K \rightarrow \mathcal{K}_\Delta(p)$ for which $T(\langle 1 : 0 \rangle) = \infty$ and

$T(\langle p^2 : pk \rangle) = k$. It is clear that T is a bijection. To show that T is an isomorphism, and to confirm that K is a group as defined, it will suffice to show that

$$T(\langle p^2 : pk \rangle \circ \langle p^2 : p\ell \rangle) = T(\langle p^2 : pk \rangle) * T(\langle p^2 : p\ell \rangle) = k * \ell$$

for all elements k and ℓ in $\mathcal{K}_\Delta(p)$ that are not equal to ∞ and are not conjugates of each other. (Compositions involving the identity element $\langle 1 : 0 \rangle$ are easily seen to satisfy the required equations. We omit those cases.)

Let k and ℓ be integers for which $P(k)$ and $P(\ell)$ are not divisible by p. Assume that ℓ is not the conjugate of k, so that $u = k + \ell + P'(0)$ is not divisible by p. We can then write $1 = ps + ut$ for some integers s and t. Let $v = k\ell - P(0)$. To calculate $\langle p^2 : pk \rangle \circ \langle p^2 : p\ell \rangle$ in G_1, we need

$$u_1 = pk + p\ell + P_1'(0) = p \cdot u \quad \text{and} \quad v_1 = (pk)(p\ell) - P_1(0) = p^2 \cdot v.$$

Here,

$$g = \gcd(p^2, p^2, u_1) = p = p^2(0) + p^2(s) + p \cdot ut,$$

where $ps + ut = 1$ as above. Therefore, $\langle p^2 : pk \rangle \circ \langle p^2 : p\ell \rangle = \langle c : m \rangle$ where

$$c = \frac{p^2}{p} \cdot \frac{p^2}{p} = p^2 \quad \text{and} \quad m = \frac{p^2 \cdot p\ell \cdot 0 + p^2 \cdot pk \cdot s + p^2 vt}{p} \equiv p(tv) \pmod{p^2}.$$

But note that $t = u^{-1}$ in $\mathbb{F}_p$, since $ut \equiv 1 \pmod{p}$. Thus, we can write

$$\langle p^2 : pk \rangle \circ \langle p^2 : p\ell \rangle = \langle p^2 : p(u^{-1}v) \rangle,$$

and so

$$T(\langle p^2 : pk \rangle \circ \langle p^2 : p\ell \rangle) = T(\langle p^2 : p(u^{-1}v) \rangle) = u^{-1}v = k * \ell,$$

as we wanted to show. □

We will refer to the set K defined in equation (3.7) as the *p-kernel of Δ in Δ_1* based on its connection to the group $\mathcal{K}_\Delta(p)$. In the next proposition, we describe the cosets of K in G_1, the p-prime subgroup of discriminant $\Delta_1 = p^2\Delta$.

Proposition 3.4.3. *Let Δ be a discriminant and let $\Delta_1 = p^2\Delta$ for some prime number p. Let G_1 be the p-prime subgroup of discriminant Δ_1 and let*

$$K = \{\langle p^e : \ell \rangle \in G_1 \mid e \le 2\},$$

the p-kernel of Δ in Δ_1. If $\langle p^{e+2} : p\ell \rangle$ is an element of G_1 not in K, then the coset of K in G_1 determined by this element consists precisely of the primitive ideal numbers of the form $\langle p^{e+2} : p\ell + p^{e+1}q \rangle$, where q is an integer.

Proof. Let $P(x)$ and $P_1(x)$ be the principal polynomials of discriminant Δ and Δ_1, respectively, as expressed in equation (3.6). If $\langle p^{e+2} : p\ell \rangle$ is an element of G_1 not in K, then e is positive. Since p^{e+2} divides $P_1(p\ell) = p^2 P(\ell)$, then p^e divides $P(\ell)$. If $\langle p^2 : pk \rangle$ is an element of K, let $u = k + \ell + P'(0)$ and $v = k\ell - P(0)$. Note that ℓ is not congruent to $\bar{k} = -k - P'(0)$ modulo p, since p divides $P(\ell)$ but not $P(k) = P(\bar{k})$. Thus, p does not divide u, and we can write $1 = p^{e+1}s + ut$ for some integers s and t.

Now consider the composite $\langle p^2 : pk \rangle \circ \langle p^{e+2} : p\ell \rangle$ in G_1, with

$$u_1 = pk + p\ell + P_1(0) = p \cdot u \quad \text{and} \quad v_1 = (pk)(p\ell) - P_1(0) = p^2 v.$$

Here, $g = \gcd(p^2, p^{e+2}, u_1) = p = p^2(0) + p^{e+2}s + u_1 t$, using the values of s and t selected in the preceding paragraph. Thus, the norm of this composite is

$$\frac{p^2}{p} \cdot \frac{p^{e+2}}{p} = p^{e+2}$$

and its character is

$$\frac{p^2 \cdot p\ell \cdot 0 + p^{e+2} \cdot pk \cdot s + v_1 \cdot t}{p} = p^{e+2}ks + pvt \equiv pvt \pmod{p^{e+2}}.$$

Now $v = u\ell - P(\ell)$, as in equation (3.2), so we find that

$$pvt = p(u\ell - P(\ell))t = p\ell \cdot ut - P(\ell) \cdot pt = p\ell(1 - p^{e+1}s) - P(\ell) \cdot pt$$
$$\equiv p\ell - P(\ell) \cdot pt \pmod{p^{e+2}},$$

here using the equation $1 = p^{e+1}s + ut$. Since p^e divides $P(\ell)$, we conclude that

$$\langle p^2 : pk \rangle \circ \langle p^{e+2} : p\ell \rangle = \langle p^{e+2} : p\ell + p^{e+1}q \rangle$$

for some integer q. We can use the counting formulas of Section 2.4, particularly as expressed in Theorem 2.4.2, to ensure that all primitive ideal numbers of this form are obtained as composites in this way. We omit the details. □

In practice, we can think of a coset of K in $G_1 = \mathcal{G}_{\Delta_1}(p^\infty)$, aside from K itself, as the collection of ideal numbers that are constructed from an individual element of $G = \mathcal{G}_\Delta(p^\infty)$ via the seeding map. We illustrate this idea with examples to conclude Section 3.4.

Example. Let $p = 5$ and $\Delta = -15$, so that $\Delta_1 = -375$. In Section 3.3, we saw that $\mathcal{K}_{-15}(5)$ is a cyclic group with five elements. Using the map described in the proof of Proposition 3.4.2, we can compile

$$K = \{\langle 1 : 0 \rangle, \langle 25 : 0 \rangle, \langle 25 : -5 \rangle, \langle 25 : 5 \rangle, \langle 25 : -10 \rangle\}$$

as the 5-kernel of Δ in Δ_1. Since 5 divides Δ, the group $G = \mathcal{G}_{-15}(5^\infty)$ contains only one nonidentity element, namely $\langle 5 : 2 \rangle$. The seeding map sends $\langle 5 : 2 \rangle$ to $\langle 25 : 10 \rangle$, which then produces $p = 5$ ideal numbers with norm 125, one of which is $\langle 125 : 10 \rangle$. The coset $\langle 125 : 10 \rangle \cdot K$ consists of these five ideal numbers, all of which are primitive. (None is the image of an element of G under the seeding map.) The 5-prime subgroup $G_1 = \mathcal{G}_{-375}(5^\infty)$ contains ten elements in total.

Example. Let $p = 3$ and $\Delta = -35$, so that $\Delta_1 = -315$. Let $P(x) = x^2 + x + 9$ and $P_1(x) = x^2 + 3x + 81$, the principal polynomials of discriminant Δ and Δ_1, respectively. Since $P(x) \equiv 0 \pmod 3$ has two solutions, $x = 0$ and $x = -1$, then the 3-kernel of Δ has $p + 1 - n_\Delta(p) = 2$ elements. The corresponding 3-kernel of Δ in Δ_1 can be written as

$$K = \{\langle 1 : 0 \rangle, \langle 9 : 3 \rangle\}.$$

In this case, each element $\langle 3^e : k \rangle$ of $G = \mathcal{G}_{-35}(3^\infty)$ produces three elements of the form $\langle 3^{e+2} : 3k + 3^{e+1}t \rangle$ in $G_1 = \mathcal{G}_{-315}(3^\infty)$. For each $e > 0$, two of these elements are primitive, while one is the image of $\langle 3^{e+1} : k + 3^e t \rangle$ in G under the seeding map. The group G_1 infinite, but its elements are partitioned into cosets of K, each with two elements.

3.5 Level subgroups

In Section 3.4, we saw how the p-prime subgroup of discriminant Δ, together with a particular subgroup of $G_1 = \mathcal{G}_{p^2\Delta}(p^\infty)$, gives us information about the entire group G_1. There is no restriction on the p-level of Δ in this process, so we could likewise use G_1 to describe the p-prime subgroup of discriminant $p^4\Delta$. In this section, we take a different approach to building from discriminants of p-level zero to those of arbitrary p-level. We fix the following notation, as we did in parts of Chapter 2. Let p be a prime number and let $\Delta(d, g)$ be a discriminant for which p does not divide g. For each integer $\ell \geq 0$, let $\Delta_\ell = \Delta(d, p^\ell g)$, so that the level of p in Δ_ℓ is ℓ. We write the principal polynomial of discriminant Δ_ℓ as $P_\ell(x)$. The following lemma compiles some useful observations about elements in prime subgroups of positive level.

Lemma 3.5.1. *Let $\Delta_\ell = \Delta(d, p^\ell g)$ be a discriminant with p-level $\ell > 0$ and let $P_\ell(x)$ be its principal polynomial. Let $\langle p^e : k \rangle$ be the congruence class of a primitive ideal number of discriminant Δ_ℓ. Then the following statements are true:*
1. *If $e > 0$, then $P_\ell(k) = p^e c$ with $\gcd(p, c) = 1$.*
2. *If $0 < e < 2\ell$, then $e = 2i$ for some integer i, and $k = p^i m$ with $\gcd(p, m) = 1$.*
3. *If $e \geq 2\ell$, then p^ℓ divides k.*

Proof. We can write $P_\ell(x) = x^2 + p^\ell s x + p^{2\ell} t$ for some integers s and t, using the definition of the principal polynomial of a discriminant. Suppose that $\langle p^e : k \rangle$ is a primitive ideal number of discriminant Δ_ℓ, so that, by definition, $P_\ell(k) = p^e c$ for some integer c. If e

is positive, then p divides $P_\ell(k)$ and, since ℓ is positive, it follows that p divides k. But then p divides $b = P'_\ell(k) = 2k + p^\ell s$. If $\langle p^e : k \rangle$ is primitive, then $g = \gcd(p^e, b, c) = 1$. Therefore, p cannot divide c. This establishes statement (1).

Assume that $0 < e < 2\ell$, so that statement (1) applies. Note that $k \neq 0$ in this case, since $P_\ell(0) = p^{2\ell} t$. So, we can write $k = p^i m$ with $\gcd(p, m) = 1$ for some $i \geq 0$. Then

$$P_\ell(k) = p^{2i} m^2 + p^{\ell+i} sm + p^{2\ell} t = p^{2i}(m^2 + p^{\ell-i} sm + p^{2(\ell-i)} t) = p^{2i} c.$$

If $i \geq \ell$, this contradicts the assumption that $e < 2\ell$. But then with $i < \ell$, we see that c is not divisible by p. We must conclude that $e = 2i$ in this case, establishing statement (2).

Finally, assume that $e \geq 2\ell$, again with statement (1) applying. If $k = 0$, then p^ℓ divides k. If $k \neq 0$, we can again write $k = p^i m$ with $\gcd(p, m) = 1$. Here, we conclude that $i \geq \ell$ since otherwise we would be forced to conclude that $e = 2i < 2\ell$ as in the preceding paragraph. This establishes statement (3). □

We define our sets of interest for this section in the following theorem.

Theorem 3.5.2. *Let p be a prime number and let $\Delta = \Delta(d, g)$ be a discriminant with p not dividing g. For each nonnegative integer ℓ, let $\Delta_\ell = \Delta(d, p^\ell g)$ and let $G = \mathcal{G}_{\Delta_\ell}$, the group of all primitive congruence classes of ideal numbers of discriminant Δ_ℓ under composition. Then for each integer $0 \leq i \leq \ell$, the set $H_i = \{\langle a : k \rangle \in G \mid a \text{ divides } p^{2i}\}$ is a subgroup of G. If $0 \leq i < \ell$, then H_i contains p^i elements. If $\ell > 0$ and $n_\Delta(p)$ is the number of elements of $\mathcal{G}_\Delta$ having norm p, then the number of elements of H_ℓ is*

$$p^{\ell-1}(p + 1 - n_\Delta(p)) = \begin{cases} p^\ell - p^{\ell-1}, & \text{if } n_\Delta(p) = 2, \\ p^\ell, & \text{if } n_\Delta(p) = 1, \\ p^\ell + p^{\ell-1}, & \text{if } n_\Delta(p) = 0. \end{cases}$$

We refer to H_ℓ as the *p-level subgroup* of discriminant Δ_ℓ. We allow ℓ to be zero, but in that case $H_0 = \{\langle 1 : 0 \rangle\}$ is the trivial subgroup and is the only new set defined in Theorem 3.5.2. In the proof, we will assume instead that ℓ is positive.

Proof. If ℓ is a positive integer, notice that $H_0 \subseteq H_1 \subseteq \cdots \subseteq H_\ell$. In the notation of Section 2.4, we then see that $|H_i| = |H_{i-1}| + m(\ell, 2i)$ for $1 \leq i \leq \ell$, with $|H_0| = 1$. Using statements (1) and (2) of Theorem 2.4.2, then

$$|H_1| = 1 + (p - 1) = p, \quad |H_2| = p + (p^2 - p) = p^2, \quad \dots,$$
$$|H_{\ell-1}| = p^{\ell-2} + (p^{\ell-1} - p^{\ell-2}) = p^{\ell-1},$$

and

$$|H_\ell| = p^{\ell-1} + p^{\ell-1}(p - n_\Delta(p)) = p^{\ell-1}(p + 1 - n_\Delta(p)).$$

Lemma 3.5.1 shows that each element of H_ℓ equals $\langle p^{2e} : p^e k \rangle$ for some $0 \le e \le \ell$, with p not dividing k if $0 < e < \ell$. To show that H_i is a subgroup of G for $0 \le i \le \ell$, it suffices to show that each such set is closed under composition. So consider

$$\langle p^{2e_1} : p^{e_1} k_1 \rangle \circ \langle p^{2e_2} : p^{e_2} k_2 \rangle,$$

where we can assume that $2e_1 \le 2e_2 \le 2i$ for some fixed i with $0 \le i \le \ell$. Let $P_\ell(x)$ be the principal polynomial of discriminant Δ_ℓ. In the formula for composition, we have $u = p^{e_1} k_1 + p^{e_2} k_2 + P'_\ell(0)$, and as we have noted previously, p^ℓ divides $P'_\ell(0)$. Therefore, p^{e_1} divides u, and is the minimal possible value of $g = \gcd(p^{2e_1}, p^{2e_2}, u)$. The maximal value of the norm in this composite is then

$$\frac{p^{2e_1}}{p^{e_1}} \cdot \frac{p^{2e_2}}{p^{e_1}} = p^{2e_2},$$

which divides p^{2i}. Thus, each set H_i is closed under composition. □

Example. Let $\Delta = \Delta(-35, 1) = -35$ and $p = 3$, so that $\Delta_1 = \Delta(-35, 3) = -315$ is a discriminant for which the p-level equals 1. From an example in Section 2.3, we see that there are two *primitive* ideal numbers in $\mathcal{Q}_{-315}$ whose norm divides $p^{2\ell} = 9$, namely $\langle 1 : 0 \rangle$ and $\langle 9 : 3 \rangle$. (The ideal numbers $\langle 3 : 0 \rangle$, $\langle 9 : 0 \rangle$, and $\langle 9 : -3 \rangle$ are not primitive. From Figure 2.2, we see that these are images of elements of $\mathcal{Q}_{-35,3}$ under the seeding map.) Thus,

$$H_0 = \{\langle 1 : 0 \rangle\} \quad \text{and} \quad H_1 = \{\langle 1 : 0 \rangle, \langle 9 : 3 \rangle\}.$$

Here, $\langle 9 : 3 \rangle$ is its own conjugate, so is its own inverse under composition, confirming that H_1 is a subgroup of $\mathcal{G}_\Delta$ in this case.

Example. Let $\Delta = \Delta(3, 1) = 12$ and $p = 2$, so that $\Delta_2 = \Delta(3, 2^2) = 192$ is a discriminant with p-level $\ell = 2$. We can read the primitive ideal numbers in $\mathcal{Q}_{192}$ whose norm divides $p^{2\ell} = 16$ from Figure 2.3, and we find that

$$H_0 = \{\langle 1 : 0 \rangle\}, \quad H_1 = \{\langle 1 : 0 \rangle, \langle 4 : 2 \rangle\},$$

and

$$H_2 = \{\langle 1 : 0 \rangle, \langle 4 : 2 \rangle, \langle 16 : 0 \rangle, \langle 16 : 8 \rangle\}.$$

Each element in each of these groups is its own inverse.

3.6 Prime subgroups of positive level

As in the preceding section, let p be a prime number and let $\Delta = \Delta(d, g)$ with g not divisible by p. For each nonnegative integer ℓ, let $\Delta_\ell = \Delta(d, p^\ell g)$. We write $P_\ell(x)$ for the principal polynomial of discriminant Δ_ℓ and we let $n_\Delta(p)$ be the number of roots of $P_0(x)$ modulo p. In this section, we use the p-level subgroups defined in Section 3.5 to help us calculate the entire p-prime subgroup of discriminant Δ_ℓ when ℓ is positive. We first note the following classification of the number of elements in this group.

Proposition 3.6.1. *Let $G_\ell = \mathcal{G}_{\Delta_\ell}(p^\infty)$ be the p-prime subgroup and let H_ℓ be the p-level subgroup of some discriminant Δ_ℓ whose p-level is ℓ. Then the following statements are true:*

1. *If $n_\Delta(p) = 0$, then $G_\ell = H_\ell$, with $p^\ell + p^{\ell-1}$ elements in total.*
2. *If $n_\Delta(p) = 1$, then G_ℓ contains twice as many elements as H_ℓ, that is, $2p^\ell$ elements in total.*
3. *If $n_\Delta(p) = 2$, then G_ℓ is infinite. For each $e > 2\ell$, there are precisely $2(p^\ell - p^{\ell-1})$ elements of norm p^e in G_ℓ, that is, twice as many as the total number of elements in H_ℓ.*

Proof. The number of elements of H_ℓ was calculated in Theorem 3.5.2, and the number of elements of G_ℓ having norm p^e with $e > 2\ell$ is counted in statements (3), (4), and (5) of Theorem 2.4.2. □

We will examine the third case more closely, using the following example to illustrate our results.

Example. Let $p = 5$ and let $\Delta_2 = \Delta(-1, 5^2 \cdot 1) = -2500$ with principal polynomial $P_2(x) = x^2 + 625$. Here, $P_0(x) = x^2 + 1$ has $n_\Delta(p) = 2$ roots modulo 5, namely 2 and -2, and we find the following $p^\ell - p^{\ell-1} = 5^2 - 5^1 = 20$ elements of the level subgroup H_2:

$$
\begin{array}{ccccc}
\langle 1:0 \rangle & \langle 25:5 \rangle & \langle 25:-5 \rangle & \langle 25:10 \rangle & \langle 25:-10 \rangle \\
\langle 625:0 \rangle & \langle 625:25 \rangle & \langle 625:-25 \rangle & \langle 625:100 \rangle & \langle 625:-100 \rangle \\
\langle 625:125 \rangle & \langle 625:-125 \rangle & \langle 625:150 \rangle & \langle 625:-150 \rangle & \langle 625:225 \rangle \\
\langle 625:-225 \rangle & \langle 625:250 \rangle & \langle 625:-250 \rangle & \langle 625:275 \rangle & \langle 625:-275 \rangle.
\end{array}
$$

For each $e \geq 5$, there are $2(p^\ell - p^{\ell-1}) = 40$ elements of norm p^e in $G_2 = \mathcal{G}_{\Delta_2}(5^\infty)$. For instance, G_2 contains $\langle 3125:k \rangle$ for the following values of k:

$$
\begin{array}{ccccccc}
50 & -75 & -200 & 300 & -325 & 425 & 550 \\
-575 & 675 & -700 & -825 & 925 & -950 & 1050 \\
1175 & -1200 & 1300 & -1325 & -1450 & 1550 & \\
-50 & 75 & 200 & -300 & 325 & -425 & -550 \\
575 & -675 & 700 & 825 & -925 & 950 & -1050 \\
-1175 & 1200 & -1300 & 1325 & 1450 & -1550. &
\end{array}
$$

These values are grouped so that those in the first three rows satisfy $k = 25q$ with $q \equiv 2$ (mod 5) while those in the last three rows have the form $k = 25q$ with $q \equiv -2$ (mod 5). A (partial) Hasse diagram for the set $Q_{-2500}(5^\infty)$, including nodes for each $\langle 5^e : k \rangle$ with $e \le 5$, appears in Figure 3.1 at the end of this section. In this diagram, the unseeded nodes are in black—these represent the elements of the prime subgroup G_2.

The following lemma explains some patterns seen in this example.

Lemma 3.6.2. Let $\Delta_\ell = \Delta(d, p^\ell g)$ and $\Delta_0 = \Delta(d, g)$, where p does not divide g. Let e be an integer larger than 2ℓ. If $\langle p^e : p^\ell k \rangle$ is an element of $G_{\Delta_\ell}(p^\infty)$, then $\langle p^{e-2\ell} : k \rangle$ is an element of $G_{\Delta_0}(p^\infty)$.

Proof. We saw in Lemma 3.5.1 that an element of $G_\ell = G_{\Delta_\ell}(p^\infty)$ having norm p^e with $e > 2\ell$ can be written as $\langle p^e : p^\ell k \rangle$ for some integer k. If $P_\ell(x)$ and $P_0(x)$ are the principal polynomials of discriminant Δ_ℓ and Δ_0, respectively, then Lemma 1.1.3 implies that $P_\ell(p^\ell k) = p^{2\ell} P_0(k)$. If $\langle p^e : p^\ell k \rangle$ is an element of G_ℓ, so that p^e divides $P_\ell(p^\ell k)$, then $p^{e-2\ell}$ divides $P_0(k)$. But then $\langle p^{e-2\ell} : k \rangle$ is an element of $G_{\Delta_0}(p^\infty)$, as claimed. □

Example. In the preceding example, as we saw, $P_0(x) = x^2 + 1$ has roots $x = 2$ and $x = -2$ modulo 5. All elements of norm $p^{2\ell+1} = 5^5 = 3125$ in G_2 have the form $\langle 3125 : 25k \rangle$ with k congruent to 2 or -2 modulo 5. Similarly, $P_0(x) \equiv 0$ (mod 25) has solutions $x = 7$ and $x = -7$. We would likewise find that all elements of norm $5^6 = 15625$ in G_2 have the form $\langle 15625 : 25k \rangle$ with k congruent to 7 or -7 modulo 25.

We can describe a prime subgroup $G_\ell = G_{\Delta_\ell}(p^\infty)$ further by considering cosets of H_ℓ in G_ℓ. We will refer to the quotient group G_ℓ/H_ℓ of all such cosets as the *level quotient* of the prime subgroup G_ℓ. When $n_\Delta(p) = 0$ is the number of roots of $P_0(x)$ modulo p, then G_ℓ/H_ℓ is trivial. If $n_\Delta(p) = 1$, then G_ℓ/H_ℓ has two elements: the set H_ℓ and the set of all ideal numbers in G_ℓ with norm $p^{2\ell+1}$, which was noted to have the same number of elements as H_ℓ. We can restrict our attention then to the case where $n_\Delta(p) = 2$.

Proposition 3.6.3. Let $\Delta_\ell = \Delta(d, p^\ell g)$ be a discriminant of p-level $\ell > 0$. Suppose that $P_0(x)$, the principal polynomial of discriminant $\Delta_0 = \Delta(d, g)$ has two roots modulo p. Let

$$G_\ell = \{ \langle p^e : k \rangle \in G_{\Delta_\ell} \mid e \ge 0 \} \quad \text{and} \quad H_\ell = \{ \langle p^e : k \rangle \in G_{\Delta_\ell} \mid 0 \le e \le 2\ell \}$$

be the p-prime subgroup and the p-level subgroup, respectively, of discriminant Δ. Then the coset of H_ℓ in G_ℓ determined by $\langle p^e : p^\ell k_1 \rangle$, where $e > 2\ell$, consists precisely of all ideal numbers of the form $\langle p^e : p^\ell k \rangle$ in G_ℓ with $k \equiv k_1$ (mod p).

Proof. Lemma 3.5.1 shows that each element of G_ℓ not in H_ℓ can be written as $\langle p^e : p^\ell k_1 \rangle$ with $e > 2\ell$ while elements of H_ℓ have the form $\langle p^{2i} : p^i k_2 \rangle$ with $i \le \ell$. We want to show that

$$\langle p^e : p^\ell k_1 \rangle \circ \langle p^{2i} : p^i k_2 \rangle = \langle p^e : p^\ell k \rangle$$

for some k with $k \equiv k_1 \pmod{p}$. This is clearly true if $i = 0$, that is, for the identity element in H_ℓ. We consider two possibilities for the remaining cases.

If $0 < i < \ell$, then by Lemma 3.5.1 we have that p does not divide k_2. In the composition formula for $\langle p^e : p^\ell k_1 \rangle \circ \langle p^{2i} : p^i k_2 \rangle$, we find that $u = p^\ell k_1 + p^i k_2 + P_\ell'(0)$ can be written as $p^i u_1$ with p not dividing u_1. Then $g = \gcd(p^e, p^{2i}, u) = p^i = p^e r + p^{2i} s + p^i u_1 t$ for some integers r, s, and t. By the congruence cancellation property, $u_1 t \equiv 1 \pmod{p}$.

Suppose on the other hand that $i = \ell$. Then by equation (1.4), we have

$$P_\ell(p^\ell k_1) - P_\ell(p^\ell k_2) = (p^\ell k_1 - p^\ell k_2)(p^\ell k_1 + p^\ell k_2 + P_\ell'(0)) = p^\ell(k_1 - k_2) \cdot u,$$

where p^ℓ divides $u = p^\ell k_1 + p^\ell k_2 + P_\ell'(0)$, as defined in the composition formula. Statement (1) of Lemma 3.5.1 shows that $P_\ell(p^\ell k_1) - P_\ell(p^\ell k_2) = p^e c_1 - p^{2\ell} c_2$ with neither c_1 nor c_2 divisible by p. With $e > 2\ell$, we conclude that the exponent of p in u can be no larger than ℓ. That is, $u = p^\ell u_1$ with p not dividing u_1. Now $g = \gcd(p^e, p^{2\ell}, u) = p^\ell = p^e r + p^{2\ell} s + p^\ell u_1 t$, with $u_1 t \equiv 1 \pmod{p}$, again by the congruence cancellation property.

Thus, for each possible value of $i > 0$, we have $g = p^i$, and it follows that the norm of the composite $\langle p^e : p^\ell k_1 \rangle \circ \langle p^{2i} : p^i k_2 \rangle$ is $\frac{p^e}{p^i} \cdot \frac{p^{2i}}{p^i} = p^e$. Hence, this composite has the form $\langle p^e : p^\ell k \rangle$ for some integer k. Furthermore, using the second equation in (3.2), we find that

$$g \cdot p^\ell k = p^{\ell+i} k = p^e \cdot p^i k_2 r + p^{2i} \cdot p^\ell k_1 s + p^i u_1 \cdot p^\ell k_1 t - P_\ell(p^\ell k_1)t.$$

As above, $P_\ell(p^\ell k_1) = p^e c_1$. With $e > 2\ell$ and $0 < i \le \ell$, we find that each of these terms is divisible by $p^{\ell+i+1}$ except for $p^{\ell+i} k_1 \cdot u_1 t$. We conclude that $k \equiv k_1 \cdot u_1 t \equiv k_1 \pmod{p}$.

Finally, we can be sure as follows that $\langle p^e : p^\ell k_1 \rangle \circ H_\ell$ consists of all ideal numbers in G_ℓ of the form $\langle p^e : p^\ell k \rangle$ with $k \equiv k_1 \pmod{p}$. We saw that for all $e > 2\ell$, there are precisely twice as many ideal numbers in G_ℓ of norm p^e as there are elements of H_ℓ. By Lemma 3.6.2, these elements have the form $\langle p^e : p^\ell k \rangle$ where k satisfies $P_0(x) \equiv 0 \pmod{p^{e-2\ell}}$ with $P_0(x)$ the principal polynomial of discriminant Δ_0. When $n_\Delta(p) = 2$, there are two conjugate possibilities for k. It is clear that there is a one-to-one correspondence between the sets

$$\{\langle p^e : p^\ell k \rangle \mid k \equiv k_1 \pmod{p}\} \quad \text{and} \quad \{\langle p^e : p^\ell k \rangle \mid k \equiv \overline{k_1} \pmod{p}\}.$$

Thus, both sets contain the same number of elements as H_ℓ, and must be cosets of H_ℓ in G_ℓ. □

Example. For $\Delta_2 = \Delta(-1, 25)$, as in the previous example, consider

$$\langle 25 : 10 \rangle \circ \langle 3125 : 300 \rangle.$$

Here, with $P_2(x) = x^2 + 625$, we have $u = 10 + 300 + 0 = 310$ and $v = 10 \cdot 300 - 625 = 2375$. We find that $g = \gcd(25, 3125, 310) = 5 = 25(25) + 3125(0) + 310(-2)$. The norm of the composite is $\frac{25}{5} \cdot \frac{3125}{5} = 3125$ and its character is

$$\frac{1}{5}(25 \cdot 300 \cdot 25 + 3125 \cdot 10 \cdot 0 + 2375 \cdot -2) = 36550 \equiv -950 \quad (\text{mod } 3125).$$

That is,

$$\langle 25 : 10 \rangle \circ \langle 3125 : 300 \rangle = \langle 3125 : -950 \rangle.$$

With $\langle 25 : 10 \rangle$ in the level subgroup H_ℓ and $\frac{300}{25} = 12$ congruent to $\frac{-950}{25} = -38$ modulo 5, this result is as predicted in Proposition 3.6.3.

Proposition 3.6.4. *Let $\Delta_\ell = \Delta(d, p^\ell g)$ be a discriminant of p-level $\ell > 0$. Suppose that $P_0(x)$, the principal polynomial of discriminant $\Delta_0 = \Delta(d, g)$ has two roots modulo p. Let G_ℓ and H_ℓ be the p-prime subgroup and the p-level subgroup, respectively, of discriminant Δ_ℓ. Then the quotient group G_ℓ/H_ℓ is an infinite cyclic group.*

Proof. We show that the coset of H_ℓ in G_ℓ determined by an element $\langle p^{2\ell+1} : p^\ell k_1 \rangle$ generates the entire group of cosets of H_ℓ in G_ℓ. By Proposition 3.6.3, this will follow by showing that if, for some positive integer e, $\langle p^{2\ell+e} : p^\ell k_2 \rangle$ is an element of G_ℓ with $k_2 \equiv k_1$ (mod p), then

$$\langle p^{2\ell+1} : p^\ell k_1 \rangle \circ \langle p^{2\ell+e} : p^\ell k_2 \rangle = \langle p^{2\ell+e+1} : p^\ell k \rangle$$

with $k \equiv k_1$ (mod p). Let $P_\ell(x)$ and $P_0(x)$ be the principal polynomials of discriminant Δ_ℓ and Δ_0, respectively. Applying the composition formula, as well as Proposition 1.1.3, we have

$$u = p^\ell k_1 + p^\ell k_2 + P_\ell'(0) = p^\ell (k_1 + k_2 + P_0'(0)).$$

Since $k_2 \equiv k_1$ (mod p), then $u_1 = k_1 + k_2 + P_0'(0) \equiv 2k_1 + P_0'(0) \equiv P_0'(k_1)$ (mod p). By Lemma 3.6.2, k_1 is one of two solutions of $P_0(x) \equiv 0$ (mod p), and we have seen that when $m(p) = 2$, then p does not divide $P_0'(k_1)$. Thus, $u = p^\ell u_1$ with $\gcd(u_1, p) = 1$. We then have

$$g = \gcd(p^{2\ell+1}, p^{2\ell+e}, u) = p^\ell = p^{2\ell+1}r + p^{2\ell+e}s + p^\ell u_1 t,$$

for some integers r, s, and t, and it follows that $u_1 t \equiv 1$ (mod p). The norm of the composite

$$\langle p^{2\ell+1} : p^\ell k_1 \rangle \circ \langle p^{2\ell+e} : p^\ell k_2 \rangle$$

is then $\frac{p^{2\ell+1}}{p^\ell} \cdot \frac{p^{2\ell+e}}{p^\ell} = p^{2\ell+e+1}$. The character of the composite has the form $p^\ell k$ where

$$g \cdot p^\ell k = p^{2\ell} k = p^{2\ell+1} p^\ell k_2 r + p^{2\ell+e} p^\ell k_1 s + p^{2\ell} k_1 (u_1 t) - P_\ell(p^\ell k_1)t.$$

(Here, we use the second equation for v in (3.2).) We find that $k \equiv k_1(u_1 t) \equiv k_1$ (mod p), as we wanted to show, since $p^{2\ell+1}$ divides $P_\ell(p^\ell k_1)$. $\qquad\square$

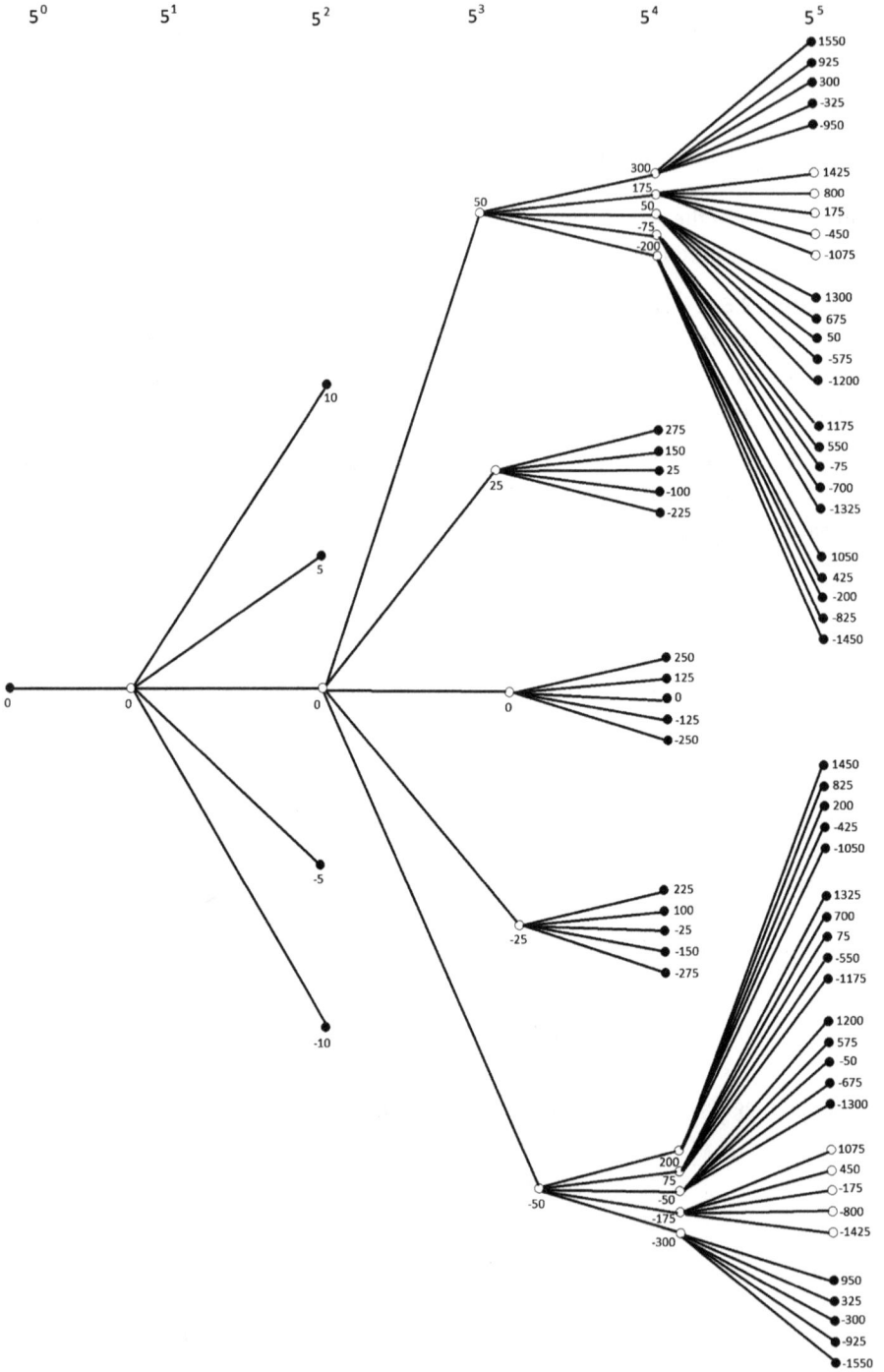

Figure 3.1: Hasse diagram for $\mathcal{Q}_{-2500}(5^{\infty})$.

Example. In our example, the coset of $\langle 3125 : 50 \rangle$ can be taken as a generator of the level quotient G_2/H_2. To compose this element with itself, we calculate $u = 50 + 50 + 0 = 100$, $v = 50 \cdot 50 - 625 = 1875$, and $g = \gcd(3125, 3125, 100) = 25 = 3125(1) + 3125(0) + 100(-31)$. The norm of the composite is $\frac{3125}{25} \cdot \frac{3125}{25} = 15625 = 5^6$ and the character is $\frac{1}{25}(3125 \cdot 50 \cdot 1 + 3125 \cdot 50 \cdot 0 + 1875 \cdot -31) = 3925$. Thus,

$$\langle 3125 : 50 \rangle^2 = \langle 15625 : 3925 \rangle.$$

Note that $\frac{3925}{25} = 157$ is congruent to $\frac{50}{25} = 2$ modulo 5. The square of $\langle 3125 : 50 \rangle \circ H_\ell$ in G_2/H_2 is represented by any element of G_2 of the form $\langle 15625 : 25k \rangle$ with $k \equiv 2 \pmod 5$.

4 Classes of ideal numbers

In Chapter 2, we considered an equivalence relation called congruence defined on the set of all ideal numbers of a fixed discriminant. We described methods of counting and listing all equivalence classes, which we called *congruence classes*, under congruence. In Chapter 3, we introduced an operation of composition on primitive congruence classes and studied the group structure produced by that operation. We now define a second equivalence relation, which we will simply call *equivalence*, on ideal numbers. We will find that the same composition operation applies to classes under this operation.

Equivalence of ideal numbers is closely related to representations of integers by quadratic forms. We will continue to focus on ideal numbers as objects in their own right, but in this chapter we will also interpret an ideal number as a quadratic form, as in Section 1.4. Recall that if $P(x)$ is the principal polynomial of discriminant Δ, then an ideal number $(a : k)$ in $\mathcal{I}_\Delta$ corresponds to $f(x,y) = ax^2 + bxy + cy^2$ where $P(k) = ac$ and $P'(k) = b$. Conversely, if $f(x,y) = ax^2 + bxy + cy^2$ is a quadratic form with $b^2 - 4ac = \Delta$, and we let $k = \frac{1}{2}(b - P'(0))$, then $(a : k)$ is an ideal number in $\mathcal{I}_\Delta$.

4.1 Equivalence of ideal numbers

We define equivalence on the set $\mathcal{I}_\Delta$ of ideal numbers of discriminant Δ in terms of the following collection of matrices.

Definition. A 2×2 matrix U is *unimodular* if its entries are integers and its discriminant equals 1. If G is the group of all unimodular matrices under matrix multiplication, then $N = \{I, -I\}$, containing the 2×2 identity matrix and its negative, is a normal subgroup of G. We define the *unimodular group* to be the quotient group $\Gamma = G/N$.

It is straightforward to show that G is a group, and that N is a normal subgroup of G, using standard properties of matrix multiplication and determinants. The effect of defining Γ as G/N is simply to identify each unimodular matrix with its negative. The characterization of Γ in the following proposition will be useful.

Proposition 4.1.1. *Let T and V be the following unimodular matrices:*

$$T = \begin{bmatrix} 1 & 1 \\ 0 & 1 \end{bmatrix} \quad and \quad V = \begin{bmatrix} 0 & -1 \\ 1 & 0 \end{bmatrix}. \tag{4.1}$$

Then every unimodular matrix can be written as a product of integer powers of T and V.

Note that $V^2 = -I$, so we can regard V as having order two in Γ, and identify V with V^{-1}. On the other hand, T has infinite order in Γ, with $T^n = \begin{bmatrix} 1 & n \\ 0 & 1 \end{bmatrix}$ for all integers n. We refer to T and V, respectively, as the *translation matrix* and the *involution matrix*.

https://doi.org/10.1515/9783111319360-004

Proof. Let $U = \left[\begin{smallmatrix} q & s \\ r & t \end{smallmatrix}\right]$ be a unimodular matrix. If $q = 0$, then $\det U = qt - rs = -rs = 1$, so we can assume that $r = 1$ and $s = -1$ (replacing U by $-U$ if necessary). Direct calculation shows that

$$\begin{bmatrix} 0 & -1 \\ 1 & t \end{bmatrix} = \begin{bmatrix} 0 & -1 \\ 1 & 0 \end{bmatrix} \cdot \begin{bmatrix} 1 & t \\ 0 & 1 \end{bmatrix} = VT^t. \tag{4.2}$$

If $q \neq 0$, we can assume that q is positive. If $q = 1$, so that $\det U = qt - rs = t - rs = 1$, then we find that

$$\begin{bmatrix} 1 & s \\ r & t \end{bmatrix} = -1 \cdot \begin{bmatrix} 0 & -1 \\ 1 & -r \end{bmatrix} \cdot \begin{bmatrix} 0 & -1 \\ 1 & s \end{bmatrix} = VT^{-r} \cdot VT^s. \tag{4.3}$$

(Here, we twice use the calculation of equation (4.2), and we again identify a matrix with its negative.) Now let q be larger than 1 and suppose that we have shown that every unimodular matrix whose upper left-hand entry is a nonnegative integer smaller than q is a product of powers of T and V. Let U be a unimodular matrix as above and write $s = qm + n$ with $0 \leq n < q$, as under the division algorithm. Then we find that

$$U \cdot T^{-m} \cdot V = \begin{bmatrix} q & s \\ r & t \end{bmatrix} \cdot \begin{bmatrix} 1 & -m \\ 0 & 1 \end{bmatrix} \cdot \begin{bmatrix} 0 & -1 \\ 1 & 0 \end{bmatrix} = \begin{bmatrix} n & -q \\ t - rm & -r \end{bmatrix} = U_1.$$

Since n, the upper left-hand entry of U_1, is nonnegative but smaller than q, we have that U_1 is a product of powers of T and V by the inductive hypothesis. But then $U = U_1 \cdot V^{-1} \cdot T^m$ has that property as well. This establishes that all unimodular matrices U are products of powers of T and V. $\qquad\qquad\square$

In practice, the effect of multiplying a unimodular matrix on the right by T^n is to add n times the first column of U to the second column. Multiplying U on the right by $\pm V$ interchanges the columns of U and changes the sign of one of those columns. Thus, we have a systematic approach to writing U as a product of powers of T and V, which we will illustrate with an example.

Example. Let $U = \left[\begin{smallmatrix} 47 & -15 \\ 22 & -7 \end{smallmatrix}\right]$, a unimodular matrix. (We can always take q, the upper left-hand entry of U, to be nonnegative, as in this example.) If necessary, multiply U by T^n for some n to ensure that the upper right-hand entry is smaller than q in absolute value. We then multiply by V to interchange columns of the resulting matrix, changing the sign of one column to make sure that the upper left-hand entry is again nonnegative. These steps are repeated until the upper left-hand entry is 1 or 0, in which case the formulas from equations (4.2) or (4.3) can be applied. The following diagram helps keep track of these steps:

$$\begin{bmatrix} 47 & -15 \\ 22 & -7 \end{bmatrix} \xrightarrow{V} \begin{bmatrix} 15 & 47 \\ 7 & 22 \end{bmatrix} \xrightarrow{T^{-3}} \begin{bmatrix} 15 & 2 \\ 7 & 1 \end{bmatrix} \xrightarrow{V} \begin{bmatrix} 2 & -15 \\ 1 & -7 \end{bmatrix} \xrightarrow{T^8} \begin{bmatrix} 2 & 1 \\ 1 & 1 \end{bmatrix} \xrightarrow{V} \begin{bmatrix} 1 & -2 \\ 1 & -1 \end{bmatrix}.$$

We could continue the process, or apply equation (4.3) to conclude that

$$U \cdot (VT^{-3}VT^8V) = VT^{-1}VT^{-2}.$$

Multiplying on the right by the inverse of the term in parentheses,

$$U = VT^{-1}VT^{-2}VT^{-8}VT^3V.$$

Other expressions are possible.

We now define *equivalence* of ideal numbers in $\mathcal{I}_\Delta$ as follows.

Theorem 4.1.2. *Let $P(x)$ be the principal polynomial of discriminant Δ. Let $(a : k)$ be an ideal number in $\mathcal{I}_\Delta$, with $P(k) = ac$ and $P'(k) = b$. If $U = \left[\begin{smallmatrix} q & s \\ r & t \end{smallmatrix}\right]$ is a unimodular matrix, let $(a : k) \cdot U = (a_1 : k_1)$, where*

$$a_1 = aq^2 + bqr + cr^2 \quad and \quad k_1 = k + (aqs + brs + crt). \tag{4.4}$$

Then $(a_1 : k_1)$ is also an element of $\mathcal{I}_\Delta$, and the following statements are true:
1. *$(a : k) \cdot I = (a : k)$, where I is the identity element of Γ.*
2. *If U_1 and U_2 are in Γ, then $(a : k) \cdot (U_1 U_2) = ((a : k) \cdot U_1) \cdot U_2$.*
3. *If we say that $(a : k) \sim (a_1 : k_1)$ if and only if $(a : k) \cdot U = (a_1 : k_1)$ for some U in Γ, then $\sim$ is an equivalence relation on $\mathcal{I}_\Delta$.*

If each of q, r, s, and t is changed in sign, then the values of a_1 and k_1 in equation (4.4) are unchanged. Thus, $(a : k) \cdot U$ is not affected by replacing U by $-U$, which are equal in the unimodular group Γ. Statements (1) and (2) in Theorem 4.1.2 verify that the transformation defined here is a *right group action* of Γ on $\mathcal{I}_\Delta$. We note two special cases of this action before proving Theorem 4.1.2.

Definition. If $U = T^n$, where T is as in equation (4.1), then $q = 1, r = 0, s = n$, and $t = 1$, so that

$$(a : k) \cdot T^n = (a : k + na). \tag{4.5}$$

We refer to this transformation, or to T^n itself, as a *translation* of $\mathcal{I}_\Delta$.

Definition. If $U = V$, as in equation (4.1), then $q = 0, r = 1, s = -1$, and $t = 0$, so that

$$(a : k) \cdot V = (c : k - b) = (c : \overline{k}). \tag{4.6}$$

(Here, $k - b = k - P'(k) = k - (2k + P'(0)) = -k - P'(0) = \overline{k}$, the complement of k with respect to Δ.) We refer to V as the *involution* of $\mathcal{I}_\Delta$.

Proposition 1.2.2 ensures that $(a : k + an)$ is an element of $\mathcal{I}_\Delta$. Recall also that $P(\overline{k}) = P(k) = ac$, so that $(c : \overline{k})$ is likewise an element of $\mathcal{I}_\Delta$. Since every unimodular matrix

U is a product of powers of T and V, it follows that $(a : k) \cdot U$, as defined in (4.4), is an element of $\mathcal{I}_\Delta$. We show this more directly in the following proof.

Proof of Theorem 4.1.2. If $(a : k)$ is an ideal number of discriminant Δ, with $P(k) = ac$ and $P'(k) = b$, define the *matrix* of $(a : k)$ to be

$$M = \begin{bmatrix} 2a & b \\ b & 2c \end{bmatrix}.$$

Note that the determinant of M is $4ac - b^2 = 4P(k) - P'(k)^2 = -\Delta$. If U is a unimodular matrix, with entries as in Theorem 4.1.2, and U' is the transpose of U, then U and U' both have determinant 1. Thus, the determinant of $U'MU = M_1$ is also $-\Delta$. Direct calculation shows that

$$U'MU = \begin{bmatrix} q & r \\ s & t \end{bmatrix} \cdot \begin{bmatrix} 2a & b \\ b & 2c \end{bmatrix} \cdot \begin{bmatrix} q & s \\ r & t \end{bmatrix}$$

$$= \begin{bmatrix} 2(aq^2 + brq + cr^2) & 2aqs + b(qt + rs) + 2crt \\ 2aqs + b(qt + rs) + 2crt & 2(as^2 + bst + ct^2) \end{bmatrix}.$$

If we let

$$a_1 = aq^2 + brq + cr^2, \quad b_1 = 2aqs + b(qt + rs) + 2crt,$$

$$c_1 = as^2 + bst + ct^2, \quad \text{and} \quad k_1 = \frac{1}{2}(b_1 - P'(0)),$$

then $a_1 x^2 + b_1 xy + c_1 y^2$ is a quadratic form of discriminant Δ, so that $(a_1 : k_1)$ is an ideal number in $\mathcal{I}_\Delta$ by Proposition 1.4.1. Since $\det(U) = qt - rs = 1$, so $qt + rs = 2rs + 1$, then

$$k_1 = \frac{b_1 - P'(0)}{2} = \frac{2aqs + b(qt + rs) + 2crt - P'(0)}{2}$$

$$= aqs + brs + crt + \frac{b - P'(0)}{2} = k + (aqs + brs + crt).$$

Now note that:

1. $I'MI = M$ if I is the 2×2 identity matrix,
2. if $U'MU = M_1$, then $(U^{-1})'M_1 U^{-1} = M$,
3. if $U_1'MU_1 = M_1$ and $U_2'M_1 U_2 = M_2$, then $(U_1 U_2)'M(U_1 U_2) = M_2$,

using properties of transpose matrices. Thus, the $\sim$ relation has the reflexive, symmetric, and transitive properties. The first and third of these observations also establish the group action claims of the theorem. □

Definition. We denote the equivalence class of an ideal number $(a : k)$ under equivalence as $[a : k]$, and call this the *class* of $(a : k)$.

If an ideal number $(a : k)$ is *congruent* to $(a : \ell)$, in that $\ell \equiv k \pmod{a}$, then it is also true that $(a : k)$ is *equivalent* to $(a : \ell)$, via a translation. Recall, however, that $(a : k)$ is also congruent to $(-a : k)$, using the definition of congruence from Chapter 2. On the other hand, it may or may not be the case that $(a : k)$ is equivalent to $(-a : k)$. (See Proposition 4.2.3 and the example that follows it.)

Proposition 4.1.3. *Equivalent ideal numbers have the same content.*

Proof. In Proposition 2.1.3, we showed that congruent ideal numbers have the same content. In particular, the content of $(a : k)$ and $(a : k) \cdot T = (a : k+a)$ are equal. Furthermore, since $\gcd(a, b, c) = \gcd(c, -b, a)$, then $(a : k)$ and $(a : k) \cdot V = (c : \overline{k})$ have the same content. Hence, in applying a sequence of translations and involutions to $(a : k)$, the content remains the same. □

Thus, we can refer to the content of a class of ideal numbers, or refer to a class of ideal numbers as being primitive or not primitive, without ambiguity. Recall from Proposition 1.5.3 that every ideal number with content m has the form $(ma : mk)$, with discriminant $m^2\Delta$, where $(a : k)$ is a primitive ideal number of discriminant Δ. Our next proposition shows that classes of primitive ideal numbers of discriminant Δ are essentially the same as classes of ideal numbers with content m in $\mathcal{I}_{m^2\Delta}$.

Proposition 4.1.4. *If $(a : k) \sim (a_1 : k_1)$ for primitive ideal numbers in $\mathcal{I}_\Delta$, then in $\mathcal{I}_{m^2\Delta}$,*

$$(ma : mk) \sim (ma_1 : mk_1).$$

Proof. Let $P(x)$ and $P_m(x)$ be the principal polynomials of discriminant Δ and $m^2\Delta$, respectively. Recall from Proposition 1.1.3 that $P_m(mx) = m^2 P(x)$ and $P'_m(mx) = mP(x)$ for all x. Thus, if $P(k) = ac$, then $P_m(mk) = (ma)(mc)$. Notice that if $\overline{k} = -k - P'(0)$, then

$$\overline{mk} = -mk - P'_m(0) = -mk - P'_m(m0) = -mk - mP'(0) = m \cdot \overline{k}.$$

Hence, if $(a : k) \cdot V = (c : \overline{k})$, then $(ma : mk) \cdot V = (mc : m \cdot \overline{k})$. Furthermore, if $(a : k) \cdot T = (a : k+a)$, then $(ma : mk) \cdot T = (ma : mk + ma) = (ma : m(k+a))$. Therefore, if a sequence of translations and involutions converts $(a : k)$ to $(a_1 : k_1)$, then the same sequence converts $(ma : mk)$ to $(ma_1 : mk_1)$. □

4.2 Proper representations by ideal numbers

We introduce the following terminology in the context of ideal number equivalence.

Definition. Let $(a : k)$ be an ideal number of discriminant Δ. We say that $(a : k)$ *properly represents* an integer m if $(a : k)$ is equivalent to an ideal number in $\mathcal{I}_\Delta$ having norm m.

It is immediate from this definition that equivalent ideal numbers in $\mathcal{I}_\Delta$ properly represent the same collection of integers. In general, we may wish to describe this collection of integers for each class of ideal numbers of a particular discriminant, a question that motivates much of our development in the remainder of this text. In this section, we note some preliminary results.

Proposition 4.2.1. *Let $P(x)$ be the principal polynomial of discriminant Δ. Let $(a : k)$ be an ideal number in $\mathcal{I}_\Delta$, with $P(k) = ac$ and $P'(k) = b$. Then $(a : k)$ properly represents an integer m if and only if there are integers q and r with $\gcd(q, r) = 1$ for which $aq^2 + bqr + cr^2 = m$.*

In this situation, we also say that (q, r) is a *proper representation* of m by $(a : k)$. Note that $(-q, -r)$ is also a proper representation of m by $(a : k)$ in this case. We will typically treat the ordered pairs (q, r) and $(-q, -r)$ as identical in this context, in the same way, and for the same reason, as a unimodular matrix and its negative are identified.

Proof. Suppose that $(a : k)$ properly represents m, so that $(a : k) \cdot U = (m : \ell)$ for some unimodular matrix U and integer ℓ. Equation (4.4) shows that

$$aq^2 + bqr + cr^2 = m$$

for some integers q and r, namely the entries in the first column of U. Furthermore, q and r must be relatively prime in this case, since the determinant of U equals 1. Conversely, if q and r are integers with $\gcd(q, r) = 1$ for which $aq^2 + bqr + cr^2 = m$, then, by the Euclidean algorithm, we can find integers s and t so that $qt - rs = 1$. If $U = \left[\begin{smallmatrix} q & s \\ r & t \end{smallmatrix}\right]$, then $(a : k) \cdot U = (m : \ell)$ for some integer ℓ, so that $(a : k)$ properly represents m. □

Recall that if $f(x, y) = ax^2 + bxy + cy^2$ is a quadratic form, and $f(q, r) = m$ for some integers q and r, then we say that $f(x, y)$ *represents* m. If $f(q, r) = m$ with $\gcd(q, r) = 1$, then $f(x, y)$ *properly represents* m. Thus, Proposition 4.2.1 states that $(a : k)$ properly represents m if and only if the quadratic form written in ideal number notation as $f = (a : k)$ properly represents m.

Our next proposition shows, among other things, that an ideal number and its conjugate properly represent the same collection of integers.

Proposition 4.2.2. *Suppose that $(a : k) \sim (a_1 : k_1)$ in $\mathcal{I}_\Delta$. Then the following are true:*
1. $(-a : \overline{k}) \sim (-a_1 : \overline{k_1})$.
2. $(a : \overline{k}) \sim (a_1 : \overline{k_1})$.
3. $(-a : k) \sim (-a_1 : k_1)$.

In other words, the negatives, the conjugates, and the negative conjugates of two equivalent ideal numbers are also equivalent.

Proof. If $U = \begin{bmatrix} q & s \\ r & t \end{bmatrix}$ is unimodular, let $\overline{U} = \begin{bmatrix} q & -s \\ -r & t \end{bmatrix}$, which is also unimodular. Let $P(x)$ be the principal polynomial of discriminant Δ, and let $P(k) = ac$ and $P'(k) = b$. Suppose that $(a : k) \cdot U = (a_1 : k_1)$, so that $a_1 = aq^2 + bqr + cr^2$ and $k_1 = k + (aqs + brs + crt)$. Note that

$$\overline{k_1} = -P'(0) - k_1 = (-P'(0) - k) - (aqs + brs + crt) = \overline{k} - (aqs + brs + crt).$$

For statement (1), recall that $P(\overline{k}) = P(k) = (-a)(-c)$ and that $P'(\overline{k}) = -P'(k) = -b$. We find that

$$-a_1 = (-a)q^2 + (-b)qr + (-c)r^2$$

and

$$\overline{k_1} = \overline{k} + ((-a)qs + (-b)rs + (-c)rt),$$

implying that $(-a : \overline{k}) \cdot U = (-a_1 : \overline{k_1})$.

For statement (2), with $P(\overline{k}) = ac$ and $P'(\overline{k}) = -b$, we observe that

$$a_1 = aq^2 + (-b)q(-r) + c(-r)^2$$

and

$$\overline{k_1} = \overline{k} + (aq(-s) + (-b)(-r)(-s) + c(-r)t).$$

This shows that $(a : \overline{k}) \cdot \overline{U} = (a_1 : \overline{k_1})$.

For statement (3), with $P(k) = (-a)(-c)$ and $P'(k) = b$, we have

$$-a_1 = (-a)q^2 + bq(-r) + (-c)(-r)^2$$

and

$$k_1 = \overline{k} + ((-a)q(-s) + b(-r)(-s) + (-c)(-r)t),$$

so that $(-a : k) \cdot \overline{U} = (-a_1 : k_1)$. $\square$

Proposition 4.2.3. *Let Δ be a negative discriminant. If an ideal number $(a : k)$ in $\mathcal{I}_\Delta$ properly represents an integer m, then a and m are either both positive or both negative.*

Proof. We want to show that if $(a : k) \cdot U = (m : \ell)$ for some U in the unimodular group Γ, then a and m have the same sign. It will suffice to show this when $U = T$, a translation, and when $U = V$, the involution. The first claim is obvious, as T leaves the norm of an ideal number unchanged. If $P(x)$ is the principal polynomial of discriminant Δ, then the norms of $(a : k)$ and $(a : k)V = (c : \overline{k})$ satisfy the equation $P(k) = ac$. Since $P'(k)^2 - 4P(k) = \Delta$, then $4ac = P'(k)^2 - \Delta$ is positive when Δ is negative. In that case, a and c are either both positive or both negative. $\square$

Definition. If $(a : k)$ is an ideal number of negative discriminant, we say that $(a : k)$ is *positive definite* if a is positive, and that $(a : k)$ is *negative definite* if a is negative. An ideal number of positive discriminant is said to be *indefinite*.

Proposition 4.2.3 shows that a positive definite ideal number cannot properly represent a negative number. In particular, $(a : k)$ is never equivalent to $(-a : k)$ when Δ is negative. However, by Proposition 4.2.2, a positive definite ideal number properly represents an integer m if and only if its negative (or negative conjugate) properly represents $-m$. Thus, when Δ is a negative discriminant, we can often restrict our attention to positive definite ideal numbers.

On the other hand, when Δ is positive, then an ideal number properly represents both positive and negative integers, as the following example illustrates.

Example. Let $\Delta = 5$, with principal polynomial $P(x) = x^2 + x - 1$. Since the ideal number $(1 : 0)$ is equivalent to itself, then $(1 : 0)$ properly represents 1. But with $P(0) = -1 = 1 \cdot -1$ and $\overline{0} = -0 - P'(0) = -1$, we also see that $(1 : 0) \cdot V = (-1 : -1)$. Thus, $(1 : 0)$ properly represents -1 also. (Applying a translation to $(-1 : -1)$ also shows that $(1 : 0) \sim (-1 : 0)$ in this case. So, it is possible for $(a : k)$ to be equivalent to $(-a : k)$ when Δ is positive.)

4.3 Reduction of ideal numbers

To help us describe the relation of equivalence on $\mathcal{I}_\Delta$, it will be useful to have a complete set of class representatives. Our next result provides a strong upper bound on possibilities. We will refine this observation in Chapter 6 and in Chapter 7 for ideal numbers of negative discriminant and positive discriminant, respectively.

Proposition 4.3.1. *If Δ is a discriminant, then every element of $\mathcal{I}_\Delta$ is equivalent to an ideal number $(a : k)$ for which the following are true:*
1. *If Δ is positive, then $4a^2 \le \Delta$.*
2. *If Δ is negative, then $3a^2 \le -\Delta$.*

Proof. Let $P(x)$ be the principal polynomial of discriminant Δ. Among all the ideal numbers of $\mathcal{I}_\Delta$ in a given class, let $(a : k)$ be a representative for which $|a|$ is as small as possible. We say that a is a *minimal norm* for the class if this is true. We can further assume that $P'(k)^2 \le a^2$. (If a is positive, e. g., we can apply a translation to ensure that

$$\frac{-a - P'(0)}{2} < k \le \frac{a - P'(0)}{2},$$

so that $P'(k) = 2k + P'(0)$ satisfies $-a < P'(k) \le a$.) If $P(k) = ac$, then $(a : k) \sim (c : \overline{k})$ by an involution. Since $|a|$ is minimal, then $|a| \le |c|$, which implies that $a^2 \le |ac| = |P(k)|$. Recalling that $\Delta = P'(k)^2 - 4P(k)$, we then consider two cases:

1. If $P(k)$ is negative, then $a^2 \le -P(k)$. It follows that

$$4a^2 \le -4P(k) \le P'(k)^2 - 4P(k) = \Delta.$$

This case can occur only when Δ is positive.

2. If $P(k)$ is positive, then $a^2 \le P(k)$. Here, note that $-P'(k)^2 \ge -a^2$, so that

$$3a^2 = 4a^2 - a^2 \le 4P(k) - P'(k)^2 = -\Delta.$$

This case can occur only when Δ is negative.

Thus, in every class of ideal numbers of discriminant Δ, a representative $(a : k)$ with minimal norm (in absolute value) has the upper bound on a^2 noted. □

Corollary 4.3.2. *For every discriminant Δ, there are finitely many classes of ideal numbers in $\mathcal{I}_\Delta$.*

Proof. For each discriminant Δ, there are finitely many integers a for which $4a^2 \le \Delta$ (if Δ is positive) or for which $3a^2 \le -\Delta$ (if Δ is negative). Given a, there are finitely many values of k with $0 \le k < |a|$, as we can assume is true, using a translation, for an ideal number of norm a. Thus, there are finitely many potential representatives of classes of ideal numbers, and so only finitely many distinct classes in $\mathcal{C}_\Delta$. □

In Section 2.1, we outlined an algorithm listing all congruence classes $\langle a : k \rangle$ of ideal numbers in $\mathcal{Q}_\Delta$ whose norms satisfy $0 < a \le u$, where u is some arbitrary upper bound. Proposition 4.3.1 implies that we can use this approach to list potential class representatives under equivalence, with $u = \lfloor \sqrt{\frac{\Delta}{4}} \rfloor$ if Δ is positive and $u = \lfloor \sqrt{\frac{-\Delta}{3}} \rfloor$ if Δ is negative. However, we must consider both $(a : k)$ and $(-a : k)$ as being in possibly distinct classes under equivalence. We illustrate this approach in the following two examples.

Example. If $\Delta = -35$, let $u = \lfloor \sqrt{\frac{35}{3}} \rfloor = 3$. In an example in Section 2.1, we listed all congruence classes in $\mathcal{Q}_{-35}$ of norm $0 < a \le 40$. From that list, we conclude that every positive definite ideal number in $\mathcal{I}_{-35}$ is equivalent to one of the following ideal numbers:

$$(1 : 0), \quad (3 : 0), \quad (3 : -1),$$

while every negative definite ideal number is equivalent to one of the following:

$$(-1 : 0), \quad (-3 : 0), \quad (-3 : -1).$$

However, we cannot assume (as yet) that the classes of these ideal numbers are in fact distinct.

Example. For $\Delta = \Delta(30, 1) = 120$, let $u = \lfloor \sqrt{\frac{120}{4}} \rfloor = 5$. The principal polynomial of discriminant 120 is $P(x) = x^2 - 30$. To find potential equivalence class representatives, it suffices to test $-2 \le x \le 2$, as in the following table:

x	0	±1	±2
$P(x)$	−30	−29	−26

We conclude that every ideal number in $\mathcal{I}_{120}$ is equivalent to at least one of the following:

$$(1:0) \quad (-1:0) \quad (2:0) \quad (-2:0) \quad (3:0) \quad (-3:0) \quad (5:0) \quad (-5:0).$$

Again, there might be other equivalences among these ideal numbers.

Proposition 4.3.1 also suggests a systematic method of *reduction* of ideal number classes that we can typically use to find a class representative with minimal norm. Before illustrating the idea with examples, we note the following proposition, which connects the reduction process to proper representations.

Proposition 4.3.3. *Let $(a : k)$ be an ideal number of discriminant Δ, and suppose that (q, r) is a proper representation of an integer m by $(a : k)$. Let T and V be the translation and involution matrices defined as in equation (4.1). Then:*
1. *$(r, -q)$ is a proper representation of m by $(a : k) \cdot V$, and*
2. *$(q - nr, r)$ is a proper representation of m by $(a : k) \cdot T^n$ for all integers n.*

Proof. Let $P(k) = ac$ and $P'(k) = b$, where $P(x)$ is the principal polynomial of discriminant Δ. Let $f(x, y) = ax^2 + bxy + cy^2$ be the quadratic form given as $f = (a : k)$ in ideal number notation. We are given that $f(q, r) = m$ for some integers q and r with $\gcd(q, r) = 1$.

(1) By equation (4.6), $(a : k) \cdot V = (c : \overline{k})$, where $\overline{k}$ is the conjugate of k with respect to Δ. Since $P(\overline{k}) = P(k) = ca$ and $P'(\overline{k}) = -P'(k) = -b$, the quadratic form with ideal number expression $(c : \overline{k})$ is $f_1(x, y) = cx^2 - bxy + ay^2$. Then

$$f_1(r, -q) = cr^2 - br(-q) + a(-q)^2 = aq^2 + bqr + cr^2 = m,$$

so that $(r, -q)$ is a proper representation of m by $(a : k) \cdot V$.

(2) By equation (4.5), $(a : k) \cdot T^n = (a : k + na)$. Using equation (1.4), we find that

$$P(k + na) = P(k) + (k + na - k)(k + na + k + P'(0)) = a(an^2 + bn + c),$$

while $P'(k + na) = 2(k + na) + P'(0) = b + 2an$. Hence,

$$f_2(x, y) = ax^2 + (b + 2an)xy + (an^2 + bn + c)y^2$$

is the quadratic form with ideal number expression $(a : k + na)$. Direct expansion and simplification shows that

$$f_2(q - nr, r) = a(q - nr)^2 + (b + 2an)(q - nr)r + (an^2 + bn + c)r^2$$
$$= aq^2 + bqr + cr^2 = m.$$

A common divisor of $q - nr$ and r also divides $q = (q - nr) + n(r)$. So, if q and r are relatively prime, then $\gcd(q - nr, r) = 1$. Thus, $(q - nr, r)$ is a proper representation of m by $(a : k) \cdot T^n$. □

Example. Let $\Delta = -35$ with principal polynomial $P(x) = x^2 + x + 9$. Since $P(47) = 2265 = 151 \cdot 15$, we have $(151 : 47)$ as an element of $\mathcal{I}_{-35}$. The following diagram illustrates a sequence of steps showing that $(151 : 47)$ is equivalent to $(1 : 0)$:

$$(151 : 47) \xrightarrow{V} (15 : -48) \xrightarrow{T^3} (15 : -3) \xrightarrow{V} (1 : 2) \xrightarrow{T^{-2}} (1 : 0).$$

Given an ideal number $(a : k)$, we apply a translation, if necessary, to ensure that $P(k)$ is as small as possible. If $P(k) = ac$ with $|c| < |a|$, we apply the involution to replace $(a : k)$ by an ideal number of norm c, and we repeat the process as needed.

An implication of this example is that $(1 : 0)$ properly represents 151, as we can verify using Proposition 4.3.3. Note that $(q, r) = (1, 0)$ is a proper representation of 151 by $(151 : 47)$. We can follow the sequence of translations and involutions that transforms $(151 : 47)$ noted above to transform $(1, 0)$ to a proper representation of 151 by $(1 : 0)$. Specifically, each application of V changes (q, r) to $(r, -q)$ (or to $(-r, q)$ as we prefer), and each application of T^n changes (q, r) to $(q - nr, r)$. We illustrate this below with a similar arrow diagram as above:

$$(1, 0) \xrightarrow{V} (0, 1) \xrightarrow{T^3} (-3, 1) \xrightarrow{V} (1, 3) \xrightarrow{T^{-2}} (7, 3).$$

The conclusion is that $(7, 3)$ is a proper representation of 151 by $(1 : 0)$. In fact, since $P(0) = 1 \cdot 9$ and $P'(0) = 1$, the quadratic form $f = (1 : 0)$ is $f(x, y) = x^2 + xy + 9y^2$. We find that

$$f(7, 3) = 7^2 + 7 \cdot 3 + 9 \cdot 3^2 = 49 + 21 + 81 = 151,$$

verifying the claim.

We introduce the following notation for later examples of representation calculations. Let Δ be a discriminant with principal polynomial $P(x)$. We write $(a : k)_r^q$ for an ideal number $(a : k)$ in $\mathcal{I}_\Delta$ together with an ordered pair (q, r) of relatively prime integers. We then denote the action of a translation T^n or an involution V on an expression of this type to be

$$(a : k)_r^q \xrightarrow{T^n} (a : k + na)_r^{q-nr} \quad \text{and} \quad (a : k)_r^q \xrightarrow{V} (c : \overline{k})_{-q}^r, \tag{4.7}$$

where $P(k) = ac$ and $\bar{k}$ is the conjugate of k with respect to Δ. (We can change the sign of both the subscript and superscript, and will typically do so to ensure that the subscript is nonnegative.) If we can transform $(m : \ell)_0^1$ to $(a : k)_r^q$ by a sequence of translations and involutions, then (q, r) is a proper representation of m by $(a : k)$. We refer to this process as a *representation reduction* of the ideal number $(m : \ell)$.

Example. Again with $\Delta = -35$, consider $(283 : 37)$, an ideal number since $P(37) = 1415 = 283 \cdot 5$. Here, we can apply the following sequence of involutions and translations:

$$(283 : 37)_0^1 \xrightarrow{V} (5 : -38)_1^0 \xrightarrow{T^8} (5 : 2)_1^{-8}$$
$$\xrightarrow{V} (3 : -3)_8^1 \xrightarrow{T} (3 : 0)_8^{-7} \xrightarrow{V} (3 : -1)_7^8.$$

We conclude that $(8, 7)$ is a proper representation of 283 by $(3 : -1)$. With $P(-1) = 3 \cdot 3$ and $P'(-1) = -1$, the quadratic form with ideal number expression $(3 : -1)$ is $f(x, y) = 3x^2 - xy + 3y^2$. We verify that

$$f(8, 7) = 3 \cdot 8^2 - 8 \cdot 7 + 3 \cdot 7^2 = 192 - 56 + 147 = 283.$$

As an accidental observation in the preceding example, note that $(3 : 0)$ is equivalent to $(3 : -1)$ in $\mathcal{I}_{-35}$. Thus, not all of the potential representatives found in a previous example belong to separate classes. We will pursue the question of finding a precise set of class representatives further for negative discriminants in Chapter 6 and for positive discriminants in Chapter 7.

4.4 Automorphs of ideal numbers

The following definition gives us a better handle on equivalence of ideal numbers.

Definition. Let $(a : k)$ be an ideal number of discriminant Δ. If a unimodular matrix U has the property that $(a : k) \cdot U = (a : k)$, then we say that U is an *automorph* of $(a : k)$.

Proposition 4.4.1. *For each ideal number $(a : k)$ in $\mathcal{I}_\Delta$, the set of all automorphs of $(a : k)$ forms a subgroup of the unimodular group Γ, which we write as* $\mathrm{Aut}(a : k)$.

Proof. Let $(a : k)$ be an ideal number of discriminant Δ.
1. It is always true that $(a : k) \cdot I = (a : k)$, so $\mathrm{Aut}(a : k)$ contains the identity matrix.
2. If $(a : k) \cdot U = (a : k)$, then $(a : k) \cdot U^{-1} = (a : k)$, so that $\mathrm{Aut}(a : k)$ contains the inverse of each of its elements.
3. If $(a : k) \cdot U_1 = (a : k)$ and $(a : k) \cdot U_2 = (a : k)$, then $(a : k) \cdot U_1 U_2 = (a : k)$. Thus, $\mathrm{Aut}(a : k)$ is closed under multiplication.

Thus, $\mathrm{Aut}(a : k)$ is a subgroup of the unimodular group. $\qquad\square$

We also have the following precise characterization of automorphs.

Proposition 4.4.2. *Let $P(x)$ be the principal polynomial of discriminant Δ. Let $(a : k)$ be an ideal number in $\mathcal{I}_\Delta$, with $P(k) = ac$ and $P'(k) = b$. Then a unimodular matrix $U = \begin{bmatrix} q & s \\ r & t \end{bmatrix}$ is an automorph of $(a : k)$ if and only if the following are true:*

1. $aq^2 + bqr + cr^2 = a.$
2. *a divides cr, and $s = -\frac{cr}{a}$.*
3. *a divides br, and $t = q + \frac{br}{a}$.*

Proof. We want to describe all U for which $(a : k) \cdot U = (a : k)$. If U has the entries above, then Theorem 4.1.2 implies that $aq^2 + bqr + cr^2 = a$ and $aqs + brs + crt = 0$. Thus, we see that

$$as = as - 0 \cdot q = (aq^2 s + bqrs + cr^2 s) - (aq^2 s + bqrs + cqrt)$$
$$= cr(rs - qt) = -cr$$

and

$$at = at - 0 \cdot r = (aq^2 t + bqrt + cr^2 t) - (aqrs + bqrs + cr^2 t)$$
$$= (aq + br)(qt - rs) = aq + br,$$

using the assumption that $qt - rs = 1$ in both equations. If a divides both cr and br, we can solve for integers s and t as given in (2) and (3). □

If $f(x,y) = ax^2 + bxy + cy^2$ is the quadratic form written as $(a : k)$ in ideal number notation, then Proposition 4.4.2 shows that we can find all automorphs of $(a : k)$ by finding all proper representations of a by $f(x,y)$. Of course, $(x,y) = (1,0)$ is an obvious solution yielding the identity automorph, but other examples might exist. We illustrate this with the following example and general statement.

Example. Let $\Delta = \Delta(2,1) = 8$. Consider the ideal number $(1 : 0)$ in $\mathcal{I}_8$, with $f(x,y) = x^2 - 2y^2$ as the corresponding quadratic form. Here, we might notice that $f(3,2) = 1$, and with $-cr = -(-2)2 = 4$ and $q + br = 3 + (0)2 = 3$, we have $U = \begin{bmatrix} 3 & 4 \\ 2 & 3 \end{bmatrix}$ as an automorph of $(1 : 0)$. Powers of U, such as

$$U^2 = \begin{bmatrix} 17 & 24 \\ 12 & 17 \end{bmatrix}, \quad U^3 = \begin{bmatrix} 99 & 140 \\ 70 & 99 \end{bmatrix}, \quad U^4 = \begin{bmatrix} 577 & 816 \\ 408 & 577 \end{bmatrix}, \quad \ldots,$$

are also automorphs of $(1 : 0)$.

We will see later that the group of automorphs of an ideal number of a positive discriminant is always infinite. On the other hand, when Δ is negative, then the group of automorphs of an ideal number is finite. We can describe the following general case in full.

Proposition 4.4.3. *Let Δ be a negative discriminant. If T and V are the unimodular matrices defined in equation (4.1), then the group of automorphs of the ideal number $(1 : 0)$ in $\mathcal{I}_\Delta$ is:*

1. $\{I\}$, *if* $\Delta < -4$,
2. $\{I, V\}$, *if* $\Delta = -4$, *and*
3. $\{I, VT, (VT)^2\}$, *if* $\Delta = -3$.

For the proof, we recall the following observation from Proposition 1.4.2. If $f(x, y) = ax^2 + bxy + cy^2$ is a quadratic form with $\Delta = b^2 - 4ac$, and $f(q, r) = m$, then

$$4am = (2aq + br)^2 - \Delta r^2. \tag{4.8}$$

Proof. Let $f(x, y) = x^2 + bxy + cy^2$ be the quadratic form of discriminant Δ having ideal number expression $(1 : 0)$. If $f(q, r) = 1$, then equation (4.8) shows that $4 = (2q + br)^2 - \Delta r^2$. We can find all solutions of this equation, depending on $\Delta < 0$. (Recall that (q, r) and $(-q, -r)$ are regarded as the same solution.)

1. If $\Delta < -4$, then $4 = (2q + br)^2 - \Delta r^2$ is possible only when $r = 0$ and $q = 1$.
2. If $\Delta = -4$, then $f(x, y) = x^2 + y^2$, and equation (4.8) becomes $4 = (2q)^2 + 4r^2$. Here, we have the solution $(q, r) = (0, 1)$, in addition to $(q, r) = (1, 0)$.
3. If $\Delta = -3$, then $f(x, y) = x^2 + xy + y^2$, and equation (4.8) implies that $4 = (2q + r)^2 + 3r^2$. In addition to $(q, r) = (1, 0)$, we can have $r = 1$ with $2q + r = \pm 1$. This gives us two more distinct solutions: $(q, r) = (0, 1)$ and $(q, r) = (-1, 1)$.

By Proposition 4.4.2, each solution of $f(q, r) = 1$ listed above leads to an automorph of $(1 : 0)$ in matrix form. For instance, in case (3), where $a = 1$, $b = 1$, and $c = 1$, the solution $(q, r) = (-1, 1)$ implies that $s = -cr = -1$ and $t = q + br = 0$, so that $U = \begin{bmatrix} -1 & -1 \\ 1 & 0 \end{bmatrix}$ is an automorph of $f(x, y) = x^2 + xy + y^2$. With similar calculations, we find that for $f(x, y)$ the quadratic form that equals $(1 : 0)$ as an ideal number expression:

1. $\mathrm{Aut}(f) = \{\begin{bmatrix} 1 & 0 \\ 0 & 1 \end{bmatrix}\} = \{I\}$, if $\Delta < -4$,
2. $\mathrm{Aut}(f) = \{\begin{bmatrix} 1 & 0 \\ 0 & 1 \end{bmatrix}, \begin{bmatrix} 0 & -1 \\ 1 & 0 \end{bmatrix}\} = \{I, V\}$, if $\Delta = -4$,
3. $\mathrm{Aut}(f) = \{\begin{bmatrix} 1 & 0 \\ 0 & 1 \end{bmatrix}, \begin{bmatrix} 0 & -1 \\ 1 & 1 \end{bmatrix}, \begin{bmatrix} -1 & -1 \\ 1 & 0 \end{bmatrix}\} = \{I, VT, (VT)^2\}$, if $\Delta = -3$,

as we wanted to show. □

We conclude Section 4.4 with the following proposition.

Proposition 4.4.4. *Let $(a : k)$ be an ideal number of discriminant Δ. Suppose that*

$$(a : k) \cdot U = (a_1 : k_1)$$

for some unimodular matrix U. Then the unimodular matrix U_1 also has the property that $(a : k) \cdot U_1 = (a_1 : k_1)$ if and only if $U_1 = WU$ for some automorph W of $(a : k)$.

In group-theoretic terms, Proposition 4.4.4 implies that if $(a : k) \sim (a_1 : k_1)$, then the collection of matrices U for which $(a : k) \cdot U = (a_1 : k_1)$ is a right coset of $\mathrm{Aut}(a : k)$ in Γ.

Proof. If $U_1 = WU$ for some automorph W of $(a : k)$, then

$$(a : k) \cdot U_1 = (a : k) \cdot (WU) = ((a : k) \cdot W) \cdot U = (a : k) \cdot U = (a_1 : k_1).$$

Conversely, suppose that $(a : k) \cdot U_1 = (a_1 : k_1)$ for some unimodular matrix U. Then

$$((a : k) \cdot U_1) \cdot U^{-1} = (a : k) \cdot (U_1 U^{-1}) = (a_1 : k_1) \cdot U^{-1} = (a : k).$$

But then $U_1 U^{-1} = W$ is an automorph of $(a : k)$, that is, $U_1 = WU$ for some automorph W of $(a : k)$. □

Proposition 4.4.4 can help us describe different representations of an integer by a given quadratic form.

Example. Let $f(x, y) = x^2 - 2y^2$, the quadratic form whose ideal number expression is $(1 : 0)$ in $\mathcal{I}_8$. In a previous example, we saw that $(1 : 0)$ has infinitely many automorphs. It follows that if $f(q, r) = m$, then there are likewise infinitely many representations of m by f. For instance, if we note that $f(17, 14) = -103$, then other representations of -103 by f include:

$$\begin{bmatrix} 3 & 4 \\ 2 & 3 \end{bmatrix} \begin{bmatrix} 17 \\ 14 \end{bmatrix} = \begin{bmatrix} 107 \\ 76 \end{bmatrix}, \quad \begin{bmatrix} 17 & 24 \\ 12 & 17 \end{bmatrix} \begin{bmatrix} 17 \\ 14 \end{bmatrix} = \begin{bmatrix} 625 \\ 442 \end{bmatrix},$$

and so forth. (Here, we are using only the first column of a matrix U for which $(1 : 0) \cdot U = (-103 : k)$ for some k. That first column is sufficient for our purposes in this case.)

4.5 The class group of a discriminant

To conclude Chapter 4, we use the relation of equivalence of ideal numbers to define a collection of groups that we will consider throughout the remainder of the text.

Definition. Let Δ be a discriminant. If Δ is positive, let $\mathcal{C}_\Delta$ be the set of all classes of primitive ideal numbers in $\mathcal{I}_\Delta$. If Δ is negative, let $\mathcal{C}_\Delta$ be the set of all classes of primitive *positive definite* ideal numbers in $\mathcal{I}_\Delta$. We refer to $\mathcal{C}_\Delta$ as the *ideal number class group*, or simply the *class group*, of discriminant Δ.

Proposition 4.1.3 ensures that the collection of primitive ideal number classes of a particular discriminant is well-defined. The restriction of $\mathcal{C}_\Delta$ to include only positive definite ideal number classes when Δ is negative is mainly for convenience. By Proposition 4.2.2, $(a : k) \sim (b : \ell)$ if and only if $(-a : k) \sim (-b : \ell)$. Thus it is easy to extend results about the set $\mathcal{C}_\Delta$ to the entire collection of primitive classes if necessary.

We can define the same operation of composition on C_Δ as we did for congruence classes of ideal numbers. As our terminology suggests, C_Δ has the structure of a group under this operation. For convenience, we repeat the definition of composition below.

Definition. Let $P(x)$ be the principal polynomial of discriminant Δ and let $[a : k]$ and $[b : \ell]$ be elements of C_Δ. Let $u = k + \ell + P'(0)$ and $v = k\ell - P(0)$, and let $g = \gcd(a, b, u) = ar + bs + ut$ for some integers r, s, and t. Then define $[a : k] \circ [b : \ell]$ to equal $[c : m]$, where

$$c = \frac{a}{g} \cdot \frac{b}{g} \quad \text{and} \quad m = \frac{a\ell r + bks + vt}{g}. \tag{4.9}$$

We refer to $[c : m]$ as the *composite* of $[a : k]$ and $[b : \ell]$ and to this operation on C_Δ as *composition*.

Proposition 4.5.1. *Let Δ be a discriminant and let C_Δ be the set of all classes of primitive ideal numbers in $\mathcal{I}_\Delta$, positive definite if Δ is negative. Then C_Δ is a finite Abelian group under the operation of composition defined above.*

Proof. In Appendix C, a relation of equivalence is defined on nontrivial ideals of a quadratic domain $D = D_\Delta$. The operation of ideal multiplication is well-defined on the set of equivalence classes of ideals of D, and the subset of classes of primitive ideals forms a group under this operation. It is further established that if $A = \langle a : k \rangle$ and $B = \langle b : \ell \rangle$ are primitive ideals, written in ideal number notation, then for the corresponding ideal numbers in $\mathcal{I}_\Delta$, either $(a : k) \sim (b : \ell)$ or $(a : k) \sim (-b : \ell)$. Ideal multiplication induces the operation of ideal number composition above, which is then well-defined on ideal number classes. The class group C_Δ can be regarded as identical to the group of ideal classes, or it may have twice as many elements as that group. In either case, C_Δ is a finite Abelian group under composition. $\square$

When Δ is negative, the group C_Δ of primitive positive definite ideal numbers is always the same as the ideal class group developed in Appendix C. (This is another reason it is convenient to restrict attention to positive definite ideal numbers.) When Δ is positive, then C_Δ equals the ideal class group if and only if the ideal numbers $(1 : 0)$ and $(-1 : 0)$ are equivalent.

The operation of composition has an important connection to representations of integers by quadratic forms. The group properties of C_Δ then give us further information about these representations.

Theorem 4.5.2. *Let f_1 and f_2 be primitive quadratic forms of discriminant Δ, written as $f_1 = (a_1 : k_1)$ and $f_2 = (a_2 : k_2)$ in ideal number notation. Let*

$$[a_1 : k_1] \circ [a_2 : k_2] = [a : k]$$

in the class group C_Δ, and let f be the quadratic form expressed as $(a : k)$ in ideal number notation. Suppose that f_1 and f_2 properly represent integers m_1 and m_2, respectively. Then f represents $m_1 m_2$. If $\gcd(m_1, m_2) = 1$, then this representation is proper.

Proof. If f_1 and f_2 properly represent integers m_1 and m_2, respectively, then $(a_1 : k_1) \sim (m_1 : \ell_1)$ and $(a_2 : k_2) \sim (m_2 : \ell_2)$ for some integers ℓ_1 and ℓ_2. But then, applying the composition formula, there is an integer ℓ so that

$$[a : k] = [a_1 : k_1] \circ [a_2 : k_2] = [m_1 : \ell_1] \circ [m_2 : \ell_2] = [m : \ell],$$

where $m = (m_1 m_2)/g^2$ for some integer g, a greatest common divisor of a collection of integers including m_1 and m_2. The quadratic form $f = (a : k)$ then properly represents $(m_1 m_2)/g^2$, and so represents $m_1 m_2$. If $\gcd(m_1, m_2) = 1$, then we can be sure that $g = 1$, so that f properly represents $m_1 m_2$. □

Example. Let $\Delta = \Delta(-35, 1) = -35$, with principal polynomial $P(x) = x^2 + x + 9$. In an example in Section 4.3, we saw that each *positive definite* ideal number in $\mathcal{I}_{-35}$ is equivalent to $(1 : 0)$, $(3 : 0)$, or $(3 : -1)$, but then saw, in a later example, that $(3 : 0)$ and $(3 : -1)$ are equivalent to each other. We will assume for now that $(1 : 0)$ and $(3 : 0)$ are not equivalent, a fact that we will prove in Chapter 5 using the concept of genus equivalence. Thus,

$$C_{-35} = \{[1 : 0], [3 : 0]\},$$

a group with two elements. The class $[1 : 0]$ serves as the identity element of every class group. Applying the definition of composition with $a = 3 = b$ and $k = 0 = \ell$, so that $u = 1$, $v = -9$, and $g = \gcd(3, 3, 1) = 1 = 3(0) + 3(0) + 1(1)$, we conclude that

$$[3 : 0] \circ [3 : 0] = [9 : -9].$$

But note that

$$(9 : -9) \xrightarrow{T} (9 : 0) \xrightarrow{V} (1 : -1) \xrightarrow{T} (1 : 0),$$

using the reduction process introduced in Section 4.3. Thus, we can say that $[3 : 0]^2 = [1 : 0]$ in C_{-35}, as must be the case in a group of order two. (We could also use the fact that $(3 : 0) \sim (3 : -1)$ to argue that $[3 : 0] = [3 : -1]$ is its own inverse in C_{-35}.)

Now let

$$f_1(x, y) = x^2 + xy + 9y^2 \quad \text{and} \quad f_2(x, y) = 3x^2 + xy + 3y^2,$$

the quadratic forms of discriminant $\Delta = -35$ having ideal number expressions $(1 : 0)$ and $(3 : 0)$, respectively. Since $[1 : 0] \circ [3 : 0] = [3 : 0]$ in the class group C_{-35}, an implication of Proposition 4.5.2 is that if f_1 and f_2 properly represent m_1 and m_2, respectively, then

f_2 represents $m_1 m_2$, though not necessarily properly. For instance, since $f_1(2, 1) = 15$ and $f_2(1, -1) = 5$, then f_2 represents $15 \cdot 5 = 75$. We might notice that $f_2(5, 0) = 75$ in fact. Here, we can be sure that no proper representation exist. Otherwise, there must be an ideal number $(75 : k)$ in $\mathcal{I}_{-35}$. But results from Chapter 2 show that this is impossible, since $p = 5$ divides the primitive discriminant $\Delta = -35$.

As another example, since $f_1(3, 2) = 51$ and $f_2(1, 1) = 7$, with $\gcd(51, 7) = 1$, then f_2 properly represents $51 \cdot 7 = 357$. To illustrate how such a representation can be constructed, note that $(1 : 0) \sim (51 : 23)$ and $(3 : 0) \sim (7 : 3)$, say by Theorem 4.1.2, using unimodular matrices whose first columns have entries 3 and 2, or 1 and 1, respectively. Now

$$[51 : 23] \circ [7 : 3] = [357 : 3083]$$

by the composition formula. We can verify as follows that this class reduces to $[3 : 0]$, and construct a representation of 357 by $(3 : 0)$, using the notation of equations (4.7):

$$(357 : 3083)_0^1 \xrightarrow{T^{-9}} (357 : -130)_0^1 \xrightarrow{V} (47 : 129)_1^0 \xrightarrow{T^{-3}} (47 : -12)_1^3$$
$$\xrightarrow{V} (3 : 11)_3^{-1} \xrightarrow{T^{-4}} (3 : -1)_3^{11} \xrightarrow{V} (3 : 0)_{11}^{-3}.$$

We verify that $f_2(-3, 11) = 3 \cdot (-3)^2 + (-3)(11) + 3(11)^2 = 27 - 33 + 363 = 357$.

5 Genera of ideal numbers

In Section 4.5, we saw that the collection of classes of primitive ideal numbers of a fixed discriminant forms a group under composition. In this chapter, we define another equivalence relation, called *genus equivalence*, on ideal numbers. We will demonstrate how this relation helps us compute the class group of a particular discriminant and describe the integers that are properly represented by various classes more fully. In this chapter, we use the notation of Legendre symbols. The reader may wish to review the definition of $(\frac{a}{p})$ and the properties of Legendre symbols listed in Section A.7 of Appendix A.

5.1 The genus group of a discriminant

In this section, as a first step in defining genus equivalence of ideal numbers, we associate a group of numbers to each discriminant as follows.

Definition. Let $\Delta = \Delta(d, g)$ be a discriminant, where $d \neq 1$ is a square-free integer and g is a positive integer. Let $G = \mathbb{Z}_{|\Delta|}^*$ be the group of units, under multiplication, in the ring of congruence classes of integers modulo $|\Delta|$. Let $G^2 = \{x^2 \mid x \in G\}$ be the set of all squares of elements of G and define H as follows:

$$H = \begin{cases} G^2 \cup (1 - \frac{\Delta}{4})G^2, & \text{if } d \text{ is even and } g \text{ is odd,} \\ G^2 \cup (1 - \frac{\Delta}{2})G^2, & \text{if } d \equiv 1 \pmod 4 \text{ and } g \equiv 2 \pmod 4, \\ G^2, & \text{in all other cases.} \end{cases}$$

Then H is a subgroup of G, called the *principal genus* of Δ. If a is an element of G, then the coset aH is called the *genus* of a with respect to Δ, pluralized as *genera*. The quotient group G/H is called the *genus group* of Δ.

We verify some claims made in this definition. Since G is a finite Abelian group, the subset G^2 is closed under multiplication ($x^2 \cdot y^2 = (xy)^2$ if multiplication is commutative), and so is a subgroup of G. Properties of the quotient group G/G^2 show that the union of G^2 with any one of its cosets is likewise closed under multiplication. (For instance, if a and b are elements of the same coset xG^2, then $aG^2 \cdot bG^2 = (ab)G^2 = x^2G^2$. But then ab is in G^2 since x^2 is in G^2 by definition.) Thus, H, as defined above, is a subgroup of G in every case. The following definition and proposition gives us an alternative characterization of genera.

Definition. Let Δ be a discriminant. Then we associate a collection of *genus symbols* to each element a in $G = \mathbb{Z}_{|\Delta|}^*$ as follows:
1.   If p is an odd prime divisor of Δ, let s_p equal the Legendre symbol $(\frac{a}{p})$.

https://doi.org/10.1515/9783111319360-005

2. If $\Delta \equiv 0$ or $12 \pmod{16}$, let

$$s_{-1} = \begin{cases} 1, & \text{if } a \equiv 1 \pmod 4, \\ -1, & \text{if } a \equiv 3 \pmod 4. \end{cases}$$

3. If $\Delta \equiv 0$ or $8 \pmod{32}$, let

$$s_2 = \begin{cases} 1, & \text{if } a \equiv 1 \text{ or } 7 \pmod 8, \\ -1, & \text{if } a \equiv 3 \text{ or } 5 \pmod 8. \end{cases}$$

4. If $\Delta \equiv 0$ or $24 \pmod{32}$, let

$$s_{-2} = \begin{cases} 1, & \text{if } a \equiv 1 \text{ or } 3 \pmod 8, \\ -1, & \text{if } a \equiv 5 \text{ or } 7 \pmod 8. \end{cases}$$

When $\Delta \equiv 0 \pmod{32}$, then each of the symbols s_{-1}, s_2, and s_{-2} is defined for a in $\mathbb{Z}_{|\Delta|}^*$. But an examination of the possibilities for a modulo 8 shows that the product of these three symbols must equal 1. (See Proposition 5.1.2 at the end of this section.) Aside from this restriction, genus symbols for a are independent of each other.

Proposition 5.1.1. *Let $G = \mathbb{Z}_{|\Delta|}^*$, where Δ is a discriminant. Then the principal genus of Δ consists precisely of the elements a in G for which every defined genus symbol equals 1.*

Proof. Let $\Delta = \Delta(d, g)$ be a discriminant and let a be an integer relatively prime to Δ. If Δ is odd, then a is congruent to a square modulo $|\Delta|$ if and only if $\left(\frac{a}{p}\right) = 1$ for every odd prime divisor of Δ. If Δ is divisible by 4 but not by 8, then $a \equiv 1 \pmod 4$ must also be true. If 8 divides Δ, then we must add the condition that $a \equiv 1 \pmod 8$. Thus, if a is an element of G^2, then every genus symbol assigned to a is 1. The additional elements in some cases of the principal genus H, depending on the values of d and g, are likewise obtained by other conditions on genus symbols. We verify this in one situation, leaving other cases to be illustrated by examples.

Suppose that $d \equiv 6 \pmod 8$ and g is odd, so that $\Delta = 4dg^2 \equiv 24 \pmod{32}$. Here, the principal genus is defined as $G^2 \cup (1 - dg^2)G^2$. If p is an odd prime divisor of Δ, then all elements a of $(1 - dg^2)G^2$ satisfy $\left(\frac{a}{p}\right) = 1$. On the other hand, since $1 - dg^2 \equiv 1 - 6 \cdot 1 \pmod 8$, then $a \equiv 3 \pmod 8$. From the definition of the symbol s_{-2}, we have that all genus symbols defined in this case equal 1. $\qquad\square$

If q is an odd prime, then $H_q = \{x^2 \bmod q \mid 1 \le x \le \frac{1}{2}(q - 1)\}$ is the set of all distinct squares in $\mathbb{Z}_q^*$. We can often use this fact, along with the Chinese remainder theorem, to compile the principal genus of a discriminant Δ without calculating all squares of elements of G.

Example. Let $\Delta = \Delta(-35, 1) = -35$, so that $G = \mathbb{Z}_{35}^*$. The squares modulo 5 and modulo 7 are

$$H_5 = \{x^2 \bmod 5 \mid 1 \le x \le 2)\} = \{1, 4\}$$

and

$$H_7 = \{x^2 \bmod 7 \mid 1 \le x \le 3\} = \{1, 4, 2\},$$

respectively. The solution of $x \equiv s \pmod 5$ and $x \equiv t \pmod 7$ is $x \equiv -14s + 15t \pmod{35}$ (using the Euclidean algorithm as in Corollary A.6.2). We find the principal genus $H = G^2$ by substituting all possibilities of $s = 1$ or 4 and $t = 1, 2,$ or 4, as in the following table:

s	t	x = −14s + 15t	x mod 35
1	1	1	1
1	2	16	16
1	4	46	11
4	1	−41	29
4	2	−26	9
4	4	4	4

For example, $11 \equiv 1 \equiv 1^2 \pmod 5$ and $11 \equiv 4 \equiv 2^2 \pmod 7$, and we can verify that $11 = x^2$ in G if $x \equiv 1 \pmod 5$ and $x \equiv 2 \pmod 7$, that is, $x = 16$ in G. We calculate the distinct cosets of H in G in the following table. The genus symbols associated to each coset are also listed as sequences of + and − signs for s_5 and s_7 in that order. It is straightforward to verify that these are well-defined. For instance, each element a listed in $2 \cdot H$ is congruent to 2 or 3 modulo 5, but to 1, 2, or 4 modulo 7:

Coset	a modulo 35	Genus
1 · H	1, 4, 9, 11, 16, 29	++
2 · H	2, 8, 18, 22, 32, 23	−+
3 · H	3, 12, 27, 33, 13, 17	−−
6 · H	6, 24, 19, 31, 26, 34	+−

Let $\phi(n)$ denote the *Euler function* evaluated at n, that is, the number of elements in $\mathbb{Z}_n^*$. There are $\phi(35) = \phi(5) \cdot \phi(7) = 24$ elements in G, partitioned into four genera each containing six elements. The genus group of $\Delta = -35$ is

$$G/H = \{1 \cdot H, 2, \cdot H, 3 \cdot H, 6 \cdot H\},$$

with the square of each element equal to $1 \cdot H$. We can also identify each genus with its ordered pair of genus symbols. In any event, G/H is isomorphic to $\mathbb{Z}_2 \times \mathbb{Z}_2$, here using a multiplicative model, $\{1, -1\}$, for the cyclic groups of order two.

Example. Let $\Delta = \Delta(30, 1) = 120$. Since $d = 30$ is even and $g = 1$ is odd, the principal genus of Δ is $H = G^2 \cup (1 - \frac{\Delta}{4})G^2 = G^2 \cup (-29 \cdot G^2)$, where $G = \mathbb{Z}_{120}^*$. Note that -29 is congruent to 1 modulo both 3 and 5, but is congruent to 3 modulo 8. We can compile the elements of H as solutions of

$$x \equiv r \pmod{3}, \quad x \equiv s \pmod{5}, \quad x \equiv t \pmod{8},$$

with $r = 1$, $s = 1$ or 4, and $t = 1$ or 3. A general solution of this system is given by $x \equiv 40r + 96s + 105t \pmod{120}$ and we find, omitting the calculations, that $H = \{1, 19, 49, 91\}$. (Note that $G^2 = \{1, 49\}$ and $(-29)G^2 = \{91, 19\}$.) Here, $\Delta \equiv 24 \pmod{32}$, so the genus symbols defined for Δ are s_{-2}, s_3, and s_5, presented in that order in the following table. For reasons explained in Section 5.3, we separate these cosets so that the product of genus symbols is +1 in the first set of columns and −1 in the second set:

Coset	a modulo 120	Genus	Coset	a modulo 120	Genus
$1 \cdot H$	1, 19, 49, 91	+ + +	$11 \cdot H$	11, 41, 59, 89	+ − +
$7 \cdot H$	7, 13, 37, 103	− + −	$23 \cdot H$	23, 47, 53, 77	− − −
$17 \cdot H$	17, 83, 107, 113	+ − −	$31 \cdot H$	31, 61, 79, 109	− + +
$29 \cdot H$	29, 71, 101, 119	− − +	$43 \cdot H$	43, 67, 73, 97	+ + −

Here, G contains $\phi(120) = \phi(2^3) \cdot \phi(3) \cdot \phi(5) = 4 \cdot 2 \cdot 4 = 32$ elements, partitioned into eight genera with four elements each. The genus group of $\Delta = 120$ is isomorphic to $\mathbb{Z}_2 \times \mathbb{Z}_2 \times \mathbb{Z}_2$.

Example. For $\Delta = \Delta(-30, 1) = -120$, the calculations are similar to the previous example but, with $1 - \frac{\Delta}{4} = 31 \equiv 7 \pmod{8}$, we let $t = 1$ or 7. Here again, $G^2 = \{1, 49\}$ is the set of squares in $G = \mathbb{Z}_{120}^*$, but $31G^2 = \{31, 79\}$. We find the following cosets of $H = \{1, 31, 49, 79\}$ in G, with genus symbols s_2, s_3, and s_5:

Coset	a modulo 120	Genus	Coset	a modulo 120	Genus
$1 \cdot H$	1, 31, 49, 79	+ + +	$7 \cdot H$	7, 73, 97, 103	+ + −
$11 \cdot H$	11, 29, 59, 101	− − +	$19 \cdot H$	19, 61, 91, 109	− + +
$13 \cdot H$	13, 37, 43, 67	− + −	$41 \cdot H$	41, 71, 89, 119	+ − +
$17 \cdot H$	17, 23, 47, 113	+ − −	$53 \cdot H$	53, 77, 83, 107	− − −

Again G/H is isomorphic to $\mathbb{Z}_2 \times \mathbb{Z}_2 \times \mathbb{Z}_2$.

Example. Let $\Delta = \Delta(-35, 3) = -315$ and let $G = \mathbb{Z}_{315}^*$. Squares of integers modulo 315 must be squares modulo 35 and modulo 9. The squares modulo 35 (of integers relatively prime to 35) were found as 1, 4, 9, 11, 16, and 29 in a previous example. The squares modulo 9 are found to be 1, 4, and 7, which are all the distinct integers modulo 9 that are squares modulo 3. The elements of $H = G^2$ are congruent to $36s - 35t$ modulo 315, where s is one of the six squares modulo 35 and t is one of the three squares modulo 9. There are eighteen possibilities in total:

$$H = \{1, 4, 16, 46, 64, 79, 106, 109, 121, 151, 169, 184, 211, 214, 226, 256, 274, 289\}.$$

The $\phi(315) = \phi(3^2) \cdot \phi(5) \cdot \phi(7) = 144$ elements of G are partitioned into eight genera, each containing eighteen elements. Below we list a representative for each coset, together with the genus symbols (in order as s_5, s_7, and s_3) for each one:

Coset	Genus	Coset	Genus
$1 \cdot H$	$+ + +$	$2 \cdot H$	$- + -$
$11 \cdot H$	$+ + -$	$19 \cdot H$	$+ - +$
$13 \cdot H$	$- - +$	$22 \cdot H$	$- + +$
$17 \cdot H$	$- - -$	$26 \cdot H$	$+ - -$

Here, we list the genera for which $s_5 \cdot s_7 = 1$ in the first columns and those for which $s_5 \cdot s_7 = -1$ in the last columns, again for reasons explained in Section 5.3.

Example. Let $\Delta = \Delta(-35, 2) = -140$ and let $G = \mathbb{Z}_{140}^*$. The subgroup G^2 can be obtained as all the elements of G that are congruent to 1, 4, 9, 11, 16, or 29 modulo 35 and to 1 modulo 4, that is, $G^2 = \{1, 9, 29, 81, 109, 121\}$. Since $d = -35 \equiv 1 \pmod 4$ and $g = 2$ is even, the principal genus is the union of this set with $(1 - \frac{\Delta}{2})G^2 = 71G^2$. Thus,

$$H = \{1, 9, 11, 29, 39, 51, 71, 79, 81, 99, 109, 121\}.$$

(The elements of H not in G^2 are most easily obtained by adding 70 to or subtracting 70 from each of the elements of G^2.) The $\phi(140) = \phi(2^2) \cdot \phi(5) \cdot \phi(7) = 48$ elements of G are partitioned into four genera, each containing twelve elements. The defined genus symbols are s_5 and s_7 only. (Note that $-140 \equiv 4 \pmod{16}$ and $-140 \equiv 20 \pmod{32}$, so none of the symbols s_{-1}, s_2, or s_{-2} is defined for this discriminant.) The genus group can be written as $G/H = \{1 \cdot H, 3 \cdot H, 19 \cdot H, 23 \cdot H\}$, among other possibilities.

Example. Let $\Delta = \Delta(30, 2) = 480$ and let $G = \mathbb{Z}_{480}^*$. The set of squares of elements of G is made up of the solutions of the system

$$x \equiv r \pmod{32}, \quad x \equiv s \pmod 3, \quad x \equiv t \pmod 5$$

where $r = 1, 9, 17$, or 25, $s = 1$, and $t = 1$ or 4. We find that $H = G^2$ contains the following eight elements:

$$H = \{1, 49, 121, 169, 241, 289, 361, 409\}.$$

The $\phi(480) = \phi(2^5) \cdot \phi(3) \cdot \phi(5) = 2^4 \cdot 2 \cdot 4 = 128$ elements of G are partitioned into sixteen genera, each containing eight elements. The defined genus symbols for $\Delta = 480$ are s_3, s_5, s_{-2}, s_{-1}, and s_2, listed in this order in the following table of genera. However, the product of the last three must be 1, as we noted earlier, and as is confirmed in Proposition 5.1.2:

Coset	Genus	Coset	Genus
$1 \cdot H$	$+ + + + +$	$61 \cdot H$	$+ + - + -$
$19 \cdot H$	$+ + + - -$	$31 \cdot H$	$+ + - - +$
$13 \cdot H$	$+ - - + -$	$73 \cdot H$	$+ - + + +$
$7 \cdot H$	$+ - - - +$	$43 \cdot H$	$+ - + - -$
$29 \cdot H$	$- + - + -$	$41 \cdot H$	$- + + + +$
$71 \cdot H$	$- + - - +$	$11 \cdot H$	$- + + - -$
$17 \cdot H$	$- - + + +$	$53 \cdot H$	$- - - + -$
$83 \cdot H$	$- - + - -$	$23 \cdot H$	$- - - - +$

In the first and second sets of columns, we list all genera for which $s_3 \cdot s_5 \cdot s_{-2}$ equals $+1$ or -1, respectively. Before explaining the significance of this distinction, in Section 5.3, we might observe that the set of genera in the first columns forms a subgroup of the genus group.

We conclude this section with a statement about the number of distinct genera of a discriminant, illustrated by these examples.

Proposition 5.1.2. *Suppose that a discriminant Δ has n distinct odd prime divisors. Let $G = \mathbb{Z}_{|\Delta|}^*$ and let H be the principal genus of Δ. Then the number of distinct genera of Δ is*

$$|G/H| = \begin{cases} 2^n, & \text{if } \Delta \text{ is odd} \quad \text{or} \quad \Delta \equiv 4 \pmod{16}, \\ 2^{n+1}, & \text{if } \Delta \equiv 8 \text{ or } 12 \pmod{16} \quad \text{or} \quad \Delta \equiv 16 \pmod{32}, \\ 2^{n+2}, & \text{if } \Delta \equiv 0 \pmod{32}. \end{cases}$$

Proof. If Δ has n distinct odd prime divisors, then there are 2^n ways of selecting the corresponding genus symbols to be $+1$ or -1. If Δ is odd, or if $\Delta \equiv 4 \pmod{16}$ (so that $\Delta \equiv 4$ or $20 \pmod{32}$), these are the only genus symbols defined. If $\Delta \equiv 8$ or $12 \pmod{16}$ or $\Delta \equiv 16 \pmod{32}$, there is one additional symbol defined: s_{-1} if $\Delta \equiv 12 \pmod{16}$ or $\Delta \equiv 16 \pmod{32}$; s_2 if $\Delta \equiv 8 \pmod{32}$; s_{-2} if $\Delta \equiv 24 \pmod{32}$. The number of possible genus symbol assignments is doubled in these cases. Finally, if $\Delta \equiv 0 \pmod{32}$, then s_{-1}, s_2, and s_{-2} are all defined. But, by examining all possibilities for an odd number modulo 8, in the following table, we see that the product of these symbols must equal 1:

$a \bmod 8$	s_{-1}	s_2	s_{-2}
1	+	+	+
3	−	−	+
5	+	−	−
7	−	+	−

Thus, in this case, we can select only two of these symbols independently, yielding a maximum of 2^{n+2} genera. Finally, in each case in which genus symbols can be selected independently, the Chinese remainder theorem ensures the existence of some integer a that satisfies a collection of congruences necessary so that a belongs to the genus in question. □

5.2 Genus equivalence of ideal numbers

We now use the genus group of a discriminant to define a relation of genus equivalence on the ideal numbers of that discriminant. We begin with the following observation.

Proposition 5.2.1. *If $(a : k)$ is a primitive ideal number of discriminant Δ, then $(a : k)$ properly represents some integer m for which $\gcd(m, \Delta) = 1$.*

Proof. Let $P(x)$ be the principal polynomial of discriminant Δ, with $P(k) = ac$ and $P'(k) = b$. By definition, $(a : k)$ is primitive if $\gcd(a, b, c) = 1$. If p is a prime common divisor of c and Δ, then p divides $b^2 = \Delta + 4ac$, so that p divides b. But then p cannot also divide a. Now let q be the product of all prime divisors of Δ that do not also divide c. The ideal number $(a : k)$ properly represents $m = f(q, 1) = aq^2 + bq + c$. If p is a prime divisor of Δ, then either p divides q or p divides c, but not both. If p divides q, then p cannot divide $m = (aq + b)q + c$. If p divides c, then p divides b, but divides neither a nor q. Again we conclude that p cannot divide m. Thus, m is an integer properly represented by $(a : k)$ that is relatively prime to Δ. □

Proposition 5.2.2. *Let $(a : k)$ be a primitive ideal number of discriminant Δ. Suppose that m and n are integers that are relatively prime to Δ, and that $(a : k)$ properly represents both m and n. Then m and n are elements of the same genus in the genus group of Δ.*

Proof. Let $P(x)$ be the principal polynomial of discriminant Δ. If $(a : k)$ properly represents m, then $(a : k)$ is equivalent to an ideal number $(m : \ell)$. If $(a : k)$ also properly represents n, then $(m : \ell)$ properly represents n. Thus, if $P(\ell) = mc$ and $P'(\ell) = b$, then there are integers q and r with $\gcd(q, r) = 1$ so that $n = mq^2 + bqr + cr^2$, as in equation (4.4). From Proposition 1.4.2, we have

$$4mn = (2mq + br)^2 - \Delta r^2. \tag{5.1}$$

Now let H be the principal genus of Δ. To show that m and n are in the same genus, that is, $mH = nH$ in the quotient group G/H, it will suffice to show that mn is an element of H. (Here, $mH = nH$ if and only if $m^2H = mnH$, but m^2 is an element of H.) We can do so by showing that each genus symbol for mn equals 1.

If p is an odd prime divisor of Δ, then equation (5.1) shows that $4mn$ is congruent to a square modulo p. Thus,

$$1 = \left(\frac{4mn}{p}\right) = \left(\frac{2}{p}\right)^2\left(\frac{mn}{p}\right) = \left(\frac{mn}{p}\right).$$

Now assume that $\Delta = b^2 - 4mc$ is even, so that $b = 2b_0$ and $\Delta = 4\Delta_0$ for some integers b_0 and Δ_0. Equation (5.1) can be rewritten as

$$mn = (mq + b_0r)^2 - \Delta_0r^2. \tag{5.2}$$

Since we assume that m and n are relatively prime to Δ, then $mq + b_0r$ and Δ_0r^2 have opposite parity.

Suppose that $\Delta \equiv 12 \pmod{16}$, so that $\Delta_0 \equiv 3 \pmod 4$. Since a square is congruent to 0 or 1 modulo 4, then

$$mn = (mq + b_0r)^2 - \Delta_0r^2 \equiv \begin{cases} 1 - 3(0) \pmod 4, & \text{if } r \text{ is even,} \\ 0 - 3(1) \pmod 4, & \text{if } r \text{ is odd.} \end{cases}$$

In either case, $mn \equiv 1 \pmod 4$.

For the remaining cases, suppose that Δ_0 is even, so that $mq + b_0r$ is odd. The square of an odd number is congruent to 1 modulo 8, and so equation (5.2) implies that

$$mn = (mq + b_0r)^2 - \Delta_0r^2 \equiv \begin{cases} 1 \pmod 8, & \text{if } r \text{ is even,} \\ 1 - \Delta_0 \pmod 8, & \text{if } r \text{ is odd.} \end{cases}$$

Now we can make the following statements:

1. If $\Delta \equiv 0 \pmod{32}$, so that $\Delta_0 \equiv 0 \pmod 8$, then $mn \equiv 1 \pmod 8$. The symbols s_{-1}, s_2, and s_{-2}, which are all defined in this case, all equal 1.
2. If $\Delta \equiv 8 \pmod{32}$, so that $\Delta_0 \equiv 2 \pmod 8$, then $mn \equiv 1$ or $7 \pmod 8$ and the symbol s_2 equals 1.
3. If $\Delta \equiv 16 \pmod{32}$, so that $\Delta_0 \equiv 4 \pmod 8$, then $mn \equiv 1$ or $5 \pmod 8$, that is, $mn \equiv 1 \pmod 4$. The symbol s_{-1} equals 1.
4. If $\Delta \equiv 24 \pmod{32}$, so that $\Delta_0 \equiv 6 \pmod 8$, then $mn \equiv 1$ or $3 \pmod 8$ and the symbol s_{-2} equals 1.

Therefore, every genus symbol for the integer mn equals 1. We conclude that mn is in the principal genus of Δ, as we wanted to show. □

We now associate a collection of genus symbols to each primitive ideal number and define a relation of genus equivalence as follows.

Definition. Let $(a : k)$ be a primitive ideal number of discriminant Δ. Let m be an integer properly represented by $(a : k)$ for which $\gcd(m, \Delta) = 1$. Then the *genus symbols* of $(a : k)$ are defined as follows:

1. If p is an odd prime divisor of Δ, let s_p equal the Legendre symbol $\left(\frac{m}{p}\right)$.
2. If $\Delta \equiv 0$ or $12 \pmod{16}$, let

$$s_{-1} = \begin{cases} 1, & \text{if } m \equiv 1 \pmod 4, \\ -1, & \text{if } m \equiv 3 \pmod 4. \end{cases}$$

3. If $\Delta \equiv 0$ or $8 \pmod{32}$, let

$$s_2 = \begin{cases} 1, & \text{if } m \equiv 1 \text{ or } 7 \pmod 8, \\ -1, & \text{if } m \equiv 3 \text{ or } 5 \pmod 8. \end{cases}$$

4. If $\Delta \equiv 0$ or $24 \pmod{32}$, let

$$s_{-2} = \begin{cases} 1, & \text{if } m \equiv 1 \text{ or } 3 \pmod 8, \\ -1, & \text{if } m \equiv 5 \text{ or } 7 \pmod 8. \end{cases}$$

5. If Δ is negative, let

$$s_\infty = \begin{cases} 1, & \text{if } m \text{ is positive}, \\ -1, & \text{if } m \text{ is negative}. \end{cases}$$

We say that primitive ideal numbers $(a : k)$ and $(a_1 : k_1)$ of the same discriminant are *genus equivalent*, and write $(a : k) \approx (a_1 : k_1)$, if they have precisely the same collection of genus symbols. Finally, if $(a : k) \approx (a_1 : k_1)$ in $\mathcal{I}_\Delta$, then we also say that $(ma : mk) \approx (ma_1 : mk_1)$ in $\mathcal{I}_{m^2\Delta}$ for every positive integer m. (These ideal numbers have content m. We do not assign genus symbols to ideal numbers that are not primitive.)

In practice, if $(a : k)$ is a primitive ideal number of discriminant Δ, and $P(k) = ac$ for the principal polynomial $P(x)$ of discriminant Δ, then we can use the integers a and c to determine the genus symbols of $(a : k)$. If $P'(k) = b$ and m is properly represented by $(a : k)$, then we have $m = aq^2 + bqr + cr^2$ for some integers q and r with $\gcd(q, r) = 1$. In this case, from Proposition 1.4.2,

$$4am = (2aq + br)^2 - \Delta r^2 \quad \text{and} \quad 4cm = (2cr + bq)^2 - \Delta q^2.$$

If a prime number p divides Δ, then p cannot divide both a and c. Otherwise, p would divide $b^2 = \Delta + 4ac$, but then p divides $\gcd(a, b, c)$, contrary to the assumption that $(a : k)$

is primitive. Now if m is relatively prime to Δ, and p is an odd prime divisor of Δ, we find that either $\left(\frac{m}{p}\right) = \left(\frac{a}{p}\right)$ or $\left(\frac{m}{p}\right) = \left(\frac{c}{p}\right)$, by the same type of argument as in the proof of Proposition 5.2.2. We draw similar conclusions for other genus symbols if Δ is even.

Example. Consider $(3 : 15)$ in $\mathcal{I}_{480}$. The principal polynomial of discriminant $\Delta = 480$ is $P(x) = x^2 - 120$. Since $P(15) = 105 = 3 \cdot 35$ and $P'(15) = 30$, with $\gcd(3, 30, 35) = 1$, we know that $(3 : 15)$ is a primitive ideal number. The defined genus symbols for primitive ideal numbers in $\mathcal{I}_{480}$ are s_3, s_5, s_{-2}, s_{-1}, and s_2, applying an example from §5.1. We can use $a = 3$ to calculate that $s_5 = -1$, $s_{-2} = 1$, $s_{-1} = -1$, and $s_2 = -1$, and we can use $c = 35$ to find that $s_3 = -1$. (We could also use a calculation such as $a(2)^2 + b(2)(1) + c(1)^2 = 107$ to verify this conclusion with a single value of m.)

Propositions 4.2.3, 5.2.1, and 5.2.2 ensure that genus symbols of a primitive ideal number are well-defined. Note that s_∞ distinguishes between positive definite and negative definite ideal numbers of negative discriminant. Genus equivalence is an equivalence relation on the set, $\mathcal{I}_\Delta$, of ideal numbers of discriminant Δ. We refer to the equivalence class of $(a : k)$ under this relation as the *genus* of $(a : k)$. If $(a : k)$ is primitive, we will typically denote this genus as a sequence of $+$ and $-$ symbols in some fixed order.

Example. Let $\Delta = -35$, with principal polynomial $P(x) = x^2 + x + 9$. The defined genus symbols for an ideal number $(a : k)$ of discriminant -35 are s_5, s_7, and s_∞, and we will use this order below. We say, for instance, that $(151 : 47)$ is in the $+ + +$ genus since $s_5 = \left(\frac{151}{5}\right) = \left(\frac{1}{5}\right) = 1$, $s_7 = \left(\frac{151}{7}\right) = \left(\frac{4}{7}\right) = 1$, and $s_\infty = 1$. On the other hand, $(283 : 37)$ is in the $- - +$ genus, as $s_5 = \left(\frac{283}{5}\right) = \left(\frac{3}{5}\right) = -1$, $s_7 = \left(\frac{283}{7}\right) = \left(\frac{3}{7}\right) = -1$, and $s_\infty = 1$. Thus, $(283 : 37)$ is not genus equivalent to $(151 : 5)$.

The proof of our main theorem for this section is now almost immediate.

Theorem 5.2.3. *For each discriminant Δ, equivalent ideal numbers in $\mathcal{I}_\Delta$ are genus equivalent.*

Proof. If $(a : k)$ is equivalent to $(a_1 : k_1)$, then these ideal numbers properly represent the same collection of integers. If they are primitive, it is immediate that their genus symbols are precisely the same, and so they are also genus equivalent.

If a primitive ideal number $f = (a : k)$ is equivalent to $f_1 = (a_1 : k_1)$, then for every positive integer m, we have $(ma : mk)$ equivalent to $(ma_1 : mk_1)$ by Proposition 4.1.4. In this case, $(a : k)$ is genus equivalent to $(a_1 : k_1)$, and thus $(ma : mk)$ is genus equivalent to $(ma_1 : mk_1)$ by definition. Since every ideal number of content m has this form, this completes the proof that equivalent ideal numbers are always genus equivalent. □

Example. We saw that $(151 : 47)$ and $(283 : 37)$ have different collections of genus symbols in $\mathcal{I}_{-35}$, so are not genus equivalent. We can now be sure that these two ideal numbers are not equivalent. In examples in Section 4.3, we found that $(151 : 47)$ is equivalent to $(1 : 0)$ and that $(283 : 37)$ is equivalent to $(3 : -1)$. It likewise follows that $(1 : 0)$ and

$(3 : -1)$ represent different classes of ideal numbers in $\mathcal{I}_{-35}$. This confirms a claim from an example concluding Section 4.5. The class group of discriminant $\Delta = -35$ is

$$\mathcal{C}_{-35} = \{[1 : 0], [3 : 0]\},$$

with two distinct classes in distinct genera.

5.3 The existence of genera of ideal numbers

In this section, we establish two theorems on the existence of primitive ideal numbers having a particular combination of genus symbols. To simplify our arguments, we will assume that in every genus of ideal numbers in $\mathcal{I}_\Delta$, there is a representative of the form $(p : k)$ or $(-p : k)$, where p is an odd prime number not dividing Δ. (We will demonstrate how such an ideal number can be found in an example.)

Theorem 5.3.1. *Let $\Delta = \Delta(d, 1)$ be a primitive discriminant. Then there is an ideal number in $\mathcal{I}_\Delta$ with a given collection of genus symbols if and only if the product of those genus symbols is 1.*

Example. Let $\Delta = \Delta(-35, 1) = -35$, for which we have seen that there are three defined genus symbols: s_5, s_7, and s_∞. There are $2^3 = 8$ ways of assigning 1 or -1 to these three symbols, but if the product is required to be 1, that number is cut in half. The following table, using the principal polynomial $P(x) = x^2 + x + 9$ to calculate $P(k) = ac$, shows that there are at least four distinct genera of ideal numbers in $\mathcal{I}_{-35}$:

f	$P(k)$	s_5	s_7	s_∞
$(1 : 0)$	$1 \cdot 9$	$+$	$+$	$+$
$(3 : 0)$	$3 \cdot 3$	$-$	$-$	$+$
$(-1 : 0)$	$-1 \cdot -9$	$+$	$-$	$-$
$(-3 : 0)$	$-3 \cdot -3$	$-$	$+$	$-$

Theorem 5.3.1 implies that every ideal number in $\mathcal{I}_{-35}$ is genus equivalent to one of these.

To prove Theorem 5.3.1, we introduce the following terminology and lemma, an application of the quadratic reciprocity theorem (Theorem A.8.6). If q is an odd prime, let

$$Q = \begin{cases} q, & \text{if } q \equiv 1 \pmod 4, \\ -q, & \text{if } q \equiv 3 \pmod 4. \end{cases}$$

We call Q the *signed prime* of q. Note that $Q \equiv 1 \pmod 4$ for every odd prime q.

Lemma 5.3.2. *Let q be an odd prime with signed prime Q. Then $(\frac{Q}{p}) = (\frac{p}{q})$ for every odd prime $p \neq q$.*

Proof. Let p be an odd prime number with $p \neq q$. If $q \equiv 1 \pmod 4$, then $(\frac{Q}{p}) = (\frac{q}{p}) = (\frac{p}{q})$ by quadratic reciprocity. On the other hand, if $q \equiv 3 \pmod 4$, then

$$\left(\frac{Q}{p}\right) = \left(\frac{-q}{p}\right) = \left(\frac{-1}{p}\right)\left(\frac{q}{p}\right) = \begin{cases} 1 \cdot (\frac{p}{q}), & \text{if } p \equiv 1 \pmod 4, \\ -1 \cdot -(\frac{p}{q}), & \text{if } p \equiv 3 \pmod 4, \end{cases}$$

again using quadratic reciprocity and properties of Legendre symbols. In either case, $(\frac{Q}{p}) = (\frac{p}{q})$. □

Proof of Theorem 5.3.1. Let $\Delta = \Delta(d, 1)$ be a primitive discriminant. Write the odd prime numbers dividing d as $q_1, q_2, \ldots, q_n$, where $q_i \equiv 3 \pmod 4$ for $1 \leq i \leq m$ and $q_i \equiv 1 \pmod 4$ for $m + 1 \leq i \leq n$. (We allow the possibility that $m = 0$ or $m = n$, or that $n = 0$.) Let Q_i be the signed prime of q_i, so that $Q_i = -q_i$ for $1 \leq i \leq m$ and $Q_i = q_i$ for $m+1 \leq i \leq n$. Since each Q_i is congruent to 1 modulo 4, and the same is true for their product, we find that

$$d = Q_1 \cdot Q_2 \cdots Q_n \cdot e \quad \text{where} \quad e = \begin{cases} 1, & \text{if } d \equiv 1 \pmod 4, \\ -1, & \text{if } d \equiv 3 \pmod 4, \\ 2, & \text{if } d \equiv 2 \pmod 8, \\ -2, & \text{if } d \equiv 6 \pmod 8. \end{cases} \tag{5.3}$$

If p is an odd prime number that does not divide d, then Lemma 5.3.2 implies that

$$\left(\frac{\Delta}{p}\right) = \left(\frac{d}{p}\right) = \left(\frac{Q_1}{p}\right)\left(\frac{Q_2}{p}\right)\cdots\left(\frac{Q_n}{p}\right)\left(\frac{e}{p}\right) = \left(\frac{p}{q_1}\right)\left(\frac{p}{q_2}\right)\cdots\left(\frac{p}{q_n}\right)\left(\frac{e}{p}\right). \tag{5.4}$$

Let $P(x)$ be the principal polynomial of discriminant Δ and let p be an odd prime number not dividing Δ, so that $(\frac{\Delta}{p})$ is 1 or -1. If there is an integer k for which p divides $P(k)$, then the equation $P'(k)^2 - 4P(k) = \Delta$ shows that Δ is congruent to a square modulo p. Conversely, if $r^2 \equiv \Delta \pmod p$, then we can select r to have the same parity as $P'(0)$, and we find that $k = \frac{1}{2}(r - P'(0))$ is an integer for which p divides $P(k)$. Thus, there are ideal numbers of the form $(p : k)$ and $(-p : k)$ in $\mathcal{I}_\Delta$ if and only if $(\frac{\Delta}{p}) = 1$.

Assuming that $(\frac{\Delta}{p}) = 1$, write the genus symbols for the ideal number $(p : k)$ as s_{q_i} for $1 \leq i \leq n$, as s_e (if $e \neq 1$), and as s_∞ (if $\Delta < 0$). The final product of Legendre symbols in equation (5.4) is precisely the product of these genus symbols, aside from s_∞, which equals 1 for $(p : k)$ in any case. (For instance, if $e = -2$, then $(\frac{e}{p}) = (\frac{-1}{p})(\frac{2}{p})$, which equals 1 if $p \equiv 1$ or 3 $\pmod 8$ and -1 if $p \equiv 5$ or 7 $\pmod 8$.) Thus, an ideal number $(p : k)$ exists if and only if the product of its genus symbols is 1.

Again assuming that $(\frac{\Delta}{p}) = 1$, write the genus symbols for the ideal number $(-p : k)$ as t_{q_i} for $1 \leq i \leq n$, as t_e (if $e \neq 1$), and as t_∞ (if $\Delta < 0$). Notice that $t_{q_i} = -s_{q_i}$ if and only if $1 \le i \le m$, that $t_e = -s_e$ if and only if e is negative, and that $t_\infty = -s_\infty$ if those symbols are defined. Equation (5.3) can be rewritten as $d = q_1 \cdot q_2 \cdots q_n \cdot e \cdot (-1)^m$, so that d is negative if m is even and e is negative, or if m is odd and e is positive. Thus, we see that, in every case, an even number of the t-symbols are changed in sign from the corresponding s-symbols. Thus, the product of the t-symbols also equals 1, completing the proof. □

Example. Let $\Delta = \Delta(-42, 1) = -168$, with principal polynomial $P(x) = x^2 + 42$. The defined genus symbols for an ideal number in $\mathcal{I}_{-168}$ are s_{-2}, s_3, s_7, and s_∞. (We will write genus symbols as sequences of plus and minus signs in this order below.) We demonstrate the existence of an ideal number in the $- - ++$ genus, for instance. Such an ideal number can have the form $(p : k)$ where p is a prime number for which $p \equiv 5$ or $7 \pmod 8$, $p \equiv 2 \pmod 3$, and $p \equiv 1, 2$, or $4 \pmod 7$. Any system of this form has a unique solution modulo 168, and Dirichlet's theorem on primes in arithmetic progressions ensures the existence of a prime number satisfying the resulting congruence. We find, for instance, $p = 53$ as a solution of $p \equiv 5 \pmod 8$, $p \equiv 2 \pmod 3$, and $p \equiv 4 \pmod 7$. By trial-and-error, we find $(53 : 8)$ as an ideal number in the desired genus. Note that $(-53 : 8)$ is also an ideal number in $\mathcal{I}_{-168}$. Here, all genus symbols are changed in sign, giving us an example of an ideal number in the $+ + --$ genus. (In the notation of the proof of Theorem 5.3.1, there are $m = 2$ prime divisors of Δ congruent to 3 modulo 4, and $e = -2$.)

Now we consider primitive ideal numbers $(a : k)$ in $\mathcal{I}_\Delta$ when the discriminant Δ is not primitive. We begin with the following lemma.

Lemma 5.3.3. *Let $\Delta = \Delta(d, 1)$ and $\Delta_g = \Delta(d, g)$ for some integer $g > 1$. Let p be a prime number that does not divide Δ_g. If $(p : k)$ is an ideal number in $\mathcal{I}_\Delta$, then $(p : gk)$ is a primitive ideal number in $\mathcal{I}_{\Delta_g}$. Conversely, if $(p : \ell)$ is a primitive ideal number in $\mathcal{I}_{\Delta_g}$, then $\ell \equiv gk \pmod p$ for some integer k, and then $(p : k)$ is an ideal number in $\mathcal{I}_\Delta$.*

Proof. Let $P(x)$ and $P_g(x)$ be the principal polynomials of discriminant Δ and Δ_g, respectively. By Proposition 1.1.3, $P_g(gx) = g^2 P(x)$ for all x. If p does not divide $\Delta_g = g^2\Delta$, and $(p : k)$ is an ideal number in $\mathcal{I}_\Delta$, then p divides $P(k)$. Hence, p divides $P_g(gk) = g^2 P(k)$, so that $(p : gk)$ is an ideal number in $\mathcal{I}_{\Delta_g}$. If $P(k) = pc$ and $P'(k) = b$, then $\gcd(p, b, c) = 1$. (Recall that every ideal number of primitive discriminant is primitive.) The content of $(p : gk)$ is $\gcd(p, gb, g^2 c)$, which also equals 1 since p does not divide g. Thus, $(p : gk)$ is primitive in $\mathcal{I}_{\Delta_g}$.

Conversely, suppose that $(p : \ell)$ is a primitive ideal number in $\mathcal{I}_{\Delta_g}$. Since p does not divide g, the linear congruence $gx \equiv \ell \pmod p$ has a solution, say k. But now $(p : gk)$ is also a primitive ideal number in $\mathcal{I}_{\Delta_g}$. Since p divides $P_g(gk) = g^2 P(k)$ and p does not divide g, then p divides $P(k)$. Thus, $(p : k)$ is an ideal number in $\mathcal{I}_\Delta$. □

Theorem 5.3.4. *Let $\Delta = \Delta(d, 1)$ and $\Delta_g = \Delta(d, g)$ for some integer $g > 1$. Then there is a primitive ideal number in $\mathcal{I}_{\Delta_g}$ having a particular collection of genus symbols if and only if the following statements are true:*

1. *The product of all of the genus symbols that are defined for an ideal number of discriminant Δ equals 1.*
2. *If $\Delta_g \equiv 0$ (mod 32), then the product of the genus symbols s_{-1}, s_2, and s_{-2} equals 1.*

Proof. We saw in the proof of Proposition 5.1.2 that if $\Delta \equiv 0$ (mod 32), then each of the symbols s_{-1}, s_2, and s_{-2} is defined for an element a in $G = \mathbb{Z}^*_{|\Delta|}$, but that their product equals 1. (This case can occur only when $\Delta = \Delta(d, g)$ with g even.) The same must be true for every primitive ideal number in Δ.

Now let $\Delta = \Delta(d, 1)$ and $\Delta_g = \Delta(d, g)$ for some integer $g > 1$. By Lemma 5.3.3, a primitive ideal number $(p : \ell)$ exists in $\mathcal{I}_{\Delta_g}$ if and only if there is an ideal number $(p : k)$ in $\mathcal{I}_\Delta$. Theorem 5.3.1 shows then that the product of the genus symbols for Δ_g that are also defined for an ideal number of discriminant Δ must equal 1. No additional symbols defined for Δ_g affect the existence of an ideal number in $\mathcal{I}_\Delta$ (aside from the condition noted in statement (2)), so these symbols can be assigned as 1 or –1 independent of the others. □

We illustrate the final claim of this proof in the next example.

Example. Let $\Delta = \Delta(-42, 1) = -168$ and $\Delta_5 = \Delta(-42, 5) = -4200$. The defined genus symbols for primitive ideal numbers in $\mathcal{I}_{-4200}$ are $s_{-2}, s_3, s_7, s_\infty$, and s_5, with the first four also defined for ideal numbers in $\mathcal{I}_{-168}$. Theorem 5.3.4 implies that the only requirement for existence of a primitive ideal number in $\mathcal{I}_{-4200}$ is that the product of the first four symbols equals 1. For instance, we saw in a previous example that $(53 : 8)$ is in the $--++$ genus of $\mathcal{I}_{-168}$. Lemma 5.3.3 shows that then $(53 : 40) \sim (53 : -13)$ is a primitive ideal number in $\mathcal{I}_{-4200}$. (To check, the principal polynomial of discriminant Δ_5 is $P_5(x) = x^2 + 1050$, and we find that $P_5(-13) = 1219 = 53 \cdot 23$.) Since $(\frac{53}{5}) = -1$, this is in the $--+-$ genus. To find an element in the $--+++$ genus, we can begin by looking for an ideal number $(a : k)$ in $\mathcal{I}_{-168}$ for which $a \equiv 53$ (mod 168) and $a \equiv 1$ (mod 5). (There are many other possibilities.) Trial-and-error calculations yield $(221 : 20)$ in $\mathcal{I}_{-168}$ and so $(221 : 100)$ as a primitive ideal number in $\mathcal{I}_{-4200}$ having the desired genus symbols. (Here, 221 is not prime, but this does not affect genus symbol calculations.)

Example. Let $\Delta = \Delta(-35, 1) = -35$ and $\Delta_3 = \Delta(-35, 3) = -315$. The genus symbols defined for a primitive ideal number f in $\mathcal{I}_{-315}$ are s_5, s_7, s_∞, and s_3. All of these except the last are also defined for an ideal number of discriminant –35. The product of the first three symbols is 1, but s_3 can be chosen arbitrarily. This gives us $2^3 = 8$ possible combinations of genus symbols in total, so eight genera of primitive ideal numbers in $\mathcal{I}_{-315}$. The following table, using $P(x) = x^2 + 3x + 81$ as the principal polynomial of discriminant –315, shows that each of these allowed genera exists:

$(a : k)$	$P(k)$	s_5	s_7	s_∞	s_3
$(1 : 0)$	$1 \cdot 81$	+	+	+	+
$(5 : 1)$	$5 \cdot 17$	–	–	+	–
$(7 : 2)$	$7 \cdot 13$	–	–	+	+
$(9 : 3)$	$9 \cdot 11$	+	+	+	–
$(-1 : 0)$	$-1 \cdot -81$	+	–	–	–
$(-5 : 1)$	$-5 \cdot -17$	–	+	–	+
$(-7 : 2)$	$-7 \cdot -13$	–	+	–	–
$(-9 : 3)$	$-9 \cdot -11$	+	–	–	+

(For the genus symbols of $(5 : 1)$, for instance, we use $c = 17$ in calculating s_5, but can use $a = 5$ for each of the others.) There are also four genera in $\mathcal{I}_{-315}$ that contain ideal numbers that are not primitive. These arise from the four genera of ideal numbers in $\mathcal{I}_{-35}$. Using the table for $\Delta = -35$ from a previous example, we could use $(3 : 0)$, $(9 : 0)$, $(-3 : 0)$, and $(-9 : 0)$ as representatives of these genera.

Example. Let $\Delta = \Delta(14, 1) = 56$ and $\Delta_2 = \Delta(14, 2) = 224$. Since 32 divides Δ_2, we find that the genus symbols for primitive ideal numbers in $\mathcal{I}_{224}$ are s_{-2}, s_7, s_{-1}, and s_2. The first two of these are also defined for $\Delta = 56$. Here, we find that there are four independent assignments of s_{-1} and s_2 possible. Once these are selected, then s_{-2} is determined, as in statement (2) of Theorem 5.3.4, and s_7 must be the same as s_{-2}, making their product 1. Thus, there are four distinct genera of primitive ideal numbers in $\mathcal{I}_{224}$. Trial-and-error calculations, with $P(x) = x^2 - 56$ the principal polynomial of discriminant 224, produce the following four genus representatives:

$(a : k)$	$P(k)$	s_{-2}	s_7	s_{-1}	s_2
$(1 : 0)$	$1 \cdot -56$	+	+	+	+
$(-1 : 0)$	$-1 \cdot 56$	–	–	–	+
$(4 : 2)$	$4 \cdot -13$	+	+	–	–
$(-4 : 2)$	$-4 \cdot 13$	–	–	+	–

Every element of $\mathcal{I}_{224}$ is genus equivalent to one of these ideal numbers.

5.4 Class groups and genera

In Chapter 4, we defined the class group of a given discriminant Δ to be the set of classes of primitive ideal numbers $(a : k)$ in $\mathcal{I}_\Delta$, with $(a : k)$ assumed to be positive definite if Δ is negative. This set, written as $\mathcal{C}_\Delta$, is a group under an operation of composition of ideal number classes. We conclude Chapter 5 with applications of genus equivalence to class groups. We will use the following two general results. The first is proved as

Theorem B.3.6 in Appendix B. The proof of the second is found in [2] and other number theory texts.

Theorem 5.4.1 (Fundamental theorem of finite Abelian groups). *If G is an Abelian group with n > 1 elements, then G is isomorphic to a unique direct product*

$$G \cong \mathbb{Z}_{n_1} \times \mathbb{Z}_{n_2} \times \cdots \times \mathbb{Z}_{n_t} \tag{5.5}$$

where each n_i is larger than 1 and n_{i+1} divides n_i for $1 \le i < t$.

Note that $n = n_1 \cdot n_2 \cdots n_t$ in this case. We allow the possibility that $t = 1$, in which case G is cyclic. If G has the form of (5.5), we say that G has *invariant factor type* $(n_1, n_2, \ldots, n_t)$. A trivial group has invariant factor type (1).

Theorem 5.4.2 (Legendre's theorem). *Let a, b, and c be integers that are squarefree, pairwise relatively prime, and neither all positive nor all negative. Then the equation*

$$ax^2 + by^2 + cz^2 = 0$$

has a nontrivial solution in integers if and only if there are integers u, v, and w satisfying

$$u^2 \equiv -bc \pmod{|a|}, \quad v^2 \equiv -ac \pmod{|b|}, \quad w^2 \equiv -ab \pmod{|c|}.$$

Since equivalent ideal numbers are genus equivalent (Theorem 5.2.3), we can view each class of ideal numbers in a class group $\mathcal{C}_\Delta$ as belonging to a particular genus. The following term is then well-defined.

Definition. If $\mathcal{C}_\Delta$ is the class group of some discriminant Δ, we define the *principal genus* of $\mathcal{C}_\Delta$ to include all classes of ideal numbers for which each defined genus symbol equals 1. More generally, we say that two classes are in the same *genus* in $\mathcal{C}_\Delta$ if they have the same collection of genus symbols.

Recall that when Δ is negative, then each ideal number $(a : k)$ has a defined genus symbol s_∞ determined by the sign of a. Since elements of $\mathcal{C}_\Delta$ are restricted to be positive definite when Δ is negative, then $s_\infty = 1$ for all classes. Thus, we can restrict our attention to the other defined genus symbols, which we will call *finite genus symbols* for comparison.

Proposition 5.4.3. *Let Δ be a discriminant. If $G = \mathcal{C}_\Delta$ is the class group of discriminant Δ and H is the principal genus of $\mathcal{C}_\Delta$, then H is a subgroup of G. The quotient group G/H can be regarded as a subgroup of the genus group of discriminant Δ.*

Proof. By Proposition 5.2.1, every class of primitive ideal numbers can be expressed as $[a : k]$ where a is relatively prime to Δ. If $[a : k] \circ [b : \ell] = [c : m]$, then $cg^2 = ab$, as in equation (4.9), for some common divisor g of a and b. It follows that the genus symbols of $[c : m]$ are the term-by-term products of the genus symbols of $[a : k]$ and of $[b : \ell]$. In particular, if all of those genus symbols are 1, then the same is true for $[c : m]$. Thus, H is a subgroup of the finite group G. More generally, in the genus group of discriminant Δ, the product of the genus of a with the genus of b equals the genus of c. Thus, we can view G/H as a subgroup of that genus group. □

Proposition 5.4.5 gives us an alternative description of the principal genus. We first state the following observation.

Lemma 5.4.4. *Let* $\Delta = \Delta(d, g)$ *be a discriminant and let* $[a : k]$ *be an element of the principal genus of the class group* C_Δ, *where we can assume that a is positive and relatively prime to* Δ. *Then the quadratic congruences* $x^2 \equiv d$ (mod a) *and* $x^2 \equiv a$ (mod d) *both have solutions.*

Proof. Let $P(x)$ be the principal polynomial of discriminant Δ. A solution of $x^2 \equiv d$ (mod a) must exist since a divides $P(k)$ for some integer k. For the second claim, the assumption that $[a : k]$ is in the principal genus is sufficient to establish that $x^2 \equiv a$ (mod p) has a solution for all prime divisors p of d, and a solution of $x^2 \equiv a$ (mod d) then exists as well. We omit further details of these claims. □

Proposition 5.4.5. *Let* Δ *be a discriminant. If* $G = C_\Delta$ *is the class group of discriminant* Δ, *then the principal genus of* C_Δ *equals* G^2, *the set of all squares of elements of G.*

Proof. Proposition 5.4.3 shows that if $[a : k]$ is an element of C_Δ, then every defined genus symbol of $[a : k]^2 = [a : k] \circ [a : k]$ is 1. Thus, G^2 is a subset of the principal genus of C_Δ. For the reverse containment, let $[a : k]$ be an element of the principal genus of C_Δ. If $\Delta = \Delta(d, g)$, then Legendre's theorem and Lemma 5.4.4 imply the existence of a nontrivial integer solution of the equation

$$x^2 - dy^2 - az^2 = 0.$$

If m is the z-coordinate in any such solution, we can use this equation to construct an ideal number class $[m : \ell]$ for which $[m : \ell]^2 = [a : k]$. (The details are omitted.) Thus, the principal genus of $G = C_\Delta$ is contained in G^2. □

We can describe the size of the principal genus of C_Δ and the number of distinct genera in terms of the invariant factor type of the class group. We state our next result for arbitrary finite Abelian groups.

Proposition 5.4.6. *Let* $(n_1, n_2, \ldots, n_t)$ *be the invariant factor type of an Abelian group G, and let s be the largest nonnegative integer so that* n_s *is even (with s = 0 if each* n_i *is odd). If* G^2 *is the subgroup of all squares of elements of G, then* $|G| = 2^s \cdot |G^2|$.

We can also describe s as the number of even terms in the invariant factor type of G, given how invariant factors are defined in terms of divisibility.

Proof. Since G is Abelian, the function $T : G \to G$ defined by $T(x) = x^2$ is a homomorphism. The image of G under T is G^2, and the kernel of T, which we write as H_2, consists of all x in G for which $x^2 = 1$. If G has invariant factor type $(n_1, n_2, \ldots, n_t)$, then Lemma B.3.5 implies that

$$|H_2| = \gcd(n_1, 2) \cdot \gcd(n_2, 2) \cdots \gcd(n_t, 2) = 2^s,$$

since $\gcd(n_i, 2)$ is 2 or 1, depending on whether n_i is even or odd. Thus, $|G| = 2^s \cdot |G^2|$ by standard properties of homomorphisms. $\qquad\square$

In particular, if $G = C_\Delta$ is the class group of discriminant Δ, with s the number of even terms in the invariant factor type of G as above, then there are 2^s distinct genera in the class group. But in Section 5.3, we also related the number of genera to the number of defined genus symbols for ideal numbers of discriminant Δ. We summarize our results as follows.

Proposition 5.4.7. *Let Δ be a discriminant and let ℓ be the number of finite genus symbols (i. e., excluding s_∞ when Δ is negative) defined for ideal numbers in $\mathcal{I}_\Delta$. Let $G = C_\Delta$ be the class group of discriminant Δ and let s be the number of even terms in the invariant factor type of G. Then*

$$s = \begin{cases} \ell - 1, & \textit{if } 32 \textit{ does not divide } \Delta, \\ \ell - 2, & \textit{if } 32 \textit{ divides } \Delta. \end{cases}$$

Proof. There are 2^ℓ ways of assigning each of the defined genus symbols to be $+1$ or -1. But from the restrictions on genus symbols noted in Theorems 5.3.1 and 5.3.4, we see that the number of possible genus symbol assignments is halved in all cases and halved again when 32 divides Δ. The formula for s follows from Proposition 5.4.6 and the observation that follows its proof. $\qquad\square$

The description of genera is highly useful in describing the structure of the class group C_Δ for a particular discriminant Δ, as we will see in the next two chapters. In Chapter 6, we will develop a practical method of listing distinct classes and describing C_Δ when Δ is negative. We do the same, with quite different methods, for positive discriminants in Chapter 7.

6 Ideal numbers of negative discriminant

At the conclusion of Chapter 4, we defined the class group of discriminant Δ. The elements of a class group, $\mathcal{C}_\Delta$, are classes $[a : k]$ of *primitive* ideal numbers in $\mathcal{I}_\Delta$. When Δ is negative, we add the condition that a is positive, so that $[a : k]$ is *positive definite*.

In Chapter 6, we describe a process of listing representatives of all distinct classes of ideal numbers of a negative discriminant, using what we define as *reduced* ideal numbers of discriminant $\Delta < 0$. By examples, we will demonstrate that it is generally possible to determine the structure of the resulting class group, $\mathcal{C}_\Delta$, specifically its *invariant factor type* (using terminology introduced in Chapter 5). We will first concentrate on class groups of negative *primitive* discriminant, in which every class of ideal numbers is an element of the class group. In Section 6.4 and Section 6.5, we consider applications of this class group structure to representations of integers by positive definite ideal numbers or quadratic forms. In Section 6.6, we extend some of our results to class groups of arbitrary, not necessarily primitive, discriminants.

6.1 Reduced ideal numbers of negative discriminant

In Section 4.3, we showed that it is possible to list potential class representatives for ideal numbers of a fixed discriminant, using an upper bound on the absolute value of the norm of a representative. We saw by example, however, that there might be equivalences among the ideal numbers produced in this way. When Δ is negative, we can refine our previous restrictions to produce a precise set of class representatives, using the following terminology.

Definition. Let $P(x)$ be the principal polynomial of some negative discriminant Δ. We say that an ideal number $(a : k)$ of discriminant Δ is *reduced* if either:
1. $a^2 < P(k)$ and $-a < P'(k) \le a$, or
2. $a^2 = P(k)$ and $0 \le P'(k) \le a$.

It is implicit in this definition that the norm of a reduced ideal number is positive, that is, $(a : k)$ is positive definite. If we let $P(k) = ac$ and $P'(k) = b$ as usual, we can also phrase the restrictions on a reduced ideal number as follows:
1. $-a < b \le a < c$, or
2. $0 \le b \le a = c$.

We also refer to the corresponding quadratic form $f(x, y) = ax^2 + bxy + cy^2$ as being reduced if these conditions hold.

In Section 6.2, we will show that every positive definite ideal number of negative discriminant Δ is equivalent to precisely one reduced ideal number. In the remainder of this section, we note some general properties of reduced ideal numbers and demon-

https://doi.org/10.1515/9783111319360-006

strate a method for listing all reduced ideal numbers of a given negative discriminant. Our first proposition implies that a congruence class of ideal numbers of negative discriminant contains no more than one reduced ideal number.

Proposition 6.1.1. *If $(a : k)$ and $(a : k_1)$ are reduced ideal numbers of negative discriminant Δ, then $|k_1 - k| < a$. In particular, if $k_1 \equiv k \pmod{a}$, then $k_1 = k$.*

Proof. We can assume without loss of generality that $k \leq k_1$, in which case $P'(k) \leq P'(k_1)$. If $(a : k)$ and $(a : k_1)$ are reduced, then

$$-a < P'(k) \leq P'(k_1) \leq a. \tag{6.1}$$

The first of these inequalities shows that $-P'(k) < a$. Then by subtracting $P'(k)$ from the other inequalities in (6.1), we see that

$$0 \leq P'(k_1) - P'(k) \leq a - P'(k) < 2a.$$

Since $P'(k_1) - P'(k) = (2k_1 + P'(0)) - (2k + P'(0)) = 2(k_1 - k)$, it follows that

$$0 \leq k_1 - k < a.$$

If a divides $k_1 - k$, then $k_1 = k$. $\qquad\square$

Proposition 6.1.1 shows that in the congruence class of a positive definite ideal number, only the representative of *minimal form*, that is, $(a : k)$ with

$$\frac{-a - P'(0)}{2} < k \leq \frac{a - P'(0)}{2},$$

has the potential of being reduced. In Section 2.1, we demonstrated an algorithm for listing all representatives for congruence classes of ideal number with the norm smaller than a given upper bound. The following proposition tells us how many of these representatives we need to compile to ensure that we obtain all reduced ideal numbers of a given negative discriminant.

Proposition 6.1.2. *Let $(a : k)$ be the ideal number in minimal form for a congruence class of positive definite ideal numbers of a given negative discriminant Δ. Then the following statements are true:*
1. *If $4a^2 < -\Delta$, then $(a : k)$ is reduced.*
2. *If $(a : k)$ is reduced, then $3a^2 \leq -\Delta$.*

Proof. If the inequality of case (1) is true, then

$$4a^2 < -\Delta = 4P(k) - P'(k)^2 \leq 4P(k),$$

using equation (1.3). Thus, $a^2 < P(k)$ and $-a < P'(k) \leq a$, using the definition of minimal form. Therefore, $(a : k)$ is reduced by definition.

Now suppose that $(a : k)$ is reduced, which implies that

$$P'(k)^2 \leq a^2 \leq P(k)$$

in every case. Then $4a^2 \leq 4P(k)$ and $-a^2 \leq -P'(k)^2$, so that

$$3a^2 = 4a^2 - a^2 \leq 4P(k) - P'(k)^2 = -\Delta,$$

establishing statement (2). □

Proposition 6.1.2 shows that if $(a : k)$ is an ideal number of discriminant $\Delta < 0$, having minimal form, then we can make the following statements:

1. If $0 < a < \sqrt{\frac{-\Delta}{4}}$, then $(a : k)$ is reduced.
2. If $\sqrt{\frac{-\Delta}{4}} \leq a \leq \sqrt{\frac{-\Delta}{3}}$, then $(a : k)$ might be reduced.
3. If $a < 0$ or $a > \sqrt{\frac{-\Delta}{3}}$, then $(a : k)$ is not reduced.

We summarize our results as an algorithm for listing the reduced ideal numbers of an arbitrary negative discriminant Δ, which we will demonstrate with examples in the remainder of this section. Note that this algorithm uses the fact that $P(k) = P(\overline{k})$, where $\overline{k} = -P'(0) - k$ is the conjugate of k with respect to Δ.

Given a discriminant $\Delta < 0$, with principal polynomial $P(x)$:
1. Calculate $u = \lfloor \sqrt{\frac{-\Delta}{3}} \rfloor$
2. Calculate $P(x)$ for $-\frac{P'(0)}{2} \leq x \leq \frac{u - P'(0)}{2}$
3. For $1 \leq a \leq u$:
   (a) For $\frac{-a - P'(0)}{2} < k \leq \frac{a - P'(0)}{2}$, test whether a divides $P(k)$
   (b) If $4a^2 \geq -\Delta$, test the other properties of a reduced ideal number
4. Return the list of $(a : k)$ that satisfy these conditions.

Example. Let $\Delta = \Delta(-155, 1) = -155$, with $P(x) = x^2 + x + 39$ as its principal polynomial. Then

$$u = \left\lfloor \sqrt{-\frac{\Delta}{3}} \right\rfloor = \lfloor \sqrt{51.7} \rfloor = 7,$$

and we calculate $P(k)$ for $-\frac{1}{2} \leq k \leq 3$ in the following table. (We label each calculation of $P(x)$ with both x and $\overline{x} = -x - 1$ to emphasize that this gives us at least seven values of $P(k)$.)

x	0, −1	1, −2	2, −3	3, −4
$P(x)$	39	41	45	51

Now for $a \leq 7$, we test whether a divides $P(k)$ for $\frac{-a-P'(0)}{2} < k \leq \frac{a-P'(0)}{2}$. (In the following list, we omit even possibilities for a, as each $P(x)$ is odd.) Note that $4a^2 < -\Delta$ for all a except 7, so no additional testing is required in most cases.

1. For $a = 1$, we look at $-1 < k \leq 0$, that is, $k = 0$ only. Here, 1 divides $P(0) = 39$, and so $(1 : 0)$ is reduced.

2. For $a = 3$, we test $-2 < k \leq 1$. We find that 3 divides $P(0) = 39 = P(-1)$, but not $P(1) = 41$. Both $(3 : 0)$ and $(3 : -1)$ are reduced.

3. For $a = 5$, we test $-3 < k \leq 2$. Here, 5 divides $P(2) = 45$, but none of the other values of $P(k)$. (Note that $k = -3$ is not considered.) Thus, $(5 : 2)$ is reduced.

4. For $a = 7$, we test $-4 < k \leq 3$, but 7 divides none of these values of $P(k)$.

Therefore, there are four reduced ideal numbers of discriminant $\Delta = -155$:

$$(1 : 0) \quad (3 : 0) \quad (3 : -1) \quad (5 : 2).$$

Example. Let $\Delta = \Delta(-271, 1) = -271$, with $P(x) = x^2 + x + 68$ as its principal polynomial. Reduced ideal numbers of discriminant Δ can have norm as large as $u = \lfloor \sqrt{-\frac{\Delta}{3}} \rfloor = 9$, and we can find all such ideal numbers by testing $P(x)$ for $-4 \leq x \leq 4$, as in the following table. (Again, we can use the fact that $P(x) = P(\overline{x})$, where $\overline{x} = -x - 1$, to reduce required calculation.)

x	0, −1	1, −2	2, −3	3, −4	4, −5
$P(x)$	68	70	74	80	88

Here, $4a^2 < -\Delta$ for $1 \leq a \leq 8$. We leave it as an exercise to verify that there are precisely eleven reduced ideal numbers of discriminant $\Delta = -271$:

$$(1 : 0) \quad (2 : 0) \quad (2 : -1) \quad (4 : 0) \quad (4 : -1)$$
$$(5 : 1) \quad (5 : -2) \quad (7 : 1) \quad (7 : -2) \quad (8 : 3) \quad (8 : -4).$$

For instance, when $a = 8$, we test $\frac{-8-1}{2} < k \leq \frac{8-1}{2}$, that is, $-4.5 < k \leq 3.5$. This includes both 3 and −4, but neither 4 nor −5.

Example. Let $\Delta = \Delta(-285, 1) = -1140$, with $P(x) = x^2 + 285$ as its principal polynomial. A reduced ideal number $(a : k)$ has $a \leq u = \lfloor \sqrt{-\frac{\Delta}{3}} \rfloor = 19$. We find all reduced ideal numbers by testing $P(x)$ for $0 \leq x \leq 9$, with $P(-k) = P(k)$:

x	0	± 1	± 2	± 3	± 4	± 5	± 6	± 7	± 8	± 9
$P(x)$	285	286	289	294	301	310	321	334	349	366

Here, $4a^2 < -\Delta$ for $1 \le a \le 16$, so for those values we need only test whether a divides $P(k)$ with $-\frac{a}{2} < k \le \frac{a}{2}$. We need to look more closely at $17 \le a \le 19$ however.

- $a = 17$ divides both $P(2)$ and $P(-2)$. But since $P(2) = 289 = 17^2$, we need the additional condition that $0 \le P'(k) \le a$. Here, $P'(2) = 4$ but $P'(-2) = -4$. Thus, $(17 : 2)$ is reduced while $(17 : -2)$ is not.
- $a = 18$ does not divide any of the $P(k)$ values listed.
- $a = 19$ divides $P(0)$ only. But $a^2 = 361 > 285 = P(0)$, so $(19 : 0)$ is not reduced.

We again leave it as an exercise to verify that there are precisely sixteen reduced ideal numbers of discriminant $\Delta = -1140$:

$$(1:0) \quad (2:1) \quad (3:0) \quad (5:0) \quad (6:3) \quad (7:3) \quad (7:-3) \quad (10:5)$$
$$(11:1) \quad (11:-1) \quad (13:1) \quad (13:-1) \quad (14:3) \quad (14:-3) \quad (15:0) \quad (17:2).$$

6.2 Reduced ideal numbers and equivalence

In this section, we show that reduced ideal numbers (and their negative conjugates) always provide a precise set of class representatives for all ideal numbers of a negative discriminant.

Proposition 6.2.1. *If Δ is a negative discriminant, then every positive definite ideal number in $\mathcal{I}_\Delta$ is equivalent to a reduced ideal number.*

Proof. Let $P(x)$ be the principal polynomial of discriminant $\Delta < 0$. In a given class of positive definite ideal numbers in $\mathcal{I}_\Delta$, select an ideal number $(a : k)$ with a as small as possible. We can apply a translation, if necessary, to assume also that $(a : k)$ is in minimal form for its congruence class, that is, with

$$\frac{-a - P'(0)}{2} < k \le \frac{a - P'(0)}{2}.$$

Since $P'(k) = 2k + P'(0)$, then $-a < P'(k) \le a$.

If $P(k) = ac$, then $(a : k) \sim (c : \bar{k})$ by an involution. Thus, $a \le c$, since otherwise the choice of a is violated. If $a < c$, then $a^2 < P(k)$ and $(a : k)$ is reduced by definition. Assume instead that $a = c$, so that $a^2 = P(k)$. If $0 \le P'(k) \le a$, then $(a : k)$ is again reduced by definition. So, assume that $-a < P'(k) < 0$. But now $(a : k) \sim (a : \bar{k})$ by an involution and, since $P'(\bar{k}) = -P'(k)$, then $0 < P'(\bar{k}) < a$. Thus, $(a : \bar{k})$ is reduced, and so $(a : k)$ is equivalent to a reduced ideal number. □

It is more difficult to prove that a positive definite ideal number is equivalent to a *unique* reduced ideal number. The following proposition is a first step in that direction.

Proposition 6.2.2. *Let* $(a : k)$ *be a reduced ideal number of negative discriminant* Δ. *If* $(a : k)$ *properly represents an integer m, then* $a \leq m$.

The proof of this proposition is by contradiction, showing that if $m < a$, then $(a : k)$ cannot properly represent m. But for later use, we will also consider the possibility that $m = a$ in this proof.

Proof. Let $P(x)$ be the principal polynomial of discriminant Δ. Let $(a : k)$ be a reduced ideal number in $\mathcal{I}_\Delta$, with $P(k) = ac$ and $P'(k) = b$. If $(a : k)$ properly represents m, then there are relatively prime integers q and r so that

$$m = aq^2 + bqr + cr^2.$$

We can assume, replacing q and r by their negatives if necessary, that r is nonnegative, with $q > 0$ if $r = 0$. From Proposition 1.4.2, we have

$$4am = (2aq + br)^2 - \Delta r^2. \tag{6.2}$$

Suppose that $m \leq a$. Since $-\Delta \geq 3a^2$ by Proposition 6.1.2, it follows that

$$4a^2 \geq 4am = (2aq + br)^2 - \Delta r^2 \geq (2aq + br)^2 + 3a^2 r^2,$$

so that

$$a^2(4 - 3r^2) \geq (2aq + br)^2.$$

Thus, r^2 is either 0 or 1, and the same is true for r (assuming as noted that r is nonnegative).

If $r = 0$, then $4a^2 \geq (2aq)^2$, which implies that $q = 1$, since $\gcd(q, r) = 1$ and q is positive if $r = 0$. Hence, $m < a$ is impossible in this case, but we obtain $m = a$ if $q = 1$ and $r = 0$.

If $r = 1$, then $a^2 \geq (2aq + b)^2$, so that

$$-a \leq 2aq + b \leq a. \tag{6.3}$$

Since $-a < b \leq a$ if $(a : k)$ is reduced, we also have that

$$a(2q - 1) < 2aq + b \leq a(2q + 1). \tag{6.4}$$

The inequalities in (6.3) and (6.4) can all hold only if $q = 0$ or $q = -1$.

1. If $q = 0$ and $r = 1$, then $m = aq^2 + bqr + cr^2 = c$. But $c \geq a$ if $(a : k)$ is reduced, so this case occurs only when $m = a = c$.

2. If $q = -1$ and $r = 1$, then $m = aq^2 + bqr + cr^2 = a - b + c$. Since $b \leq a$ if $(a : k)$ is reduced, we find that $a - b + c \geq c$. This case occurs only when $m = a = b = c$.

We conclude that no integer $m < a$ can be properly represented by a reduced ideal number $(a : k)$. □

Proposition 6.2.3. *If $(a : k)$ and $(a_1 : k_1)$ are reduced ideal numbers of negative discriminant Δ, and $(a : k)$ is equivalent to $(a_1 : k_1)$, then $(a : k) = (a_1 : k_1)$.*

Proof. If $(a : k)$ is equivalent to $(a_1 : k_1)$, then $(a : k)$ properly represents a_1, and $(a_1 : k_1)$ properly represents a. Proposition 6.2.2 then implies that $a \leq a_1$ and $a_1 \leq a$, so that $a = a_1$. Now with $(a : k)$ and $(a : k_1)$ both reduced, we have $|k_1 - k| < a$ by Proposition 6.1.1. Let $(a : k) \cdot U = (a : k_1)$ for some $U = \left[\begin{smallmatrix} q & s \\ r & t \end{smallmatrix}\right]$ in the unimodular group Γ, so that $qt - rs = 1$. Since we identify U with $-U$, we can further assume that r is positive, or that $r = 0$ and q is positive. If $P(x)$ is the principal polynomial of discriminant Δ, with $P(k) = ac$ and $P'(k) = b$, then

$$a = aq^2 + bqr + cr^2 \quad \text{and} \quad k_1 = k + (aqs + brs + crt), \tag{6.5}$$

by Theorem 4.1.2. The first equation can be rewritten as

$$4a^2 = (2aq + br)^2 - \Delta r^2,$$

as in equation (6.2). In the proof of Proposition 6.2.2, we saw that these equations can hold only in the following cases:
1. If $r = 0$, then $qt - rs = qt = 1$, so we can assume that $q = 1 = t$. But then $k_1 = k + as$ from equation (6.5). Thus, $k_1 = k$ by Proposition 6.1.1.
2. We can have $r = 1$ and $q = 0$, so that $s = -1$, but only if $a = c$. Here, $k_1 = k + (-b + at)$, implying that $|at - b| < a$ by Proposition 6.1.1. Since $(a : k)$ is reduced with $a = c$, we also know that $0 \leq b \leq a$, from which it follows that $a(t-1) \leq at - b \leq at$. Combining these inequalities, we find that $t = 0$ and $t = 1$ are the only possibilities.
   (a) If $t = 0$, then $U = V$ is the involution and $k_1 = k - b = \bar{k}$, the conjugate of k with respect to Δ. But with $P(\bar{k}) = P(k) = ac$ with $c = a$, we must also have $0 \leq P'(\bar{k}) \leq a$ for $(a : \bar{k})$ to be reduced. Recalling that $P'(\bar{k}) = -P'(k) = -b$, we find that this is possible only when $b = 0$. Thus, $k_1 = k$.
   (b) If $t = 1$, then $k_1 = k + (-b + a)$. Note that $P'(k_1) = 2k_1 + P'(0) = P'(k) + 2(a - b) = 2a - b$. If $(a : k_1)$ is reduced, we must have $-a < 2a - b \leq a$. The second of these inequalities shows that $a \leq b$, but we must also have $b \leq a$ if $(a : k)$ is reduced. The only possibility is that $a = b$, and then $k_1 = k + (-b + a) = k$.
3. We can have $r = 1$ and $q = -1$, but only if $a = b = c$. Here,

$$k_1 = k + (aqs + brs + crt) = k + a(-s + s + t) = k + at.$$

Since $k_1 \equiv k \pmod{a}$, then $k_1 = k$.

Therefore, if $(a : k)$ and $(a_1 : k_1)$ are reduced and equivalent to each other, then $a = a_1$ and $k = k_1$. That is, the two equivalent reduced ideal numbers of negative discriminant are in fact equal. □

We summarize our results from this section in the following theorem, omitting further proof.

Theorem 6.2.4. *Let Δ be a negative discriminant. Then every positive definite ideal number in $\mathcal{I}_\Delta$ is equivalent to one and only one reduced ideal number of discriminant Δ. Likewise, every negative definite ideal number in $\mathcal{I}_\Delta$ is equivalent to the negative conjugate of one and only one reduced ideal number of discriminant Δ. The class group $\mathcal{C}_\Delta$ of discriminant Δ consists precisely of the classes of all primitive reduced ideal numbers in $\mathcal{I}_\Delta$.*

6.3 Class groups of primitive negative discriminant

If d is a negative square-free integer, then $\Delta = \Delta(d, 1)$ is a primitive discriminant, and every ideal number in $\mathcal{I}_\Delta$ is primitive. Theorem 6.2.4 implies that the classes of reduced ideal numbers of discriminant Δ are the distinct elements in the class group $\mathcal{C}_\Delta$. As we saw in Section 6.1, it is straightforward to list all these reduced ideal numbers. We can perform calculations in $\mathcal{C}_\Delta$ through a combination of the composition formula of Section 4.5 and the reduction process, introduced in Section 4.3, to determined the reduced ideal number to which a given composite is equivalent. We showed in Proposition 5.4.7 that the number of finite genus symbols defined for ideal numbers of discriminant Δ determines one aspect of the resulting class group, namely the number of even terms in its invariant factor type. (See Section 5.4 or Appendix B for this terminology.) We illustrate with examples that it is typically not difficult to compute the group structure of a class group, using these facts. Our first three examples are discriminants for which we have already compiled all reduced ideal numbers, in Section 6.1.

Example. Let $\Delta = \Delta(-155, 1) = -155 = -1 \cdot 5 \cdot 31$. In an example in Section 6.1, we found four reduced ideal numbers of this discriminant, which we list below as representatives for all elements of the corresponding class group:

$$\mathcal{C}_{-155} = \{[1 : 0], [3 : 0], [3 : -1], [5 : 2]\}.$$

An Abelian group with four elements can be isomorphic to either $\mathbb{Z}_4$ or $\mathbb{Z}_2 \times \mathbb{Z}_2$. But here s_5 and s_{31} are the only finite genus symbols defined for elements of $\mathcal{I}_{-155}$. This implies that there is $\ell - 1 = 1$ even invariant factor. Thus, $\mathcal{C}_{-155}$ is isomorphic to $\mathbb{Z}_4$. We can confirm this by observing that $[3 : 0]$ and $[3 : -1]$ are conjugates, and hence inverses of each other in the class group. In $\mathbb{Z}_2 \times \mathbb{Z}_2$, on the other hand, every element is its own inverse. To illustrate computation in a class group, we verify more directly that $\mathcal{C}_{-155}$ is cyclic with $[3 : 0]$ as a generator. Composites are calculated directly from the formula of Section 4.5 (along with translations to obtain a simpler expression in most cases). In the first equation, we use an involution to reduce $[9 : -3]$ to the equivalent ideal number class $[5 : 2]$. In the second equation, an involution and translation reduce $[15 : -3]$ to $[3 : -1]$.

- $[3 : 0]^2 = [3 : 0] \circ [3 : 0] = [9 : -3] = [5 : 2]$
- $[3 : 0]^3 = [3 : 0] \circ [5 : 2] = [15 : -3] = [3 : 2] = [3 : -1]$
- $[3 : 0]^4 = [3 : 0] \circ [3 : -1] = [1 : 0]$

Note that $[5 : 2]$ and $[1 : 0]$ are the only squares of elements of C_{-155}, and are also the classes for which both finite genus symbols are +1.

Example. In a second example in Section 6.1, we saw that there are precisely eleven reduced ideal number of discriminant $\Delta = -271$. A group of prime order must be cyclic, so C_{-271} has invariant factor type (11), with any nonidentity element as a generator. Note that 271 is prime, and so there is only one finite genus symbol defined for ideal numbers of discriminant $\Delta = -271$, namely s_{271}. As predicted, there are $\ell - 1 = 0$ even invariant factors for C_{-271}. We verify with the following calculations that $[2 : 0]$ generates the entire class group:

- $[2 : 0]^2 = [2 : 0] \circ [2 : 0] = [4 : 0]$
- $[2 : 0]^3 = [4 : 0] \circ [2 : 0] = [8 : -4]$
- $[2 : 0]^4 = [8 : -4] \circ [2 : 0] = [16 : -4] = [5 : 3] = [5 : -2]$
- $[2 : 0]^5 = [5 : -2] \circ [2 : 0] = [10 : -2] = [7 : 1]$
- $[2 : 0]^6 = [7 : 1] \circ [2 : 0] = [14 : -6] = [7 : 5] = [7 : -2]$
- $[2 : 0]^7 = [7 : -2] \circ [2 : 0] = [14 : -2] = [5 : 1]$
- $[2 : 0]^8 = [5 : 1] \circ [2 : 0] = [10 : -4] = [8 : 3]$
- $[2 : 0]^9 = [8 : 3] \circ [2 : 0] = [4 : -1]$
- $[2 : 0]^{10} = [4 : -1] \circ [2 : 0] = [2 : -1]$
- $[2 : 0]^{11} = [2 : -1] \circ [2 : 0] = [1 : 0]$

For the final three calculations, we use the fact that $[8 : 3] = [2 : -1]^3$ and that $[2 : -1]$ is the inverse of $[2 : 0]$.

Example. Let $\Delta = \Delta(-285, 1) = -1140$. Here, $-285 \equiv 3 \pmod 4$ and $-285 = -1 \cdot 3 \cdot 5 \cdot 19$, with three distinct prime factors. An ideal number in $\mathcal{I}_{-1140}$ has four defined genus symbols (not counting s_∞): s_{-1}, s_3, s_5, and s_{19}. Since 32 does not divide Δ, there are $\ell - 1 = 3$ even terms in the invariant factor type of C_{-1140}.

In a third example in Section 6.1, we found precisely sixteen reduced ideal numbers of discriminant $\Delta = -1140$. The invariant factor type of an Abelian group with sixteen elements is one and only one of the following:

$$(16), \quad (8, 2), \quad (4, 4), \quad (4, 2, 2), \quad (2, 2, 2, 2).$$

Only one of these has precisely three even terms, hence it must be that C_{-1140} is isomorphic to $\mathbb{Z}_4 \times \mathbb{Z}_2 \times \mathbb{Z}_2$.

To verify this claim, we list the elements of C_{-1140} in the following table, with the classes grouped into eight distinct genera:

s_{-1}	s_3	s_5	s_{19}	Class	Class
+	+	+	+	$[1:0]$	$[6:3]$
−	−	−	−	$[2:1]$	$[3:0]$
+	−	−	+	$[5:0]$	$[17:2]$
−	+	+	−	$[10:5]$	$[15:0]$
−	+	−	+	$[7:3]$	$[7:-3]$
−	−	+	+	$[11:1]$	$[11:-1]$
+	+	−	−	$[13:1]$	$[13:-1]$
+	−	+	−	$[14:3]$	$[14:-3]$

We verify below that $A = [7 : 3]$, $B = [2 : 1]$, and $C = [5 : 0]$ can be used as generators of the class group. The entries in the first column of the following array are A^r for $0 \le r < 4$; those in the first row are $B^s \circ C^t$ for $(s, t) = (0, 0), (1, 0), (0, 1)$, and $(1, 1)$ in order; the remaining entries are composites of the row and column headers:

$$
\begin{array}{cccc}
[1:0] & [2:1] & [5:0] & [10:5] \\
[7:3] & [14:3] & [11:1] & [13:-1] \\
[6:3] & [3:0] & [17:1] & [15:0] \\
[7:-3] & [14:-3] & [11:-1] & [13:1].
\end{array}
$$

For example,

$$A \circ B \circ C = [7:3] \circ [10:5] = [70:-25] = [13:25] = [13:-1],$$

using the facts that $k = -25$ satisfies $k \equiv 3 \pmod 7$ and $k \equiv 5 \pmod{10}$, and $P(-25) = 910 = 70 \cdot 13$.

Example. Let $\Delta = \Delta(-185, 1) = -740$, with $P(x) = x^2 + 185$ as its principal polynomial. Since $-185 \equiv 3 \pmod 4$ with $185 = -1 \cdot 5 \cdot 37$, there are three finite genus symbols defined for ideal numbers in $\mathcal{I}_{-740}$: s_{-1}, s_5, and s_{37}. The invariant factor type of C_{-740} has $\ell - 1 = 2$ even terms. Using the methods of Section 6.1, we find that there are precisely sixteen reduced ideal numbers in $\mathcal{I}_{-740}$. These are listed in the following table, grouped by genus:

s_{-1}	s_5	s_{37}	Class	Class	Class	Class
+	+	+	$[1:0]$	$[9:2]$	$[9:-2]$	$[10:5]$
+	−	−	$[2:1]$	$[5:0]$	$[13:6]$	$[13:-6]$
−	−	+	$[3:1]$	$[3:-1]$	$[7:2]$	$[7:-2]$
−	+	−	$[6:1]$	$[6:-1]$	$[14:5]$	$[14:-5]$

As seen in the preceding example, there are two invariant factor types for Abelian groups with sixteen elements that have precisely two even terms: $(8, 2)$ and $(4, 4)$. In this case, trial-and-error shows that C_{-740} contains an element of order eight, such as $[3 : 1]$, ruling out the possibility that C_{-740} is isomorphic to $\mathbb{Z}_4 \times \mathbb{Z}_4$. In the following table, the first row includes $[3 : 1]^r$ and the second row contains $[3 : 1]^r \circ [2 : 1]$, each for $0 \le r < 8$. Since $[2 : 1]^2 = [1 : 0]$, and each class appears in this list, then C_{-740} is isomorphic to $\mathbb{Z}_8 \times \mathbb{Z}_2$:

$$[1 : 0] \quad [3 : 1] \quad [9 : -2] \quad [7 : 2] \quad [10 : 5] \quad [7 : -2] \quad [9 : 2] \quad [3 : -1]$$
$$[2 : 1] \quad [6 : 1] \quad [13 : 6] \quad [14 : -5] \quad [5 : 0] \quad [14 : 5] \quad [13 : -6] \quad [6 : -1].$$

For instance, $[3 : 1]^2 \circ [2 : 1] = [18 : 7] = [13 : -7] = [13 : 6]$, since $P(7) = 234 = 18 \cdot 13$.

6.4 Class groups and representations

Throughout this section, let Δ be a primitive negative discriminant and let C_Δ be the class group of discriminant Δ. Recall that an ideal number $(a : k)$ in $\mathcal{I}_\Delta$ *properly represents* an integer m if and only if $(a : k)$ is equivalent to an ideal number $(m : \ell)$ in $\mathcal{I}_\Delta$. An ordered pair (q, r) is called a *proper representation* of m by $(a : k)$ if q and r are the entries in the first column of a unimodular matrix U for which $(a : k) \cdot U = (m : \ell)$. (We can also phrase these definitions in terms of the quadratic with $(a : k)$ as its ideal number expression.) It is sufficient to describe these representations by the *reduced* ideal numbers of a given negative discriminant Δ, as equivalent ideal numbers properly represent the same collection of integers. The algebraic properties of the class group C_Δ (its invariant factor type and genera) give us additional information about the integers properly represented by a reduced ideal number, as we will demonstrate in this section.

The following terminology will be useful.

Definition. If $(a : k)$ is a reduced ideal number of negative discriminant Δ, then the *reduced conjugate* of $(a : k)$ is the reduced ideal number to which $(a : \bar{k})$ is equivalent. We may also regard the reduced conjugate of $(a : k)$ as the inverse of $[a : k]$ in the class group C_Δ.

Example. In $\mathcal{I}_{-155}$, the ideal number $(5 : 2)$ is reduced, as we have seen in previous examples. The conjugate of $(5 : 2)$ is $(5 : -3)$ but, since $(5 : -3) \sim (5 : 2)$, we can say that $(5 : 2)$ is its own reduced conjugate in C_{-155}. (In other words, $[5 : 2]$ is its own inverse in C_{-155}.)

Let $P(x)$ be the principal polynomial of discriminant Δ. For each positive integer m, let $n_\Delta(m)$ denote the number of solutions of $P(x) \equiv 0 \pmod{m}$. Recall that if $m = p$ is a prime number, then $n_\Delta(p)$ is no larger than 2.

Proposition 6.4.1. *Let* Δ *be a primitive negative discriminant and let* p *be a prime number. Then the following statements are true:*

1. *If* $n_\Delta(p) = 0$, *then* p *is not properly represented by any reduced ideal number of discriminant* Δ.
2. *If* $n_\Delta(p) = 1$, *then there is precisely one reduced ideal number* $(a : k)$ *that properly represents* p. *Here,* $(a : k)$ *must be its own reduced conjugate.*
3. *If* $n_\Delta(p) = 2$, *then* p *is properly represented by a reduced ideal number and by its reduced conjugate (which may or may not be equal to each other), but by no other reduced ideal numbers.*

Proof. If ℓ is a solution of $P(x) \equiv 0 \pmod p$, then $\overline{\ell}$, the conjugate of ℓ with respect to Δ, is also a solution. Any other solution must be congruent modulo p to one of these. Congruent ideal numbers (with positive norm) are equivalent, so can be regarded as identical for this consideration. Thus, there are at most two classes of ideal numbers with norm p.

(1) When $P(x) \equiv 0 \pmod p$ has no solutions, then there is no ideal number $(p : \ell)$ in $\mathcal{I}_\Delta$, and p is not properly represented by a reduced ideal number of discriminant Δ.

(2) When $P(x) \equiv 0 \pmod p$ has exactly one solution ℓ, then $\overline{\ell}$ is congruent to ℓ modulo p. If $(p : \ell)$ is equivalent to a reduced ideal number $(a : k)$, then $(p : \overline{\ell})$ is equivalent to $(a : \overline{k})$, which in turn is equivalent to the reduced conjugate of $(a : k)$. But if $[p : \ell] = [p : \overline{\ell}]$, then $[a : k] = [a : \overline{k}]$. Therefore, there is only one reduced ideal number that properly represents p, and that ideal number equals its reduced conjugate.

(3) Suppose finally that $P(x) \equiv 0 \pmod p$ has two solutions, ℓ and $\overline{\ell}$. In this case, if $(p : \ell)$ is equivalent to a reduced ideal number $(a : k)$, then $(p : \overline{\ell})$ is equivalent to the reduced conjugate of $(a : k)$. These reduced ideal numbers might be equal, but there are no other reduced ideal numbers in $\mathcal{I}_\Delta$ that can properly represent p. $\square$

For composite values of m, we can use the group structure of $\mathcal{C}_\Delta$ to describe the proper representations of m by reduced ideal numbers of discriminant Δ. We summarize our main result in the following theorem.

Theorem 6.4.2. *Let* $m = p_1^{e_1} \cdot p_2^{e_2} \cdots p_t^{e_t}$ *where each* p_i *is a distinct prime number and each* e_i *is a positive integer. Let* $\mathcal{C}_\Delta$ *be the class group of some negative primitive discriminant* Δ. *Then* m *is properly represented by a reduced ideal number of discriminant* Δ *if and only if the following are true:*

1. $n_\Delta(p_i) > 0$ *for* $1 \le i \le t$.
2. *If* $n_\Delta(p_i) = 1$ *for some* i, *then* $e_i = 1$.

Assuming that these conditions hold, then let $(p_i : \ell_i)$ *be an ideal number of norm* p_i *for* $1 \le i \le t$, *and let* $(a_i : k_i)$ *be the reduced ideal number to which* $(p_i : \ell_i)$ *is equivalent. Then* m *is properly represented by the reduced ideal number in the following class in* $\mathcal{C}_\Delta$:

$$[a_1 : k_1]^{e_1} \circ [a_2 : k_2]^{e_2} \circ \cdots \circ [a_t : k_t]^{e_t}.$$

Proof. Conditions (1) and (2) are necessary and sufficient for an ideal number in $\mathcal{I}_\Delta$ having norm m to exist. (See the propositions in Section 2.2.) Any such ideal number $(m : \ell)$ is equivalent to some reduced ideal number $(a : k)$, and so is properly represented by a reduced ideal number. The final claim is then a consequence of Theorem 4.5.2. $\qquad\square$

We illustrate the implications of Theorem 6.4.2 with an extended example in the remainder of this section.

Example. Let $\Delta = -155 = -1 \cdot 5 \cdot 31$, with $P(x) = x^2 + x + 39$. Here, $n_\Delta(p) = 1$ for $p = 5$ and $p = 31$ only, with $(5 : 2)$ and $(31 : 15)$ ideal numbers of these norms. There are no ideal numbers of norm $p = 2$ since $\Delta \equiv 5 \pmod 8$. The existence of ideal numbers of odd prime norm $p \neq 5, 31$ is determined by the Legendre symbol $(\frac{-155}{p})$. Note that

$$\left(\frac{-155}{p}\right) = \left(\frac{5}{p}\right)\left(\frac{-31}{p}\right) = \left(\frac{p}{5}\right)\left(\frac{p}{31}\right),$$

as in Lemma 5.3.2. Ideal numbers of norm p exist if the product of $(\frac{p}{5})$ and $(\frac{p}{31})$ is 1. (We abbreviate these Legendre symbols as s_5 and s_{31} for a fixed ideal number.) We say that an ideal number $(p : \ell)$ is in the ++ genus or $--$ genus if $s_5 = +1 = s_{31}$ or if $s_5 = -1 = s_{31}$, respectively.

In an example in Section 6.3, we found that $\mathcal{C}_{-155}$ contains four elements $[a : k]$. We list these values of a and k in the following table, along with $b = P'(k)$ and $c = \frac{1}{a} \cdot P(k)$. The genus symbols of these ideal numbers, s_5 and s_{31}, are listed as well:

a	k	b	c	s_5	s_{31}
1	0	1	39	+	+
3	0	1	13	−	−
3	−1	−1	13	−	−
5	2	5	9	+	+

An ideal number $(p : \ell)$ in the ++ genus is equivalent to either $(1 : 0)$ or $(5 : 2)$; one in the $--$ genus is equivalent to either $(3 : 0)$ or $(3 : -1)$. The following sequence of translations and involutions, using the representation reduction notation introduced in Section 4.3, shows that $(31 : 15)$ is equivalent to $(5 : 2)$, with $(q, r) = (-1, 2)$ a proper representation of 31 by $(5 : 2)$:

$$(31 : 15)_0^1 \xrightarrow{V} (9 : -16)_1^0 \xrightarrow{T^2} (9 : 2)_1^{-2} \xrightarrow{V} (5 : -3)_2^1 \xrightarrow{T} (5 : 2)_2^{-1}.$$

For $p \neq 5, 31$, if conjugate ideal numbers $(p : \ell)$ and $(p : \bar{\ell})$ exist, then they are equivalent to conjugate ideal numbers. Since $(1 : 0)$ and $(5 : 2)$ are their own reduced conjugates, while $(3 : 0)$ and $(3 : -1)$ are conjugates of each other, we can make the following statements:

1. If $(\frac{p}{5}) = 1 = (\frac{p}{31})$, then p is properly represented by either $(1:0)$ or $(5:2)$, but not by both.

2. If $(\frac{p}{5}) = -1 = (\frac{p}{31})$, then p is properly represented by both $(3:0)$ and $(3:-1)$.

We look at some specific examples below.

If $p = 59$, then $(\frac{p}{5}) = 1 = (\frac{p}{31})$, and we can verify that $(59 : 4)$ and $(59 : -5)$ are conjugate ideal numbers in $\mathcal{I}_{-155}$. Here, we find that

$$(59 : 4)_0^1 \xrightarrow{V} (1 : -5)_1^0 \xrightarrow{T^5} (1 : 0)_1^{-5}$$

and

$$(59 : -5)_0^1 \xrightarrow{V} (1 : 4)_1^0 \xrightarrow{T^{-4}} (1 : 0)_1^4,$$

so that $(-5, 1)$ and $(4, 1)$ are proper representations of 59 by $(1 : 0)$. (We can verify this claim by substituting these values for x and y in $f(x,y) = x^2 + xy + 39y^2$, the quadratic form given by $(1 : 0)$ in ideal number notation.)

If $p = 71$, then $(\frac{p}{5}) = 1 = (\frac{p}{31})$. Directly solving $P(x) \equiv 0 \pmod{71}$ shows that $(71 : 24)$ and $(71 : -25)$ are conjugate ideal numbers in $\mathcal{I}_{-155}$. Here, we find that

$$(71 : 24)_0^1 \xrightarrow{V} (9 : -25)_1^0 \xrightarrow{T^3} (9 : 2)_1^{-3} \xrightarrow{V} (5 : -3)_3^1 \xrightarrow{T} (5 : 2)_3^{-2},$$

and conclude that $(-2, 3)$ is a proper representation of 71 by $(5 : 2)$. A similar reduction of $(71 : -25)$ shows that $(-1, 3)$ is a proper representation of 71 by $(5 : 2)$. Both claims can be verified using the quadratic form $f(x,y) = 5x^2 + 5xy + 9y^2$.

If $p = 73$, then $(\frac{p}{5}) = -1 = (\frac{p}{31})$, and we find that $(73 : 32)$ and $(73 : -33)$ are conjugate ideal numbers in $\mathcal{I}_{-155}$. The following sequence shows that $(73 : 32)$ is equivalent to $(3 : -1)$, with $(3, 2)$ a proper representation of 73 by $(3 : -1)$:

$$(73 : 32)_0^1 \xrightarrow{V} (15 : -33)_1^0 \xrightarrow{T^2} (15 : -3)_1^{-2} \xrightarrow{V} (3 : 2)_2^1 \xrightarrow{T^{-1}} (3 : -1)_2^3.$$

A similar calculation, or an application of Proposition 4.2.2, shows that $(3, -2)$ is a proper representation of 73 by $(3 : 0)$.

To illustrate the implications of Theorem 6.4.2 for composite values, we begin with $m = p^e$, a power of a prime. Since the composite of an ideal number $(p : \ell)$ with its conjugate $(p : \bar{\ell})$ equals $(1 : 0)$, we do not obtain a proper representation of p^2 in this way. (We might view this as indicating an *improper* representation of p^2 by $(1 : 0)$, namely with the ordered pair $(q, r) = (p, 0)$.) When $n_\Delta(p) = 1$, then there are no ideal numbers $(p^e : \ell)$ in $\mathcal{I}_\Delta$, and so no proper representations of p^e by reduced ideal numbers, for $e > 1$. When $n_\Delta(p) = 2$, then $n_\Delta(p^e) = 2$ for all $e \geq 1$, and so two proper representations of p^e by reduced ideal numbers exist.

As an example for $\Delta = -155$, consider $p = 3$, for which $(3 : 0)$ and $(3 : -1)$ are reduced ideal numbers. Powers of these ideal numbers (or rather their congruence classes) can be calculated via the algorithm of Theorem 2.2.6. We obtain the following representatives for the powers of $(3 : 0)$, for instance:

$$(3:0), \quad (9:-3), \quad (27:6), \quad (81:6), \quad (243:-75), \quad \dots.$$

The conjugates of these ideal numbers are likewise powers of $(3 : -1)$. Theorem 6.4.2 implies that the ideal numbers listed above are equivalent to powers of the class $[3 : 0]$ as follows:

$$[3:0], \quad [5:2], \quad [3:-1], \quad [1:0], \quad [3:0], \quad \dots.$$

Considering the conjugates in the same way, we can draw the following conclusions about proper representations of 3^e for $e \geq 1$:

1. If e is odd, then 3^e has proper representations by $(3 : 0)$ and by $(3 : -1)$.
2. If $e \equiv 2 \pmod 4$, then 3^e has two proper representations by $(5 : 2)$.
3. If 4 divides e, then 3^e has two proper representations by $(1 : 0)$.

We apply a sequence of translations and involutions to $(243 : -75)$ as a specific example:

$$(243:-75)_0^1 \xrightarrow{V} (23:74)_1^0 \xrightarrow{T^{-3}} (23:5)_1^3 \xrightarrow{V} (3:-6)_3^{-1} \xrightarrow{T^2} (3:0)_3^{-7}.$$

We can confirm that 243 is properly represented by $(3 : 0)$, with $(q, r) = (-7, 3)$ as a representation.

Similar statements are true for other primes, such as $p = 73$, for which there is an ℓ with $(p : \ell) \sim (3 : 0)$. On the other hand, since we saw above that $(71 : 24) \sim (5 : 2)$, then every odd power of 71 has two proper representations by $(5 : 2)$ and every even power has two proper representations by $(1 : 0)$. Since $(59 : 4) \sim (1 : 0)$, as a third possibility, we can say that every positive power of 59 has two proper representations by $(1 : 0)$.

Finally, we consider an example of a positive integer m with more than one prime divisor, each satisfying the conditions of Theorem 6.4.2. Let $m = 391 = 17 \cdot 23$, where $(\frac{17}{5}) = -1 = (\frac{17}{31})$ and $(\frac{23}{5}) = -1 = (\frac{23}{31})$. An ideal number of norm 17 or 23 is in the $--$ genus, so is equivalent to either $(3 : 0)$ or $(3 : -1)$. Direct calculation shows that

$$(17:-4) \sim (3:0) \sim (23:5) \quad \text{and} \quad (17:3) \sim (3:-1) \sim (23:-6).$$

There are four ideal numbers with norm 391 that we can compute as composites of ideal numbers of norm 17 and 23. For instance, $(17 : -4) \circ (23 : 5) = (391 : 166)$, since $166 \equiv -4 \pmod{17}$ and $166 \equiv 5 \pmod{23}$. This composite is in the same class as $[3 : 0]^2 = [5 : 2]$, so produces a proper representation of 391 by $(5 : 2)$. Specifically, from the sequence

$$(391 : 166)_0^1 \xrightarrow{V} (71 : -167)_1^0 \xrightarrow{T^2} (71 : -25)_1^{-2}$$
$$\xrightarrow{V} (9 : 24)_2^1 \xrightarrow{T^{-3}} (9 : -3)_2^7 \xrightarrow{V} (5 : 2)_7^{-2},$$

we find that $(q, r) = (-2, 7)$ is a proper representation of 391 by $(5 : 2)$. Calculation of all composites of norm 391 appear in the following table:

Composite	Class	Representation
$(17 : -4) \circ (23 : 5) = (391 : 166)$	$[3 : 0] \circ [3 : 0] = [5 : 2]$	$(-2, 7)$
$(17 : -4) \circ (23 : -6) = (391 : 132)$	$[3 : 0] \circ [3 : -1] = [1 : 0]$	$(-8, 3)$
$(17 : 3) \circ (23 : 5) = (391 : -133)$	$[3 : -1] \circ [3 : 0] = [1 : 0]$	$(5, 3)$
$(17 : 3) \circ (23 : -6) = (391 : -167)$	$[3 : -1] \circ [3 : -1] = [5 : 2]$	$(-5, 7)$

To verify two of these claims, note that

$$5(-2)^2 + 5(-2)(7) + 9(7)^2 = 20 - 70 + 441 = 391$$

and

$$(-8)^2 + (-8)(3) + 39(3)^2 = 64 - 24 + 351 = 391,$$

using the quadratic forms with ideal number expression $(5 : 2)$ and $(1 : 0)$.

As this example illustrates, we can typically obtain some information about the integers properly represented by reduced ideal numbers of a negative primitive discriminant using a combination of genus considerations and the invariant factor type of its class group. In this example, however, there was no obvious distinction that we could make between the prime numbers represented by the two ideal numbers, $(1 : 0)$ and $(5 : 2)$, in the principal genus. In Section 6.5, we describe a situation in which a complete categorization of the integers properly represented by each reduced ideal number is possible.

6.5 Negative Boolean discriminants

A group G, written multiplicatively, is called a *Boolean group* if each element of G is its own inverse or, equivalently, if the square of each element of G is the identity element.

Proposition 6.5.1. *Every Boolean group is Abelian. A finite group G is Boolean if and only if the invariant factor type of G is (1) or $(2, 2, \ldots, 2)$ with n copies of 2.*

Proof. In every group G, it is always true that $(ab)^{-1} = b^{-1}a^{-1}$ for every a and b in G. But then if every element of G is its own inverse, we have

$$ab = (ab)^{-1} = b^{-1}a^{-1} = ba$$

for every a and b in G.

In Appendix B, we define the *exponent* of a finite group G to be the smallest positive integer t so that $a^t = 1$ for every a in G. Note that a finite group is Boolean if and only if its exponent is less than or equal to 2. If a finite Abelian group G has invariant factors $(n_1, n_2, \ldots, n_k)$, then Corollary B.3.7 shows that n_1 is the exponent of G. Thus, if G is Boolean, then either $n_1 = 1$, so that G is trivial, or $n_1 = 2$, in which case, $n_2, \ldots n_k$ must also equal 2. □

Definition. We say that a group G is *Boolean of type n* if G is isomorphic to a direct product of n copies of $\mathbb{Z}_2$. A trivial group is said to be *Boolean of type zero*.

Suppose that the class group C_Δ of a discriminant Δ is Boolean. (We will refer to Δ as a *Boolean discriminant* in this case.) Then the principal genus, being the set of squares of each of the classes in C_Δ, contains only the identity element. In that case, each coset of the principal genus contains only one class. But these cosets are the distinct genera of ideal numbers. It follows that we can completely describe the collection of integers that are properly represented by each of the reduced ideal numbers of a Boolean discriminant, using genera. We illustrate this claim with the following example.

Example. Let $\Delta = \Delta(-273, 1) = -1092$, with $P(x) = x^2 + 273$ as its principal polynomial. The maximum norm for a reduced ideal number of this discriminant is $\lfloor \sqrt{\frac{1092}{3}} \rfloor = 19$. One can compile all reduced ideals by calculating $P(x)$ for $-9 \le x \le 9$. We leave it as an exercise to verify that there are precisely the eight reduced ideal numbers listed in the table below, along with their corresponding quadratic forms. (Note that $(17 : -4)$ is not reduced since $P(-4) = 17^2$.) Since $-273 = -1 \cdot -3 \cdot -7 \cdot 13$, using *signed* primes, we find that four finite genus symbols are defined for each ideal number in $\mathcal{I}_{-1092}$. These are calculated for each reduced ideal number in the table:

$(a : k)$	$f(x, y)$	s_{-1}	s_3	s_7	s_{13}
$(1 : 0)$	$x^2 + 273y^2$	+	+	+	+
$(2 : 1)$	$2x^2 + 2xy + 137y^2$	+	−	+	−
$(3 : 0)$	$3x^2 + 91y^2$	−	+	−	+
$(6 : 3)$	$6x^2 + 6xy + 47y^2$	−	−	−	−
$(7 : 0)$	$7x^2 + 39y^2$	−	+	+	−
$(13 : 0)$	$13x^2 + 21y^2$	+	+	−	−
$(14 : 7)$	$14x^2 + 14xy + 23y^2$	−	−	+	+
$(17 : 4)$	$17x^2 + 8xy + 17y^2$	+	−	−	+

Each of the reduced forms in this list equals its own reduced conjugate, and thus is its own inverse in the class group C_{-1092}. Therefore, C_{-1092} is Boolean of type three.

Using Theorem 6.4.2 and other methods introduced in Section 6.4, we can determine whether a positive integer m is properly represented by a reduced ideal number of discriminant $\Delta = -1092$ using the prime factorization of m. Specifically, m has this property if and only if $n_\Delta(p) > 0$ for every prime divisor p of m and m is not divisible by p^2 if $p = 2$, 3, 7, or 13 (the prime divisors of Δ, for which $n_\Delta(p) = 1$). Assuming that m has these properties, we can be sure of the reduced ideal number that properly represents m purely from genus symbol calculations. We illustrate this claim with the following value of m.

Let $m = 3149 = 47 \cdot 67$. We find that $n_\Delta(p) = 2$ for both $p = 47$ and $p = 67$, since the product of the genus symbols defined for $\Delta = -1092$ is 1 for each of these primes. Specifically, each symbol is -1 for $p = 47$, since 47 is congruent to 3, 2, 5, and 8 modulo 4, 3, 7, and 13, respectively. Thus, 47 is properly represented by $(6 : 3)$, the only reduced ideal number in the $-\,-\,-\,-$ genus. (This is apparent from the quadratic form associated with $(6 : 3)$ in the table above.) On the other hand, since 67 is congruent to 3, 1, 4, and 2 modulo 4, 3, 7, and 13, respectively, we find that 67 is represented by the reduced ideal number in the $-\,+\,+\,-$ genus, that is, $(7 : 0)$. Since

$$(6 : 3) \circ (7 : 0) = (42 : 21) \sim (17 : -21) \sim (17 : -4) \sim (17 : 4),$$

it follows that 3149 is properly represented by $(17 : 4)$. (We could also determine this simply by multiplying genus symbols, to note that the product is in the $+\,-\,-\,+$ genus, for which $(17 : 4)$ is the only reduced representative.)

To verify these claims, one can check that $(47 : 3)$ and $(47 : -3)$ are incongruent ideal numbers in $\mathcal{I}_{-1092}$ as are $(67 : 14)$ and $(67 : -14)$. Composing pairs of these ideal numbers, we find that $(3149 : \ell)$ is an ideal number for $\ell = \pm 1125$ and for $\ell = \pm 1460$. We leave it as an exercise to verify that each of these ideal numbers reduces to $(17 : 4)$, producing the following proper representations of 3149 by that ideal number and its quadratic form $17x^2 + 8xy + 17y^2$: $(-14, 3)$, $(-3, 14)$, $(13, 2)$, and $(2, 13)$.

We conclude this section by considering the existence of (primitive) negative Boolean discriminants. This question is connected to the problem of finding all discriminants Δ for which the order of the class group $\mathcal{C}_\Delta$, also called the *class number* of Δ and written as h_Δ, takes on a particular value. The special case of class number $h_\Delta = 1$ has a long history. (Note that $\mathcal{C}_\Delta$ is Boolean of type 0 in this case.) Gauss conjectured, in terms of quadratic forms, a complete list of negative primitive discriminants Δ for which $\mathcal{C}_\Delta$ is trivial. The conjecture was first proved in 1967, independently by Baker and Stark. We state this result as follows.

Fact. If $\Delta = \Delta(d, 1)$ is negative, then the class group $\mathcal{C}_\Delta$ is trivial if and only if d is one of the following nine values:

$$-1 \quad -2 \quad -3 \quad -7 \quad -11 \quad -19 \quad -43 \quad -67 \quad -163.$$

More recently, extensive results have been obtained on larger class numbers, using rational points on elliptic curves, Dirichlet L-functions, and other methods that are well beyond the scope of this text. Watkins [3], in particular, compiled a table for $n \leq 100$ that included the largest (in absolute value) negative discriminant Δ for which $h_\Delta = n$ and the number of discriminants for which this occurs. This table does not include information about the group structure (invariant factor type) of class groups, but direct calculation of reduced ideal numbers for relatively small values of Δ yields the following three results.

Fact. If $\Delta = \Delta(d, 1)$ is negative, then C_Δ is isomorphic to $\mathbb{Z}_2$ if and only if d is one of the following eighteen values:

$$-5 \quad -6 \quad -10 \quad -13 \quad -15 \quad -22 \quad -35 \quad -37 \quad -51$$
$$-58 \quad -91 \quad -115 \quad -123 \quad -187 \quad -235 \quad -267 \quad -403 \quad -427.$$

Fact. If $\Delta = \Delta(d, 1)$ is negative, then C_Δ is isomorphic to $\mathbb{Z}_2 \times \mathbb{Z}_2$ if and only if d is one of the following twenty-four values:

$$-21 \quad -30 \quad -33 \quad -42 \quad -57 \quad -70 \quad -78 \quad -85$$
$$-93 \quad -102 \quad -130 \quad -133 \quad -177 \quad -190 \quad -195 \quad -253$$
$$-435 \quad -483 \quad -555 \quad -595 \quad -627 \quad -715 \quad -795 \quad -1435.$$

Fact. If $\Delta = \Delta(d, 1)$ is negative, then C_Δ is isomorphic to $\mathbb{Z}_2 \times \mathbb{Z}_2 \times \mathbb{Z}_2$ if and only if d is one of the following thirteen values:

$$-105 \quad -165 \quad -210 \quad -273 \quad -330 \quad -345 \quad -357$$
$$-385 \quad -462 \quad -1155 \quad -1995 \quad -3003 \quad -3315.$$

For Boolean groups of type greater than 3, we can use the following approach to lower the necessary amount of class group computations. In Proposition 5.4.7, we saw that if Δ is a primitive discriminant and there are ℓ finite genus symbols defined for ideal numbers in $\mathcal{I}_\Delta$, then there are $\ell - 1$ even terms in the invariant factor type of the class group C_Δ. Thus, for example, C_Δ can be Boolean of type four only if precisely five genus symbols (excluding s_∞) are defined. Since genus symbols are determined by the prime factorization of Δ, it is relatively easy to search first for the needed number of genus symbols. We conclude Section 6.5 with three computational results based on this approach and the Watkins table described above.

Proposition 6.5.2. *There is only one primitive negative discriminant Δ for which the class group C_Δ is Boolean of type four, namely $\Delta = \Delta(d, 1)$ for $d = -1365$.*

Proof. According to the table in [3], the largest (in absolute value) d for which $\Delta = \Delta(d, 1)$ has class number $h_\Delta = 16$ is $d = -31243$. Testing all square-free integers d with $-31243 \leq d < 0$ to count the number of genus symbols of $\Delta = \Delta(d, 1)$, we find that there are 375 of them with precisely five finite genus symbols defined. We can construct the collection of reduced ideal number of each of these discriminants, and determine that there is only one of them with sixteen elements, namely for $d = -1365$, as noted above. □

As an aside, for all of the values of d tested in this way, the total number of reduced ideal numbers in $\mathcal{I}_\Delta$ is a *multiple* of 16. This is as must be the case, since the invariant factor type of C_Δ has four even terms in each of these cases.

Proposition 6.5.3. *There are no primitive negative discriminants Δ for which the class group C_Δ is Boolean of type five.*

Proof. Using the table in [3], when $\Delta = \Delta(d,1)$, a class number $h_\Delta = 32$ can occur only if $-164083 \leq d < 0$. In this range of values, we find 154 possible d values for which Δ has six finite genus symbols. Constructing the collection of reduced ideal numbers for each one, we find no cases in which there are precisely 32 such ideal numbers (although each possibility is a multiple of 32). □

Proposition 6.5.4. *There are no primitive negative discriminants Δ for which the class group C_Δ is Boolean of type six.*

Proof. Following the same approach as in the preceding proof, we must test $-693067 \leq d < 0$. There are only 13 of these for which seven finite genus symbols are defined, and in no case are there precisely 64 reduced ideal numbers. (The number of reduced ideal numbers is a multiple of 64 each time.) □

The search for negative Boolean discriminants of type five was originally part of an honors project conducted by Christopher Hudert under the direction of the author. Each of the searches and calculations described in the previous three propositions took less than a minute to run on a personal computer, using Python programs written by the author. Since each increase in the number of genus symbols defined requires an additional distinct prime divisor of d, it seems reasonable that the trend demonstrated above of fewer and fewer potential Boolean discriminants of larger types will continue.

6.6 Class groups of arbitrary negative discriminant

To conclude Chapter 6, we consider class groups of negative discriminants that are not primitive. We will mainly state our results without proof and illustrate them with examples. Complete proofs of the claims made in this section appear in [1].

Throughout this section, let Δ be a discriminant (not assumed to be primitive or negative), let p be a prime number, and let $\Delta_1 = p^2\Delta$. We write the principal polynomials of discriminant Δ and Δ_1 as $P(x)$ and $P_1(x)$, respectively. We will demonstrate that we can use the class group $G = C_\Delta$ to construct $G_1 = C_{\Delta_1}$. We use the following example throughout this section.

Example. Let $\Delta = -23$ with $P(x) = x^2 + x + 6$ as its principal polynomial. The following values of $P(x)$ allow us to compute the class group $G = C_{-23}$. (We calculate more terms of $P(x)$ than necessary, based on additional need for this data in later examples.)

x	0, −1	1, −2	2, −3	3, −4	4, −5	5, −6	6, −7
$P(x)$	6	8	12	18	26	36	48

Here, $u = \lfloor \sqrt{\frac{23}{3}} \rfloor = 2$ is the maximum norm of a reduced ideal number of discriminant $\Delta = -23$. Since Δ is primitive, each ideal number in $\mathcal{I}_{-23}$ is primitive, and we find that

$$G = \{[1:0], [2:0], [2:-1]\}.$$

Now let $p = 5$, so that $\Delta_1 = p^2\Delta = -575$, with $P_1(x) = x^2 + 5x + 150$ as principal polynomial. We can compile the reduced ideal numbers of discriminant Δ_1 directly by testing for ideal numbers of maximum norm $u = \lfloor \sqrt{\frac{575}{3}} \rfloor = 14$. We find 21 reduced ideal numbers, using the following values of $P_1(x)$:

x	−2, −3	−1, −4	0, −5	1, −6	2, −7	3, −8	4, −9
$P_1(x)$	144	146	150	156	164	174	186

Three of these classes, $[5:0]$, $[10:0]$, and $[10:-5]$, are not primitive. The remaining eighteen, listed below, comprise the class group $G_1 = C_{-575}$:

$$\begin{array}{cccccc}
[1:-2] & [2:-2] & [2:-3] & [3:-2] & [3:-3] & [4:-2] \\
[4:-3] & [6:-2] & [6:-3] & [6:0] & [6:-5] & [8:-2] \\
[8:-3] & [9:-2] & [9:-3] & [12:-2] & [12:1] & [12:-6].
\end{array}$$

Note that $[12:-3]$ is not reduced since $P_1(-3) = 12^2$ but $P_1'(-3) = -1$ is negative.

Our first proposition shows that reduced ideal numbers in C_{Δ_1} that are not primitive arise directly from the elements of C_Δ.

Proposition 6.6.1. *Let Δ be a discriminant, let p be a prime number, and let $\Delta_1 = p^2\Delta$. If $(a:k)$ is a reduced ideal number of discriminant Δ, then $(pa:pk)$ is a reduced ideal number of discriminant Δ_1.*

Proof. Let $P(x)$ and $P_1(x)$ be the principal polynomials of discriminant Δ and Δ_1, respectively. If $(a:k)$ is an element of $\mathcal{I}_\Delta$, so that $P(k) = ac$ for some integer c, then

$P_1(pk) = p^2 P(k) = (pa)(pc)$. This implies that $(pa : pk)$ is an element of $\mathcal{I}_{\Delta_1}$. Suppose that $(a : k)$ is reduced in $\mathcal{I}_\Delta$. If

$$a^2 < P(k) \quad \text{and} \quad -a < P'(k) \leq a,$$

then

$$(pa)^2 < P_1(pk) = p^2 P(k) \quad \text{and} \quad -pa < P'_1(pk) = pP'(k) \leq pa.$$

Likewise, if

$$a^2 = P(k) \quad \text{and} \quad 0 \leq P'(k) \leq a,$$

then

$$(pa)^2 = P_1(pk) = p^2 P(k) \quad \text{and} \quad 0 \leq P_1(pk) = pP(k) \leq pa.$$

Thus if $(a : k)$ is reduced in $\mathcal{I}_\Delta$, then $(pa : pk)$ is reduced in $\mathcal{I}_{\Delta_1}$. □

Proposition 6.6.2. *Let p be a prime number and let G and G_1 be the class groups of discriminant Δ and $\Delta_1 = p^2\Delta$, respectively. Then every element of G_1 can be written as $[a : pk]$ where a is an integer not divisible by p and k is an integer. In this case, the function $T : G_1 \to G$ defined by $T([a : pk]) = [a : k]$ is a well-defined surjective homomorphism, which we call the projection of G_1 onto G.*

Proof. Recall from Proposition 5.2.1 that a *primitive* ideal number in $\mathcal{I}_\Delta$ properly represents an integer relatively prime to Δ. But then an element of $[a : \ell]$ of G_1 properly represents an integer not divisible by p, and we can assume that this is true for a itself. Now the linear congruence $px \equiv \ell \pmod{a}$ has a solution, which we can label as k. Thus, $[a : \ell] = [a : pk]$ in G_1.

Let $P(x)$ and $P_1(x)$ be the principal polynomials for Δ and $\Delta_1 = p^2\Delta$, respectively. If $(a : pk)$ is an ideal number in $\mathcal{I}_{\Delta_1}$, then a divides $P_1(pk) = p^2 P(k)$. Since $\gcd(a, p) = 1$, then a divides $P(k)$, so that $(a : k)$ is an element of $\mathcal{I}_\Delta$. Furthermore, again using Proposition 5.2.1, we can assume that every element of G can be written as $(a : k)$ where p does not divide a. Thus, T is surjective.

That T is well-defined and a homomorphism between the groups G_1 and G under composition is a consequence of results from Chapter 8 in [1]. (See, in particular, Theorem 8.2.2 in that text.) □

Example. Returning to the preceding example, we consider the projection $T : C_{-575} \to C_{-23}$ defined in Proposition 6.6.2. Each primitive reduced ideal number of discriminant $\Delta_1 = -575$ is first rewritten in the form $[a : pk]$. The arrow following that ideal number indicates the effect of the homomorphism T, taking $[a : pk]$ to $[a : k]$. The resulting element of $G = C_{-23}$ is replaced by its equivalent reduced ideal number:

$$
\begin{array}{rcllrcl}
[1 : -2] &=& [1 : 0] & \longrightarrow & [1 : 0] &=& [1 : 0] \\
[2 : -2] &=& [2 : 0] & \longrightarrow & [2 : 0] &=& [2 : 0] \\
[2 : -3] &=& [2 : -5] & \longrightarrow & [2 : -1] &=& [2 : -1] \\
[3 : -2] &=& [3 : -5] & \longrightarrow & [3 : -1] &=& [2 : 0] \\
[3 : -3] &=& [3 : 0] & \longrightarrow & [3 : 0] &=& [2 : -1] \\
[4 : -2] &=& [4 : -10] & \longrightarrow & [4 : -2] &=& [2 : -1] \\
[4 : -3] &=& [4 : 5] & \longrightarrow & [4 : 1] &=& [2 : 0] \\
[6 : -2] &=& [6 : 10] & \longrightarrow & [6 : 2] &=& [2 : -1] \\
[6 : -3] &=& [6 : -15] & \longrightarrow & [6 : -3] &=& [2 : 0] \\
[6 : 0] &=& [6 : 0] & \longrightarrow & [6 : 0] &=& [1 : 0] \\
[6 : -5] &=& [6 : -5] & \longrightarrow & [6 : -1] &=& [1 : 0] \\
[8 : -2] &=& [8 : -10] & \longrightarrow & [8 : -2] &=& [1 : 0] \\
[8 : -3] &=& [8 : 5] & \longrightarrow & [8 : 1] &=& [1 : 0] \\
[9 : -2] &=& [9 : -20] & \longrightarrow & [9 : -4] &=& [2 : -1] \\
[9 : -3] &=& [9 : -15] & \longrightarrow & [9 : -3] &=& [2 : 0] \\
[12 : -2] &=& [12 : 10] & \longrightarrow & [12 : 2] &=& [1 : 0] \\
[12 : 1] &=& [12 : 25] & \longrightarrow & [12 : 5] &=& [2 : -1] \\
[12 : -6] &=& [12 : -30] & \longrightarrow & [12 : -6] &=& [2 : 0].
\end{array}
$$

As one example, -30 is a multiple of 5 that is congruent to -6 modulo 12, so that

$$[12 : -6] = [12 : -30]$$

in G_1. The homomorphism T can be applied to this expression:

$$T([12 : 5(-6)]) = [12 : -6].$$

Now since $P(-6) = 36 = 12 \cdot 3$, then $[12 : -6] = [3 : 5]$ using an involution. But then $[3 : 5] = [3 : -1]$ and, since $P(-1) = 6 = 3 \cdot 2$, then $[3 : -1] = [2 : 0]$ by an involution. Therefore, $T : G_1 \to G$ takes $[12 : -6]$ to $[2 : 0]$.

The kernel of T, written as ker T, consists of all elements of G_1 taken to the identity element of G, that is, $[1 : 0]$. Using the revised expressions for elements of G_1 to which T applies, we find that in this example,

$$\ker T = \{[1 : 0], [6 : 0], [6 : -5], [8 : -10], [8 : 5], [12 : 10]\}.$$

Proposition 6.6.3. *Let $G = C_\Delta$ for some discriminant Δ, let p be a prime number, and let $G_1 = C_{p^2\Delta}$. Let $P(x)$ be the principal polynomial of discriminant Δ. Let $T : G_1 \to G$ be the projection defined as in Proposition 6.6.2. Then the kernel of T consists of $[1 : 0]$ and the distinct classes of the form $[P(k) : pk]$, where k is an integer for which p does not divide $P(k)$. These classes can also be expressed as $[p^2 : pk]$ with the same restriction on k.*

Proof. Let $P(x)$ and $P_1(x)$ be the principal polynomials for Δ and $\Delta_1 = p^2\Delta$, respectively. Since $P_1(pk) = p^2 P(k)$, then $(P(k) : pk)$ is an ideal number in $\mathcal{I}_{\Delta_1}$. The content of this ideal number is $\gcd(P(k), pP'(k), p^2)$, which is 1 if p does not divide $P(k)$. Thus, $[P(k) : pk]$ is an element of G_1 that is in the form to which T applies. Then $T([P(k) : pk]) = [P(k) : k]$ equals the identity element of G, since $P(k) = P(k) \cdot 1$. Thus, all classes $[P(k) : pk]$ are in the kernel of T, as is $[1 : 0]$. Note that $[P(k) : pk] = [p^2 : \overline{pk}] = [p^2 : p\overline{k}]$, using an involution. As k varies over all integers for which p does not divide $P(k)$, so does its conjugate with respect to Δ, $\overline{k}$. Thus, we can also say that all classes of the form $[p^2 : pk]$, with p not dividing $P(k)$, are also in the kernel of T. We omit the proof that all elements of the kernel have this form. (See Proposition 8.2.6 in [1].) $\qquad\square$

Since $T : G_1 \to G$ is surjective, and we can describe the kernel of T, we can list all elements of G_1 without first finding all reduced ideal numbers of discriminant $p^2\Delta$. If we express all elements of G as $[a : k]$ with a not divisible by p, and write

$$\ker T = \{[p^2 : p\ell] \mid p \text{ does not divide } P(\ell)\},$$

then we can express the elements of G_1 as composites of pairs of elements $[a : pk]$ and $[p^2 : p\ell]$. We illustrate this in the following example.

Example. Let $\Delta = -23$ as above and let $p = 13$. Let $G = C_{-23}$ and $G_1 = C_{-3887}$, where $-3887 = 13^2(-23)$. Let $P(x) = x^2 + x + 6$, the principal polynomial of discriminant Δ. Since 13 divides $P(\ell)$ for $\ell = 4$ and $\ell = -5$, we find that the kernel of the projection $T : G_1 \to G$ has $(p + 1) - 2 = 12$ elements, which we can express as

$$[1 : 0] \quad [169 : 0] \quad [169 : -13] \quad [169 : 13] \quad [169 : -26] \quad [169 : 26]$$
$$[169 : -39] \quad [169 : 39] \quad [169 : -52] \quad [169 : 65] \quad [169 : -78] \quad [169 : 78].$$

The classes $[1 : 0]$, $[2 : 0]$, and $[2 : -13]$ are sent to the elements of G by T. Representatives of all classes in G_1 can then be computed as composites of one of these three classes with any of the twelve classes in the kernel of T, as listed above. Thus, G_1 has 36 distinct elements in total. (We omit further calculation in this example, but it can be confirmed directly that C_{-3887} has 36 primitive reduced ideal numbers, as well as three, namely $(13 : 0)$, $(26 : 0)$, and $(26 : -13)$, with content 13.)

Readers may recognize a similarity between our alternative expression for the kernel of the T projection and the kernel of a prime subgroup of positive level, as this was defined in Section 3.4, which is in turn related to a group of matrices we defined as the prime kernel of a discriminant in Section 3.3. In general, the kernel of a p-projection T can be regarded as a quotient group of the prime kernel of a discriminant.

---

7 Ideal numbers of positive discriminant

We now turn our attention to ideal numbers of positive discriminant. As we did with negative discriminants in Chapter 6, we begin by defining a collection of *reduced* ideal numbers of positive discriminant, demonstrating how these ideal numbers can be constructed in practice, and showing that every ideal number is equivalent to one that is reduced. In this case, however, there are typically equivalences among the reduced ideal numbers. A *reduction map*, defined for each positive discriminant, allows us to describe these equivalences in full, and thus construct the class group of an arbitrary positive discriminant. We will also see that the reduction process allows us to compute representations of integers by ideal numbers of positive discriminant and their corresponding indefinite quadratic forms.

The concept of *continued fractions* of real numbers, particularly real quadratic numbers, turns out to have important connections and applications to ideal numbers of positive discriminant. Basic definitions and results concerning continued fractions appear in Appendix D. We will invoke those facts as needed in this chapter.

7.1 Reduced ideal numbers of positive discriminant

When Δ is positive, we can define a set of ideal numbers that contains representatives of every class of ideal numbers in $\mathcal{I}_\Delta$. Let $P(x)$ be the principal polynomial of discriminant Δ. We fix the following notation for the two roots of $P(x)$:

$$w = \frac{-P'(0) + \sqrt{\Delta}}{2} \quad \text{and} \quad \overline{w} = \frac{-P'(0) - \sqrt{\Delta}}{2}.$$

When Δ is positive, these roots are real numbers and $P(x)$ is negative (for real numbers x) precisely when $\overline{w} < x < w$. Note that $w - \overline{w} = \sqrt{\Delta}$ and $w + \overline{w} = -P'(0)$.

Definition. Let $P(x)$ be the principal polynomial of some positive discriminant Δ, with roots $w > \overline{w}$. We say that an ideal number $(a : k)$ in $\mathcal{I}_\Delta$ is *reduced* if

$$k - a < \overline{w} < k < k + a < w.$$

It is implicit that the norm, a, of a reduced ideal number $(a : k)$ of positive discriminant is positive. There is a unique value in the congruence class of k modulo a for which $k - a < \overline{w} < k$. Thus, there is at most one reduced ideal number in the congruence class of a particular ideal number $(a : k)$. Our first proposition shows that reduced ideal numbers tend to arise in pairs.

Proposition 7.1.1. *Let $(a : k)$ be a reduced ideal number of positive discriminant Δ. Let $w > \overline{w}$ be the roots of the principal polynomial, $P(x)$, of discriminant Δ. If $\overline{k}$ is the conjugate*

https://doi.org/10.1515/9783111319360-007

of k with respect to Δ, let ℓ be the unique integer for which $\ell \equiv \overline{k}$ (mod a) and $\ell - a < \overline{w} < \ell$. Then $(a : \ell)$ is also a reduced ideal number.

We will refer to $(a : \ell)$ as the *reduced conjugate* of $(a : k)$, using this terminology only when $(a : k)$ is reduced. It is possible that $(a : \ell) = (a : k)$, but only when $(a : k)$ is congruent to its conjugate.

Proof. We are given that $k - a < \overline{w} < k < k + a < w$. Let $\ell \equiv \overline{k}$ (mod a) be selected so that $\ell - a < \overline{w} < \ell$. Note that $P(\ell) \equiv P(\overline{k})$ (mod a) and that a divides $P(\overline{k}) = P(k)$, in which case $(a : \ell)$ is an ideal number of discriminant Δ. Recall that $\overline{k} = -k - P'(0)$, so that $k + \overline{k} = -P'(0) = w + \overline{w}$. Thus, $k - \overline{w} = w - \overline{k}$ and, since $0 < k - \overline{w}$, then $0 < w - \overline{k}$, that is, $\overline{k} < w$. We also have $w - k = \overline{k} - \overline{w}$ and, since $a < w - k$, then $a < \overline{k} - \overline{w}$, or $\overline{w} < \overline{k} - a$. The final inequality shows that $\ell \leq \overline{k} - a$. Therefore, we now have $\ell - a < \overline{w} < \ell < \ell + a \leq \overline{k} < w$, so that $(a : \ell)$ is reduced. $\qquad\square$

The following proposition gives alternative descriptions of reduced ideal numbers. Recall that if w and $\overline{w}$ are defined as above and $(a : k)$ is an ideal number in $\mathcal{I}_\Delta$, then $v = \frac{w-k}{a}$ and $\overline{v} = \frac{\overline{w}-k}{a}$ are called the major and minor roots of $(a : k)$, respectively.

Proposition 7.1.2. *Let $P(x)$ be the principal polynomial of some positive discriminant Δ and let $(a : k)$ be an ideal number in $\mathcal{I}_\Delta$. Let $P(k) = ac$ and $P'(k) = b$, and let v and $\overline{v}$ be the major and minor roots of $(a : k)$. Then $(a : k)$ is reduced if and only if any of the following statements are true:*

1. *a and $P(k - a)$ are positive, while $P(k)$ and $P(k + a)$ are negative.*
2. *a is positive, c is negative, $a + b + c$ is negative, and $a - b + c$ is positive.*
3. *$v > 1$ and $-1 < \overline{v} < 0$.*

Proof. As noted, $P(x)$ is negative if and only if $\overline{w} < x < w$. Thus, if $a > 0$, in which case $k - a < k < k + a$, then $P(k - a)$ is positive, $P(k)$ is negative, and $P(k + a)$ is negative if and only if $k - a < \overline{w} < k < k + a < w$. This establishes that statement (1) is equivalent to the definition of a reduced ideal number.

If $f(x) = ax^2 + bx + c$ has ideal number expression $(a : k)$, then $P(ax + k) = a \cdot f(x)$ for all x by Proposition 1.3.3. Thus, if a is positive, then $P(k)$, $P(k + a)$, and $P(k - a)$ have the same sign as $f(0) = c$, $f(1) = a + b + c$, and $f(-1) = a - b + c$, respectively. Hence, statements (1) and (2) are equivalent.

Finally, since $f(x)$ has two real roots, and assuming that a is positive, the conditions that $f(-1)$ is positive while $f(0)$ and $f(1)$ are negative are necessary and sufficient so that $f(x)$ has a root between -1 and 0, and a second root greater than 1. Thus, statements (2) and (3) are equivalent. $\qquad\square$

Proposition 7.1.3. *For each positive discriminant Δ, the set of reduced ideal numbers is finite.*

Proof. If $\overline{w} < k < k + a < w$, then $0 < a = (k + a) - k < w - \overline{w} = \sqrt{\Delta}$. For each a, there are precisely a values of k for which $k - a < \overline{w} < k$. With finitely possibilities for a

and k, there are finitely many possibilities for reduced ideal numbers of a given positive discriminant Δ. ☐

We can construct the set of reduced ideal numbers in $\mathcal{Q}_\Delta$ in practice by the following proposition.

Proposition 7.1.4. *Let Δ be a positive discriminant with $w > \overline{w}$ the roots of the principal polynomial $P(x)$ of discriminant Δ. Suppose that a and k satisfy the inequalities $0 < a < \frac{\sqrt{\Delta}}{2}$ and $\overline{w} < k < \overline{w} + a$, and that a divides $P(k)$, say with $P(k) = ac$. Then the ideal numbers $(a : k)$ and $(-c : k)$ are both reduced. All reduced ideal numbers in $\mathcal{I}_\Delta$ have one or both of these forms.*

Proof. When $0 < a < \frac{\sqrt{\Delta}}{2}$ and $\overline{w} < k < \overline{w} + a$, then we find that

$$k - a < \overline{w} < k < k + a < w,$$

the final inequality due to the fact that $w - \overline{w} = \sqrt{\Delta}$. If a divides $P(k)$, say with $P(k) = ac$ and $P'(k) = b$, then $(a : k)$ is an ideal number in $\mathcal{I}_\Delta$, and is reduced by definition. Note that $(-c : k)$ is an ideal number in $\mathcal{I}_\Delta$ since $P(k) = (-c)(-a)$. Now a is positive, c is negative, $a + b + c$ is negative, and $a - b + c$ is positive if and only if $-c$ is positive, $-a$ is negative, $-c + b + (-a) = -(a - b + c)$ is negative, and $-c - b + (-a) = -(a + b + c)$ is positive. Hence, $(-c : k)$ is also reduced by Proposition 7.1.2.

The inequalities $a + b + c < 0 < a - b + c$ can be rewritten as $b < a + c < -b$. Thus, $(a+c)^2 < b^2$, and then $(a-c)^2 = (a+c)^2 - 4ac < b^2 - 4ac = \Delta$. Hence, if $(a : k)$ is reduced, so that a and $-c$ are both positive, then a and $-c$ cannot both exceed $\frac{\sqrt{\Delta}}{2}$. Therefore, if we find all reduced ideal numbers $(a : k)$ with $0 < a < \frac{\sqrt{\Delta}}{2}$, as above, then those together with the corresponding $(-c : k)$ must include all reduced ideal numbers in $\mathcal{I}_\Delta$. (These lists are not necessarily disjoint.) ☐

Example. Let $\Delta = 41$, whose principal polynomial $P(x) = x^2 + x - 10$ has roots $w = \frac{-1+\sqrt{41}}{2} \approx 2.7$ and $\overline{w} = \frac{-1-\sqrt{41}}{2} \approx -3.7$. For each positive $a \le \lfloor \frac{\sqrt{41}}{2} \rfloor = 3$, we test the a smallest values of k exceeding $\overline{w}$ in the following table. If a divides $P(k)$, we also list $-c = -\frac{P(k)}{a}$:

a	k	$P(k)$	$-c$
1	-3	-4	4
2	-3	-4	2
2	-2	-8	4
3	-3	-4	—
3	-2	-8	—
3	-1	-10	—

Listing both $(a : k)$ and $(-c : k)$, we find the following complete list of reduced ideal numbers in $\mathcal{I}_{41}$:

$$(1 : -3) \quad (2 : -3) \quad (2 : -2) \quad (4 : -3) \quad (4 : -2).$$

Note that $(2 : -3)$ arises both as $(a : k)$ and $(-c : k)$, but needs to be listed only once. Here, $(1 : -3)$ is its own reduced conjugate, whereas $(2 : -3)$ and $(2 : -2)$ are reduced conjugates of each other, as are $(4 : -3)$ and $(4 : -2)$. To verify that each ideal number is reduced, as in Proposition 7.1.2, we list a, b, c, $a + b + c$, and $a - b + c$ for each of these ideal numbers (with $b = P'(k)$), together with an approximation of the major and minor roots of $(a : k)$, that is, $v = \frac{w-k}{a} = \frac{-b+\sqrt{\Delta}}{2a}$ and $\overline{v} = \frac{\overline{w}-k}{a} = \frac{-b-\sqrt{\Delta}}{2a}$:

a	b	c	$a + b + c$	$a - b + c$	v		$\overline{v}$
1	−5	−4	−8		2	5.70	−0.70
2	−5	−2	−5		5	2.85	−0.35
2	−3	−4	−5		1	2.35	−0.85
4	−5	−1	−2		8	1.43	−0.18
4	−3	−2	−1		5	1.18	−0.43

For example, in the fifth line, $v = \frac{3+\sqrt{41}}{8} \approx 1.18$ and $\overline{v} = \frac{3-\sqrt{41}}{8} \approx -0.43$. In each case, the major root of $(a : k)$ exceeds 1 while the minor root is between −1 and 0.

Example. Let $\Delta = \Delta(115, 1) = 460$, with principal polynomial $P(x) = x^2 - 115$. Here, w and $\overline{w}$ are $\pm\sqrt{115} \approx \pm10.7$, and we need to test $0 < a \leq 10$ in the algorithm of Proposition 7.1.4. It is convenient to calculate $P(x)$ for the smallest ten values of $x \geq -10$, as in the following table:

x	−10	−9	−8	−7	−6	−5	−4	−3	−2	−1
$P(x)$	−15	−34	−51	−66	−79	−90	−99	−106	−111	−114

For $1 \leq a \leq 10$, we then test whether a divides any of the first a values in the $P(x)$ row. In the following table, we list these values of a and $x = k$, together with $-c = -P(k)/a$:

a	1	2	3	3	5	6	6	9	9	10
k	−10	−9	−10	−8	−10	−7	−5	−5	−4	−5
$-c$	15	17	5	17	3	11	15	10	11	9

Eliminating the overlap between $(a : k)$ and $(-c : k)$, which is all cases where $-c \leq 10$, we find a total of sixteen reduced ideal numbers in $\mathcal{I}_{460}$:

$$(1 : -10) \quad (2 : -9) \quad (3 : -10) \quad (3 : -8) \quad (5 : -10)$$
$$(6 : -7) \quad (6 : -5) \quad (9 : -5) \quad (9 : -4) \quad (10 : -5)$$
$$(11 : -4) \quad (11 : -7) \quad (15 : -5) \quad (15 : -10) \quad (17 : -8) \quad (17 : -9).$$

To confirm, for instance, that $(17 : -9)$ is reduced, we can note that $P(-9) = -34 = 17 \cdot -2 = a \cdot c$ and $P'(-9) = -18 = b$, so that a is positive, c is negative, $a + b + c = -3$ is negative, and $a - b + c = 33$ is positive. Note that $(a : k)$ is its own reduced conjugate for exactly four of the sixteen reduced ideal numbers.

Proposition 7.1.5. *If Δ is a positive discriminant, then every ideal number in $\mathcal{I}_\Delta$ is equivalent to a reduced ideal number.*

Proof. Let $w > \overline{w}$ be the roots of the principal polynomial $P(x)$ of discriminant Δ. In each class of ideal numbers in $\mathcal{I}_\Delta$, we can select a representative $(a : k)$ with a as small as possible in absolute value. Proposition 4.3.1 shows that $4a^2 \le \Delta$ in this case, that is, $|a| < \frac{\sqrt{\Delta}}{2}$. (Equality cannot hold here since Δ is not a square.) If a is positive, we can use a translation to assume further that $k - a < \overline{w} < k$. Since $w - \overline{w} = \sqrt{\Delta}$, then $k - a < \overline{w} < k < k + a < w$, so that $(a : k)$ is reduced.

If a is negative, then we can select k so that $\overline{w} < k + a < k < w < k - a$, which implies that $P(k+a)$ and $P(k)$ are negative while $P(k-a)$ is positive. If $f = (a : k)$ is written as the quadratic polynomial $f(x) = ax^2 + bx + c$, then with a negative we see that $f(1) = a + b + c$, $f(0) = c$, and $f(-1) = a - b + c$ have the opposite signs from $P(k + a)$, $P(k)$, and $P(k - a)$, respectively, by Proposition 1.3.3. Now consider the involution $(a : k) \cdot V = (c : \overline{k})$, that is, $g(x) = cx^2 - bx + a$ as a quadratic polynomial. Since c is positive, $g(0) = a$ is negative, $g(1) = c + (-b) + a$ is negative, and $g(-1) = c - (-b) + a$ is positive, then $(c : \overline{k})$ is reduced. An ideal number is equivalent to its involution, so the class of $(a : k)$ contains a reduced ideal number in this case as well. □

Unlike in the case where Δ is negative, there are typically equivalences among the reduced elements of $\mathcal{I}_\Delta$ when Δ is positive. We will explore this further in the next few sections via a function defined on $\mathcal{I}_\Delta$.

7.2 Reduction maps

When Δ is a positive discriminant, then we associate an integer to each ideal number in $\mathcal{I}_\Delta$ as follows.

Definition. Let Δ be a positive discriminant with principal polynomial $P(x)$, and let $w = \frac{-P'(0) + \sqrt{\Delta}}{2}$. If $(a : k)$ is an ideal number in $\mathcal{I}_\Delta$, then define the *reduction number* of $(a : k)$ to be

$$q = q(a : k) = \left\lfloor \frac{w - k}{a} \right\rfloor, \tag{7.1}$$

that is, the largest integer smaller than the major root of $(a : k)$.

Proposition 7.2.1. *Let* Δ *be a positive discriminant with principal polynomial* $P(x)$*. For an ideal number* $(a : k)$ *in* $\mathcal{I}_\Delta$ *with reduction number* q*, let*

$$k' = \overline{aq + k} \quad and \quad a' = -\frac{P(k')}{a}, \tag{7.2}$$

where $\overline{aq + k}$ *denotes the conjugate of* $aq + k$ *with respect to* Δ*. Then* $(a' : k')$ *is an ideal number of discriminant* Δ*.*

Proof. Note that $P(k') = P(\overline{aq + k}) \equiv P(k) \equiv 0 \pmod{|a|}$, since the principal polynomial applied to an integer and its conjugate are equal, and a divides $P(k)$ by the definition of an ideal number. Thus, a' is an integer. Furthermore, a' divides $P(k') = -aa'$, so that $(a' : k')$ is an ideal number. $\qquad\square$

Definition. If $(a : k)$ is an ideal number of positive discriminant Δ, let

$$R(a : k) = (a' : k'),$$

where a' and k' are given as in equation (7.2). We can view R as a function from $\mathcal{I}_\Delta$ to $\mathcal{I}_\Delta$, and we call R the *reduction map* on $\mathcal{I}_\Delta$.

Example. Let $\Delta = 41$, so that $P(x) = x^2 + x - 10$ and $w = \frac{-1+\sqrt{41}}{2} \approx 2.7$. Since $P(9) = 80$, then $(8 : 9)$ is an ideal number in $\mathcal{I}_{41}$. The reduction number of $(8 : 9)$ is $q(8 : 9) = \lfloor \frac{2.7-9}{8} \rfloor = -1$. Then $aq + k = 8(-1) + 9 = 1$, and we find that $k' = \overline{1} = -2$ and $a' = -\frac{P(-2)}{8} = 1$. Thus, $R(8 : 9) = (1 : -2)$.

Example. Again with $\Delta = 41$, consider the ideal number $(-8 : -10)$. (Here, -10 is the conjugate of 9 with respect to Δ, so that $P(-10) = P(9) = 80$ as above.) The reduction number of $(-8 : -10)$ is $q(-8 : -10) = \lfloor \frac{2.7+10}{-8} \rfloor = -2$. Now $aq + k = (-8)(-2) - 10 = 6$, and we find that $k' = \overline{6} = -7$ and $a' = -\frac{P(-7)}{-8} = -\frac{32}{-8} = 4$. Thus, $R(-8 : -10) = (4 : -7)$.

We compile several important properties of reduction maps in the remainder of Section 7.2.

Proposition 7.2.2. *Let* $(a_1 : k_1)$ *and* $(a_2 : k_2)$ *be ideal numbers of positive discriminant* Δ *and let* $R : \mathcal{I}_\Delta \to \mathcal{I}_\Delta$ *be the reduction map. Then* $R(a_1 : k_1) = R(a_2 : k_2)$ *if and only if* $a_1 = a_2$ *and* $k_2 = k_1 + a_1 n$ *for some integer* n*.*

Proof. Suppose that $R(a_1 : k_1) = R(a_2 : k_2) = (a' : k')$. By equation (7.2), then

$$-a_1 a' = P(k') = -a_2 a',$$

where $P(x)$ is the principal polynomial of discriminant Δ. Thus, $a_1 = a_2$, and we call that common value a. If $q_1 = q(a_1 : k_1)$ and $q_2 = q(a_2 : k_2)$, then equation (7.2) also shows that

$$aq_1 + k_1 = \overline{k'} = aq_2 + k_2.$$

But then $k_2 = k_1 + a(q_1 - q_2) = k_1 + a_1 n$, with $n = q_1 - q_2$.

Conversely, suppose that $(a : k)$ and $(a : k + an)$ are ideal numbers in $\mathcal{I}_\Delta$ for some positive discriminant Δ. If $q_1 = q(a : k)$ and $q_2 = q(a : k + an)$, then

$$q_2 = \left\lfloor \frac{w - (k + an)}{a} \right\rfloor = \left\lfloor \frac{w - k}{a} - n \right\rfloor = \left\lfloor \frac{w - k}{a} \right\rfloor - n = q_1 - n.$$

But then $aq_2 + (k + an) = a(q_1 - n) + (k + an) = aq_1 + k$, and so $R(a : k) = R(a : k + an)$. $\quad\square$

In other words, if an ideal number is in the image of the reduction map, then its inverse image under R consists of a congruence class of ideal numbers. We can also describe the image of $\mathcal{I}_\Delta$ under the reduction map, which we do after the following lemma.

Lemma 7.2.3. *Let $P(x)$ be the principal polynomial of positive discriminant Δ and let $w = \frac{-P'(0) + \sqrt{\Delta}}{2}$, the larger root of $P(x)$. For each integer k, let $\overline{k} = -k - P'(0)$ be the conjugate of k with respect to Δ. Then $(w - k)(w - \overline{k}) = -P(k)$ for every integer k.*

Proof. Note that $w + P'(0) = \frac{-P'(0) + \sqrt{\Delta}}{2} + P'(0) = \frac{P'(0) + \sqrt{\Delta}}{2} = -\overline{w}$, where $\overline{w}$ is the smaller root of $P(x)$. Thus, $w - \overline{k} = w + k + P'(0) = k + w + P'(0) = k - \overline{w}$, and it follows that

$$(w - k)(w - \overline{k}) = -(k - w)(k - \overline{w}) = -P(k),$$

since w and $\overline{w}$ are the roots of $P(x)$. $\quad\square$

Proposition 7.2.4. *Let R be the reduction map on $\mathcal{I}_\Delta$ for some positive discriminant Δ. Then an ideal number $(a' : k')$ in $\mathcal{I}_\Delta$ is in the image of $\mathcal{I}_\Delta$ under R if and only if the reduction number $q(a' : k')$ is positive.*

Proof. Suppose that $R(a : k) = (a' : k')$ for some $(a : k)$ in $\mathcal{I}_\Delta$. If we let $q = q(a : k) = \lfloor \frac{w-k}{a} \rfloor$, then

$$\frac{w - k}{a} - q = \frac{w - (aq + k)}{a}$$

is between 0 and 1. But $aq + k$ is the conjugate of k', and we have

$$\frac{w - k'}{a'} \cdot \frac{w - \overline{k'}}{a} = -\frac{P(k')}{a'a} = 1,$$

using Lemma 7.2.3 and the fact that $P(k') = -aa'$. With $0 < \frac{w - \overline{k'}}{a} < 1$, it follows that $\frac{w-k'}{a'} > 1$ and so $q' = q(a' : k') = \lfloor \frac{w-k'}{a'} \rfloor \geq 1$.

Conversely, let $(a' : k')$ be an ideal number in $\mathcal{I}_\Delta$ with $q' = q(a' : k')$ positive. Let $a = -\frac{P(k')}{a'}$, an integer, and let k be the conjugate of k' with respect to Δ. Since $P(k) = P(k') = -aa'$, then a divides $P(k)$ and $f = (a : k)$ is an element of $\mathcal{I}_\Delta$. We show as follows that $R(a : k) = (a' : k')$. First, note that

$$\frac{w-k}{a} \cdot \frac{w-k'}{a'} = \frac{(w-\overline{k'})(w-k')}{aa'} = -\frac{P(k')}{aa'} = 1.$$

Since w is irrational and q' is positive, so that $\frac{w-k'}{a'} > 1$, then $0 < \frac{w-k}{a} < 1$ and so $q = q(a : k) = 0$. Then $R(a : k) = (a' : k')$ since $k' = \overline{k} = \overline{aq+k}$ and $-\frac{P(k')}{a} = -\frac{P(k)}{a} = a'$. (By Proposition 7.2.2, we could use any ideal number $(a : k+an)$ in place of $(a : k)$ here.) □

When the reduction map is restricted to *reduced* ideal numbers, it permutes those elements of $\mathcal{I}_\Delta$ among themselves. To show this, we begin as follows.

Lemma 7.2.5. *Suppose that $R(a : k) = (a' : k')$, where $(a : k)$ is an ideal number of positive discriminant Δ and $R : \mathcal{I}_\Delta \to \mathcal{I}_\Delta$ is the reduction map. Let $f(x) = ax^2 + bx + c$ be the polynomial having ideal number expression $(a : k)$ and let $q = q(a : k)$ be the reduction number of $(a : k)$. Then $g(x) = -f(q)x^2 - f'(q)x - a$ is the polynomial having ideal number expression $(a' : k')$.*

Proof. Let $P(x)$ be the principal polynomial of discriminant Δ. If $f(x) = ax^2 + bx + c$ is the polynomial having ideal number expression $(a : k)$, then Proposition 1.3.3 implies that $P(ax + k) = a \cdot f(x)$ for all x. Let q be the reduction number of $(a : k)$, and let $R(a : k) = (a' : k')$, so that $k' = \overline{aq+k}$, the conjugate of $aq + k$ with respect to Δ, and $P(k') = -aa'$. Let $g(x)$ be the polynomial having ideal number expression $(a' : k')$. Then we see the following, using properties of a principal polynomial and its derivative applied to an integer and its conjugate:

1. $-aa' = P(k') = P(aq + k) = a \cdot f(q)$, so that $a' = -f(q)$ is the quadratic coefficient of $g(x)$.
2. $P'(k') = -P'(aq + k) = -(2(aq + k) + P'(0)) = -2aq - P'(k) = -2aq - b = -f'(q)$ is the linear coefficient of $g(x)$.
3. Since $P(k') = a'(-a)$, then $-a$ is the constant coefficient of $g(x)$.

Thus, $g(x) = -f(q)x^2 - f'(q)x - a$, as we wanted to show. □

Proposition 7.2.6. *Let R be the reduction map on $\mathcal{I}_\Delta$ for some positive discriminant Δ. If $(a : k)$ is a reduced ideal number in $\mathcal{I}_\Delta$, then $R(a : k)$ is also reduced.*

Proof. Let $(a : k)$ be reduced, let $R(a : k) = (a' : k')$, and let $f(x) = ax^2 + bx + c$ and $g(x) = -f(q)x^2 - f'(q)x - a$ be the polynomials having ideal number expressions $f = (a : k)$ and $g = (a' : k')$, respectively, as in Lemma 7.2.5. If $v > \overline{v}$ are the two roots of $f(x)$, then $-1 < \overline{v} < 0$ and $1 < v$. With a positive, we know that $f(x)$ is negative only when $\overline{v} < x < v$. Since q is the largest integer less than v, by definition of the reduction number of $(a : k)$, we see that $f(q)$ and $f(q - 1)$ are both negative, while $f(q + 1)$ is positive. But

$$f(q + 1) = a(q + 1)^2 + b(q + 1) + c = (aq^2 + bq + c) + (2aq + b) + a$$
$$= f(q) + f'(q) + a = -g(1),$$

and

$$f(q-1) = a(q-1)^2 + b(q-1) + c = (aq^2 + bq + c) - (2aq + b) + a$$
$$= f(q) - f'(q) + a = -g(-1).$$

So, now we have that the quadratic coefficient of $g(x)$ is positive, with $g(-1) = -f(q-1)$ positive, $g(0) = -a$ negative, and $g(1) = -f(q+1)$ negative. It follows that $g(x)$ has a root between -1 and 0, and another root larger than 1. Thus, $(a' : k')$ is reduced by Proposition 7.1.2. □

Corollary 7.2.7. *Let R be the reduction map on $\mathcal{I}_\Delta$ for some positive discriminant Δ. If the domain of R is restricted to reduced ideal numbers, then R is a permutation.*

Proof. Let S be the subset of reduced ideal numbers in $\mathcal{I}_\Delta$. Proposition 7.2.6 shows that R takes S into S. Since S is finite by Proposition 7.1.3, we can show that $R : S \to S$ is a permutation by showing that R is injective when restricted to the domain S. By Proposition 7.2.2, if $R(a_1 : k_1) = R(a_2 : k_2)$, then $a_1 = a_2$ and $k_2 = k_1 + an$ for some integer n. But we also saw that the congruence class of an ideal number contains at most one reduced element. Thus, $(a_1 : k_1) = (a_2 : k_2)$, and $R : S \to S$ is a permutation. □

We will look at examples of this permutation in Section 7.3. We conclude Section 7.2 with a useful connection between the reduction map and reduced conjugates. We again begin with a lemma.

Lemma 7.2.8. *Let $(a : k)$ be a reduced ideal number of positive discriminant. Let $R : \mathcal{I}_\Delta \to \mathcal{I}_\Delta$ be the reduction map, let $R(a : k) = (a' : k')$, and let $(a : \ell)$ be the reduced conjugate of $(a : k)$. Then $\ell = k'$. Furthermore, $(a : k)$ and $(a : \ell)$ have the same reduction number.*

Proof. Let $P(x)$ be the principal polynomial of discriminant Δ, with roots $w > \overline{w}$. Let $\overline{k}$ be the conjugate of k with respect to Δ, and let $q = \lfloor \frac{w-k}{a} \rfloor$, the reduction number of $(a : k)$. We noted previously that $w + \overline{w} = -P'(0) = k + \overline{k}$, which implies that $w - k = \overline{k} - \overline{w}$. Then the largest integer q less than $\frac{w-k}{a}$ is also the largest integer for which $\overline{k} - aq$ is larger than $\overline{w}$. But now $\ell = \overline{k} - aq$ by definition. Thus, we have

$$k' = \overline{aq + k} = -P'(0) - (aq + k) = (-P'(0) - k) - aq = \overline{k} - aq = \ell.$$

For the second claim, note that $aq + k < w < (aq + k) + a$ by the definition of q. We want to show that $aq + \ell < w < (aq + \ell) + a$ as well. Since $(a : k)$ is reduced, we know that $k - a < \overline{w} < k$. Multiplying each term by -1 and adding $-P'(0) - aq$ throughout, we find that $(\overline{k} - aq) + a > w - aq > \overline{k} - aq$. But as noted above, $\ell = \overline{k} - aq$. Thus, $\ell < w - aq < \ell + a$, equivalent to what we wanted. □

Proposition 7.2.9. *Let $(a : k)$ be a reduced ideal number of positive discriminant. Let $R(a : k) = (a' : k')$, where $R : \mathcal{I}_\Delta \to \mathcal{I}_\Delta$ is the reduction map, and let $(a : \ell)$ and $(a' : \ell')$ be the reduced conjugates of $(a : k)$ and $(a' : k')$, respectively. Then $R(a' : \ell') = (a : \ell)$.*

Proof. By Lemma 7.2.8, $\ell = k'$, and so $P(\ell) = -aa'$ by equation (7.2). From the proof of Proposition 7.2.4, an ideal number sent to $(a : \ell)$ by the reduction map must have norm $-\frac{P(\ell)}{a} = a'$. The character of such an ideal number is congruent to the conjugate of ℓ, which is the conjugate of k'. By definition, ℓ' is the unique integer of this form for which $(a' : \ell')$ is reduced. Thus, $R(a' : \ell') = (a : \ell)$, as claimed. $\qquad\square$

7.3 Periodicity of reduction maps

A typical situation, which we will see applied throughout the remainder of Chapter 7, is that we need to apply the reduction map repeatedly to an ideal number, forming a sequence of ideal numbers. We will write the result of n applications of R to $(a : k)$ as $R^n(a : k) = (a_n : k_n)$. We illustrate an algorithm for this process, which we will call the *reduction algorithm* in the following examples, which also illustrate the results of Corollary 7.2.7 and Proposition 7.2.9.

Example. Consider $(8 : 9)$ in $\mathcal{I}_{41}$, as in an example in Section 7.2. We will illustrate the calculation of $R^n(8 : 9)$ for all $n \geq 0$. It is convenient to start with a table for the principal polynomial of discriminant Δ, in this case $P(x) = x^2 + x - 10$, at least for all values where $P(x)$ is negative. In the following table, we pair each x with its conjugate with respect to Δ:

x	$-3, 2$	$-2, 1$	$-1, 0$
$P(x)$	-4	-8	-10

Let $w = \frac{-1+\sqrt{41}}{2} \approx 2.7$ and $\overline{w} = \frac{-1-\sqrt{41}}{2} = -3.7$, the two roots of $P(x)$.

Now for the given ideal number $(a : k)$, let $a_0 = a$ and $k_0 = k$ and perform the following steps in order for $n \geq 0$:

1. Calculate the reduction number $q_n = q(a_n : k_n) = \lfloor \frac{w-k_n}{a_n} \rfloor$.
2. Calculate $k_{n+1} = \overline{a_n q_n + k_n}$.
3. Calculate $a_{n+1} = -P(k_{n+1})/a_n$.

We can repeat these steps indefinitely, producing $R^n(a : k) = (a_n : k_n)$ for all $n \geq 0$. The following table keeps track of these calculations in our example:

n	0	1	2	3	4	5	6	7
a	8	1	4	2	2	4	1	4
k	9	-2	-3	-2	-3	-2	-3	-3
q	-1	4	1	2	2	1	5	1

For instance, $q_0 = \lfloor \frac{2.7-9}{8} \rfloor = -1$, then $a_0 q_0 + k_0 = -8 + 9 = 1$ and $k_1 = \bar{1} = -2$, then $a_1 = -P(k_1)/a_0 = 8/8 = 1$, using the table of $P(x)$ values. Now $q_1 = \lfloor \frac{w-k_1}{a_1} \rfloor = \lfloor \frac{2.7+2}{1} \rfloor = 4$, $a_1 q_1 + k_1 = 1 \cdot 4 - 2 = 2$, so that $k_2 = \bar{2} = -3$, and $a_2 = -P(k_2)/a_1 = -P(-3)/1 = 4$.

In this example, we find eventually that $a_7 = a_2$ and $k_7 = k_2$. At that point, continued calculations simply repeat the previous five steps indefinitely. Thus, in that sense, we have calculated $R^n(8:9)$ for every $n \geq 0$.

As another observation in this example, we might recognize (from an example in Section 7.1) that the ideal numbers $(a_n : k_n)$ for $2 \leq n \leq 6$ are the *reduced* ideal numbers of discriminant $\Delta = 41$. As predicted in Proposition 7.2.6 and Corollary 7.2.7, if any $(a_n : k_n)$ is reduced, then all subsequent ideal numbers are also reduced. In this example, we see that R permutes all five reduced ideal numbers among themselves. In the terminology of permutation groups, R is a cycle of length 5 when restricted to the subset of reduced ideal numbers in $\mathcal{I}_{41}$, specifically

$$R = ((4:-3),(2:-2),(2:-3),(4:-2),(1:-3)).$$

(We can, as usual, select any of the five reduced ideal numbers as the starting point for this cycle.)

When $(a:k)$ is a reduced ideal number, we can revise the algorithm of the preceding example in a way that is somewhat easier for calculation by hand. We again illustrate this approach with an example, which shows that the permutation R is not always a single cycle.

Example. In an example in Section 7.1, we found sixteen reduced ideal numbers of discriminant $\Delta = 460$. In the following table, we list each of these paired with its reduced conjugate, along with the reduction number $q = \lfloor \frac{w-k}{a} \rfloor$ assigned to each reduced ideal number $(a:k)$. (As noted in Lemma 7.2.8, these pairs have the same reduction number.)

Reduced conjugates		q	Reduced conjugates		q
$(1:-10)$		20	$(9:-5)$	$(9:-4)$	1
$(2:-9)$		9	$(10:-5)$		1
$(3:-10)$	$(3:-8)$	6	$(11:-7)$	$(11:-4)$	1
$(5:-10)$		4	$(15:-10)$	$(15:-5)$	1
$(6:-7)$	$(6:-5)$	2	$(17:-9)$	$(17:-8)$	1

We also repeat the calculation of $P(x) = x^2 - 115$ for convenience:

x	± 10	± 9	± 8	± 7	± 6	± 5	± 4	± 3	± 2	± 1
$P(x)$	-15	-34	-51	-66	-79	-90	-99	-106	-111	-114

Now given any reduced ideal number $(a : k)$, we calculate $R(a : k) = (a' : k')$ as follows. Let k' be the character of the reduced conjugate of $(a : k)$. (Lemma 7.2.8 implies that $k' = \ell$ if the reduced conjugate of $(a : k)$ is $(a : \ell)$. Note that $\ell = k$ for some of the reduced ideal numbers listed above.) Then let $a' = -\frac{P(k')}{a}$, using the table of $P(x)$ values. For instance, if $(a : k) = (1 : -10)$, which is its own reduced conjugate, then $k' = -10$ and $a' = -\frac{P(-10)}{1} = 15$. Now with $(a : k) = (15 : -10)$, we have that $k' = -5$, the character of the reduced conjugate of $(15 : -10)$, and $a' = -\frac{P(-5)}{15} = 6$. Our first table shows that if we continue this process, we return to $(1 : -10)$ after ten steps, with ten distinct ideal numbers appearing before that occurs. In this procedure, we do not need to use the reduction number q as such, but in the following tables, we list these as well for future use:

n	0	1	2	3	4	5	6	7	8	9	10
a	1	15	6	11	9	10	9	11	6	15	1
k	-10	-10	-5	-7	-4	-5	-5	-4	-7	-5	-10
q	20	1	2	1	1	1	1	1	2	1	20

Applying the same procedure to $(2 : -9)$, which did not appear in the first list, we find that it takes six steps before $(2 : -9)$ recurs. The other six reduced ideal numbers arise in the process:

n	0	1	2	3	4	5	6
a	2	17	3	5	3	17	2
k	-9	-9	-8	-10	-10	-8	-9
q	9	1	6	4	6	1	9

We conclude that, as a permutation of the reduced ideal numbers, the reduction map R is a product of two disjoint cycles, one of length ten and the other of length six.

Definition. Let R be the reduction map on $\mathcal{I}_\Delta$ for some positive discriminant Δ. We say that an ideal number $(a : k)$ in $\mathcal{I}_\Delta$ is *periodic* if there is a nonnegative integer m and a positive integer ℓ such that $R^{m+\ell}(a : k) = R^m(a : k)$. We say that $(a : k)$ is *purely periodic* if we can take m to be 0 in this equation, that is, if $R^\ell(a : k) = (a : k)$ for some positive integer ℓ. In either case, the smallest such positive integer ℓ is called the *period* of $(a : k)$.

In the remainder of this section, we show that every ideal number in $\mathcal{I}_\Delta$ is periodic, and that the purely periodic ideal numbers are precisely the *reduced* ideal numbers in $\mathcal{I}_\Delta$. We begin with two lemmas.

Lemma 7.3.1. *Let R be the reduction map on $\mathcal{I}_\Delta$ for some positive discriminant Δ. Let $(a : k)$ be an element of $\mathcal{I}_\Delta$ and let $R(a : k) = (a' : k')$. Then the following statements are true.*

1. *If $0 < a < \frac{\sqrt{\Delta}}{2}$, then $(a' : k')$ is reduced.*
2. *If $\frac{\sqrt{\Delta}}{2} < a < \sqrt{\Delta}$, then $0 < a' < \frac{\sqrt{\Delta}}{2}$.*
3. *If $a > \sqrt{\Delta}$, then either $0 < a' < \frac{a}{4}$ or $0 < -a' < a - \sqrt{\Delta}$.*
4. *If $a < 0$, then $0 < a' < -a + \sqrt{\Delta}$.*

Proof. Let $w > \overline{w}$ be the two roots of the principal polynomial $P(x)$ of discriminant Δ. Recall that $w - \overline{w} = \sqrt{\Delta}$ and that $P(x) < 0$ if and only if $\overline{w} < x < w$. If $0 < a < \frac{\sqrt{\Delta}}{2}$, let k_1 be the smallest integer so that $k_1 \equiv k \pmod{a}$ and $k_1 > \overline{w}$, and consider the ideal number $(a : k_1)$. Since $w - \overline{w} = \sqrt{\Delta} > 2a$, then $k_1 - a < \overline{w} < k_1 < k_1 + a < w$. Thus, $(a : k_1)$ is reduced and $R(a : k_1)$ must likewise be reduced. But $R(a : k_1) = R(a : k)$ by Proposition 7.2.2, so $(a' : k')$ is reduced. This establishes statement (1).

For the remaining statements, we need some additional properties of the principal polynomial. Since $P(x) = (x - w)(x - \overline{w})$ for all x, we find that the minimum value of $P(x)$ is $-\frac{\Delta}{4}$, taken on at $x = \frac{w + \overline{w}}{2}$. Furthermore, $P(w + x) = x(x + w - \overline{w}) = x^2 + x\sqrt{\Delta}$ for all real numbers x. Let $q = \lfloor \frac{w - k}{a} \rfloor$ be the reduction number of $(a : k)$, so that $q < \frac{w - k}{a} < q + 1$. Note that $aq + k$ is between $w - a$ and w in every case. If a is negative, so that $w < aq + k < w - a$, then $0 < P(aq + k) < P(w - a) = a^2 - a\sqrt{\Delta}$. If $0 < a < \sqrt{\Delta}$, then $\overline{w} = w - \sqrt{\Delta} < w - a < aq + k < w$, and so $-\frac{\Delta}{4} < P(aq + k) < 0$. Finally, if $a > \sqrt{\Delta}$, then $P(aq + k)$ might be either positive or negative, but must satisfy $-\frac{\Delta}{4} < P(aq + k) < P(w - a) = a^2 - a\sqrt{\Delta}$. We can now prove statements (2)–(4), using the fact that $P(aq + k) = P(k') = -aa'$.

(2) If $\frac{\sqrt{\Delta}}{2} < a < \sqrt{\Delta}$, then $-\frac{\Delta}{4} < -aa' < 0$, that is, $0 < aa' < \frac{\Delta}{4}$. It follows that $0 < a' < \frac{\sqrt{\Delta}}{2}$ since otherwise aa' exceeds this upper bound.

(3) If a is larger than $\sqrt{\Delta}$, then either $-\frac{\Delta}{4} < -aa' < 0$ or $0 < -aa' < a^2 - a\sqrt{\Delta}$. In the first case, then $0 < a' < \frac{\Delta}{4a} < \frac{a^2}{4a} = \frac{a}{4}$. In the second case, $0 < -a' < a - \sqrt{\Delta}$.

(4) If a is negative, then $0 < -aa' < a^2 - a\sqrt{\Delta}$. Dividing through by the positive number $-a$, then $0 < a' < -a + \sqrt{\Delta}$. $\qquad\square$

Lemma 7.3.2. *Let R be the reduction map on $\mathcal{I}_\Delta$ for some positive discriminant Δ. Let $(a : k)$ be an element of $\mathcal{I}_\Delta$ and let*

$$R(a : k) = (a' : k') \quad and \quad R(a' : k') = (a'' : k'').$$

If a and a' are both positive, then a'' must also be positive.

Proof. Let $P(x)$ be the principal polynomial of discriminant Δ, with roots $w > \overline{w}$. Suppose that a and a' are positive, but that a'' is negative. From Lemma 7.3.1, we see that $a' > \sqrt{\Delta}$ in this case. Since $P(k') = -aa'$ is negative, we must have $\overline{w} < k' < w$. With $(a' : k')$ the image of an element of $\mathcal{I}_\Delta$ under R, we further have that $q' = q(a' : k') \geq 1$ by Proposition 7.2.4. Thus, with $a' > \sqrt{\Delta}$, $q' \geq 1$ and $k' > \overline{w}$, we see that $a'q' + k' > \sqrt{\Delta} \cdot 1 + \overline{w} =$

w. But $q' = \lfloor \frac{w-k'}{a'} \rfloor \geq 1$, with a' positive, forces w to be larger than $a'q' + k'$. This is a contradiction, so we cannot have a'' negative. □

Proposition 7.3.3. *Let R be the reduction map on $\mathcal{I}_\Delta$ for some positive discriminant Δ. If $(a : k)$ is an ideal number in $\mathcal{I}_\Delta$, then there is a nonnegative integer m so that $R^m(a : k)$ is reduced.*

Proof. For each nonnegative integer i, write $R^i(a : k)$ as $(a_i : k_i)$. Suppose that there is some i for which a_i is negative. Lemma 7.3.1 shows that a_{i+1} is then positive but that a_{i+2} might be negative. In that case, we have that $0 < a_{i+1} < -a_i + \sqrt{\Delta}$ and $0 < -a_{i+2} < a_{i+1} - \sqrt{\Delta}$. But then $0 < -a_{i+2} < -a_i$, so that this alternation of signs cannot continue indefinitely. From Lemma 7.3.2, it follows that there is a nonnegative integer n so that a_i is positive for $i \geq n$. But now we see from Lemma 7.3.1 that a_i is strictly decreasing for $i \geq n$, until we obtain $a_j < \frac{\sqrt{\Delta}}{2}$ for some $j \geq n$. Finally, if $m = j + 1$, then Lemma 7.3.1 implies that $R^m(a : k)$ is reduced. □

Corollary 7.3.4. *Let $\mathcal{I}_\Delta$ be the set of all ideal numbers of some positive discriminant Δ. Then every element $(a : k)$ of $\mathcal{I}_\Delta$ is periodic, and $(a : k)$ is purely periodic if and only if $(a : k)$ is reduced.*

Proof. Let $R : \mathcal{I}_\Delta \to \mathcal{I}_\Delta$ be the reduction map. Since the set of reduced ideal numbers in $\mathcal{I}_\Delta$ is finite, Corollary 7.2.7 shows that every reduced ideal number is purely periodic. Conversely, suppose that $(a : k)$ is purely periodic in $\mathcal{I}_\Delta$, so that there is a smallest positive integer ℓ so that $R^\ell(a : k) = (a : k)$. Note that then $R^\ell(R^\ell(a : k)) = (a : k)$, that is, $R^{2\ell}(a : k) = (a : k)$. More generally, $R^{q\ell}(a : k) = (a : k)$ for all $q \geq 0$. Now Proposition 7.3.3 implies that there is some $m \geq 0$ so that $R^m(a : k)$ is reduced. We can write $m = q\ell - r$ with $0 \leq r < \ell$, as a variation on the division algorithm. But then

$$(a : k) = R^{q\ell}(a : k) = R^r(R^m(a : k)).$$

Since R takes reduced ideal numbers to reduced ideal numbers, then $(a : k)$ must itself be reduced.

Finally, if $(a : k)$ is an arbitrary element of $\mathcal{I}_\Delta$, then there is a nonnegative integer m so that $R^m(a : k)$ is reduced. In that case, there is an $\ell > 0$ so that $R^\ell(R^m(a : k)) = R^m(a : k)$ since a reduced ideal number is purely periodic. By definition, then $(a : k)$ is periodic. □

7.4 Reduction and equivalence

When Δ is a positive discriminant, we have a practical method of constructing the collection of reduced ideal numbers in $\mathcal{I}_\Delta$. Every ideal number of positive discriminant is equivalent to at least one reduced ideal number, but we have not claimed that this equivalence is unique. In fact, there are typically equivalences among these reduced ideal numbers that allow us to cut down the list of class representatives much further. The reduction map introduced in the preceding sections, which permutes reduced ideal numbers among themselves, is a method of recognizing these equivalences.

Proposition 7.4.1. *Let Δ be a positive discriminant, with principal polynomial $P(x)$ having roots $w > \overline{w}$. Let $(a : k)$ be an ideal number in $\mathcal{I}_\Delta$, with $q = q(a : k) = \lfloor \frac{w-k}{a} \rfloor$ the corresponding reduction number. Let k' be the conjugate of $aq + k$ with respect to Δ, let $P(k') = -aa'$, so that $R(a : k) = (a' : k')$, where R is the reduction map on $\mathcal{I}_\Delta$. Let T and V be the translation and involution matrices, as defined in equation (4.1). Then*

$$(a : k) \cdot (T^q V) = (-a' : k') \quad and \quad (-a : k) \cdot (T^{-q} V) = (a' : k'). \tag{7.3}$$

Proof. Note that

$$T^q V = \begin{bmatrix} 1 & q \\ 0 & 1 \end{bmatrix} \cdot \begin{bmatrix} 0 & -1 \\ 1 & 0 \end{bmatrix} = \begin{bmatrix} q & -1 \\ 1 & 0 \end{bmatrix}$$

and

$$T^{-q} V = \begin{bmatrix} 1 & -q \\ 0 & 1 \end{bmatrix} \cdot \begin{bmatrix} 0 & -1 \\ 1 & 0 \end{bmatrix} = \begin{bmatrix} -q & -1 \\ 1 & 0 \end{bmatrix}.$$

Let $P(k) = ac$ and $P'(k) = b$, so that $f(x) = ax^2 + bx + c$ is the quadratic polynomial written as $(a : k)$ in ideal number notation. Recall from Proposition 1.3.3 that $P(ax + k) = a \cdot f(x)$ for all x. Now by Theorem 4.1.2, if we write $(a : k) \cdot (T^q V) = (m : \ell)$, then

$$m = a(q)^2 + b(q)(1) + c(1)^2 = f(q) = \frac{1}{a}P(aq + k) = \frac{1}{a}P(k') = -a',$$

since k' is the conjugate of $aq + k$ and $P(k') = -aa'$, while

$$\ell = k + (a(q)(-1) + b(-1)(1) + c(1)(0)) = k - aq - b = -(aq + k) - P'(0) = \overline{aq + k} = k',$$

using the fact that $b = P'(k) = 2k + P'(0)$. Thus, $(a : k) \cdot (T^q V) = (-a' : k')$, as we wanted to show.

For the second claim, note that $P(k) = ac = (-a)(-c)$. If $(-a : k) \cdot (T^{-q} V) = (m : \ell)$, then

$$m = -a(-q)^2 + b(-q)(1) - c(1)^2 = -f(q) = a',$$

and

$$\ell = k + (-a(-q)(-1) + b(-1)(1) - c(1)(0)) = k - aq - b = k',$$

by the same calculations as above. Thus, $(-a : k) \cdot (T^{-q} V) = (a' : k')$. $\quad\square$

We have already seen (Proposition 7.1.5) that every ideal number of positive discriminant Δ is equivalent to a reduced ideal. Proposition 7.4.1 shows that typically there are equivalences among the reduced forms themselves. We illustrate this by returning to two examples that we have considered earlier in this chapter.

Example. For $\Delta = 41$, we saw that there are five reduced ideal numbers in total. The following table compiles the effect of applying the reduction map repeatedly to one of them, namely $(1 : -3)$. (While the starting point is different, the calculations of this table are the same as in an example in Section 7.3.)

i	0	1	2	3	4	5
a	1	4	2	2	4	1
k	-3	-3	-2	-3	-2	-3
q	5	1	2	2	1	5

The columns for $i = 0$ and $i = 5$ are identical, indicating that $(1 : -3)$ is purely periodic with period $\ell = 5$. Each reduced ideal number in $\mathcal{I}_{41}$ appears as $R^n(1 : -3)$ with $0 \leq n < 5$. We can now apply Proposition 7.4.1 repeatedly to establish the following list of equivalences. (As in previous examples, we also note the unimodular matrix that establishes the equivalence, from equations (7.3).)

$$(1 : -3) \xrightarrow{T^5 V} (-4 : -3) \xrightarrow{T^{-1} V} (2 : -2) \xrightarrow{T^2 V} (-2 : -3) \xrightarrow{T^{-2} V} (4 : -2)$$

$$\xrightarrow{TV} (-1 : -3) \xrightarrow{T^{-5} V} (4 : -3) \xrightarrow{TV} (-2 : -2) \xrightarrow{T^{-2} V} (2 : -3)$$

$$\xrightarrow{T^2 V} (-4 : -2) \xrightarrow{T^{-1} V} (1 : -3).$$

We see in this case that $(1 : -3)$ is equivalent to each of the reduced ideal numbers of discriminant 41, as well as the negative conjugate of each one. Since every ideal number in $\mathcal{I}_{41}$ is equivalent to a reduced ideal number, we conclude that every ideal number in $\mathcal{I}_{41}$ is equivalent to $(1 : -3)$. That is, $\mathcal{C}_{41} = \{[1 : -3]\}$, a trivial group. (We could use $[1 : 0]$, or the class of any other ideal number, as the representative.)

As an aside for this example, which we will say more about later, notice that our calculations show that

$$U = T^5 V \cdot T^{-1} V \; T^2 V \cdot T^{-2} V \cdot TV \cdot T^{-5} V \cdot TV \cdot T^{-2} V \cdot T^2 V \cdot T^{-1} V$$

is an automorph of the ideal number $(1 : -3)$, in that $(1 : -3) \cdot U = (1 : -3)$.

Example. For $\Delta = 460$, we found sixteen reduced ideal numbers in $\mathcal{I}_\Delta$ and saw that the reduction map partitions them into two cycles, one of length ten and the other of length six. We repeat the data from an example in Section 7.3 below:

i	0	1	2	3	4	5	6	7	8	9	10
a	1	15	6	11	9	10	9	11	6	15	1
k	−10	−10	−5	−7	−4	−5	−5	−4	−7	−5	−10
q	20	1	2	1	1	1	1	1	2	1	20

i	0	1	2	3	4	5	6
a	2	17	3	5	3	17	2
k	−9	−9	−8	−10	−10	−8	−9
q	9	1	6	4	6	1	9

If we apply Proposition 7.4.1 to $(2 : -9)$, for instance, we obtain the following equivalences:

$$(2 : -9) \xrightarrow{T^9 V} (-17 : -9) \xrightarrow{T^{-1}V} (3 : -8) \xrightarrow{T^6 V} (-5 : -10)$$
$$\xrightarrow{T^{-4}V} (3 : -10) \xrightarrow{T^6 V} (-17 : -8) \xrightarrow{T^{-1}V} (2 : -9).$$

This time, we obtain only three of the reduced ideal numbers in the cycle of $(2 : -9)$, along with the negative conjugates of the other three. Repeated application of Proposition 7.4.1 to this case will only repeat this pattern indefinitely. But by the same process, we find that $(17 : -9)$ is equivalent to three reduced ideal numbers (itself included) and that $(1 : -10)$ and $(15 : -10)$ are equivalent to five each:

$$(1 : -10) \sim (6 : -5) \sim (9 : -4) \sim (9 : -5) \sim (6 : -7)$$
$$(15 : -10) \sim (11 : -7) \sim (10 : -5) \sim (11 : -4) \sim (15 : -5)$$
$$(2 : -9) \sim (3 : -8) \sim (3 : -10) \quad \text{and} \quad (17 : -9) \sim (5 : -10) \sim (17 : -8).$$

Since every ideal number in $\mathcal{I}_{460}$ is equivalent to a reduced ideal number, we conclude that there are at most four distinct classes in $\mathcal{C}_{460}$.

In this example, we can be sure that these classes are, in fact, distinct by considering genus equivalence. For $\Delta = 460 = \Delta(115, 1)$, since $115 \equiv 3 \pmod 4$ and $115 = 5 \cdot 23$, we find that three genus symbols are defined for each ideal number: s_{-1}, s_5, and s_{23}. In each of the collections of equivalent ideal numbers above, there is at least one $(a : k)$ with a relatively prime to Δ, which we can use to calculate these genus symbols:

$(a : k)$	s_{-1}	s_5	s_{23}
$(1 : -10)$	+	+	+
$(3 : -8)$	−	−	+
$(11 : -7)$	−	+	−
$(17 : -9)$	+	−	−

Since, for instance, $(1 : -10)$ is not genus equivalent to $(11 : -7)$, they cannot be equivalent, even though $R^3(1 : -10) = (11 : -7)$. (Note, however, that $(1 : -10) \sim (-11 : -7)$ as an application of Proposition 7.4.1. One can verify that each genus symbol of $(-11 : -7)$ is +1.) Our conclusion is that $\mathcal{C}_{460}$ has four elements:

$$\mathcal{C}_{460} = \{[1 : -10], [2 : -9], [5 : -10], [10 : -5]\},$$

here using the reduced ideal number of smallest norm within each class. Other expressions are possible. For instance, we could also write

$$\mathcal{C}_{460} = \{[1 : -10], [2 : -9], [-2 : -9], [-1 : -10]\}$$

(with $[-2 : -9] = [5 : -10]$ and $[-1 : -10] = [10 : -5]$), so that the norm of each representative is as small as possible in absolute value. In any event, we can say that $\mathcal{C}_{460} \cong \mathbb{Z}_2 \times \mathbb{Z}_2$, since the square of each class is in the principal genus.

We summarize the implications of Proposition 7.4.1, as illustrated by these examples, in the following corollary, omitting further proof.

Corollary 7.4.2. *Let Δ be a positive discriminant and let R be the reduction map on $\mathcal{I}_\Delta$. If $(a : k)$ is an ideal number in $\mathcal{I}_\Delta$, then $(a : k)$ is equivalent to $R^{2n}(a : k)$ for every integer $n \geq 0$. Suppose that $(a : k)$ is a reduced ideal number, with period ℓ under the reduction map. If ℓ is odd, then there are (at least) ℓ reduced ideal numbers to which $(a : k)$ is equivalent, namely $R^{2n}(a : k)$ for $0 \leq n < \ell$. If ℓ is even, then $(a : k)$ is equivalent (at least) to each of the $\ell/2$ reduced ideal numbers $R^{2n}(a : k)$ for $0 \leq n < \ell/2$.*

It is a fact that the only equivalences among reduced ideal numbers of positive determinant are those described here. That is, if $(a : k)$ and $(a' : k')$ are reduced ideal numbers in $\mathcal{I}_\Delta$ and $(a : k) \sim (a' : k')$, then $(a' : k') = R^{2n}(a : k)$ for some integer $n \geq 0$. To prove this, we will, in the remainder of Chapter 7, introduce a numerical topic with important applications to indefinite ideal numbers.

7.5 The continued fraction of an indefinite ideal number

The continued fraction expansion of a real number v is an important computational method, particularly for approximating v by rational numbers. (Basic definitions and facts about continued fractions are compiled in Appendix D. We will invoke these results as needed.) When v is a real *quadratic number*, that is, a root of a quadratic polynomial with integer coefficients, the continued fraction expansion of v has several other characteristics and applications. In Section 1.5, we established a correspondence between irrational quadratic numbers and primitive ideal numbers. We now associate a continued fraction expansion to every ideal number of positive discriminant, using the reduction map on $\mathcal{I}_\Delta$. This gives us a method of computing the continued fraction expansion of a real quadratic number, as well as providing an approach to applications, especially to representations of integers by indefinite quadratic forms.

Definition. Let Δ be a positive discriminant and let $R : \mathcal{I}_\Delta \to \mathcal{I}_\Delta$ be the reduction map. If $(a : k)$ is an ideal number in $\mathcal{I}_\Delta$, let $R^n(a : k) = (a_n : k_n)$ for every nonnegative integer n. Let q_n be the reduction number of $(a_n : k_n)$, that is, $q_n = \lfloor \frac{w - k_n}{a_n} \rfloor$, where w is the larger root of $P(x)$, the principal polynomial of discriminant Δ. Then we define the *continued fraction* of $(a : k)$ to be the sequence $\langle q_0, q_1, q_2, q_3, \dots \rangle$. We associate two additional sequences s_i and t_i, for $i \geq -2$, to this continued fraction by the following recursive definitions:

$$s_{-2} = 0, \quad s_{-1} = 1, \quad s_i = s_{i-2} + s_{i-1} q_i \quad \text{for } i \geq 0 \tag{7.4}$$

and

$$t_{-2} = 1, \quad t_{-1} = 0, \quad t_i = t_{i-2} + t_{i-1} q_i \quad \text{for } i \geq 0. \tag{7.5}$$

We refer to s_i/t_i, for $i \geq 0$, as the sequence of *convergents* of this continued fraction. We call s_i and t_i the *numerator* and *denominator* sequences, respectively, of these convergents. We may also refer to these more simply, although less precisely, as the *convergent sequences* of $(a : k)$.

Proposition 7.5.1. *Let $(a : k)$ be an ideal number of positive discriminant Δ and let $\langle q_0, q_1, q_2, \dots \rangle$ be the continued fraction of $(a : k)$. Then q_i is a positive integer for all $i > 0$. There is a smallest nonnegative integer m and a smallest positive integer ℓ so that $q_i = q_{i+\ell}$ for all $i \geq m$. If $m = 0$, then q_0 is positive.*

Proof. The first claim is a consequence of Proposition 7.2.4, since every ideal number $(a_i : k_i)$ with $i > 0$ (in the notation of the preceding definition) is the image of an ideal number under the reduction map. The periodicity of each ideal number in $\mathcal{I}_\Delta$ under the reduction map, as noted in Corollary 7.3.4, establishes the second result. If $m = 0$, then $q_0 = q_\ell$, which must be positive since $\ell > 0$. □

In Section 7.3, we illustrated repeated application of the reduction map via tables in several examples. We repeat one such table in the following example, with additional rows for the convergent sequence calculations.

Example. Consider $(8 : 9)$ in $\mathcal{I}_{41}$, where $P(x) = x^2 + x - 10$ is the principal polynomial. The following table compiles information from repeated application of the reduction map to this ideal number:

i	0	1	2	3	4	5	6	7	8	9	10
a	8	1	4	2	2	4	1	4	2	2	4
k	9	−2	−3	−2	−3	−2	−3	−3	−2	−3	−2
q	−1	4	1	2	2	1	5	1	2	2	1
s	−1	−3	−4	−11	−26	−37	−211	−248	−707	−1662	−2369
t	1	4	5	14	33	47	268	315	898	2111	3009

We obtain the entries in the s row by multiplying the value of q immediately above the entry of interest by the value of s one place to the left and adding the value of s two places to the left. We will typically omit the initial terms, that is, for $i = -2$ and $i = -1$. For the numerator sequence, these are 0 and 1. The t row is calculated similarly, but with initial terms 1 and 0 instead.

As we saw previously, the rows for a and k begin to repeat a pattern eventually—in this case, the entries of columns 2 through 6 repeat indefinitely. The same is true for the q row, so that the continued fraction of $(8 : 9)$ similarly repeats. We write $\langle -1, 4, \overline{1, 2, 2, 1, 5} \rangle$ for the continued fraction of $(8 : 9)$ in $\mathcal{I}_{41}$ following the convention of placing a bar over the repeating pattern of reduction numbers. Notice, however, that the s and t values continue to grow in absolute value. (Although we refer to them as the convergent sequences of $(8 : 9)$ for simplicity, they clearly do not converge in the usual sense.)

With $w = \frac{-1 + \sqrt{41}}{2}$ the larger root of $P(x)$, we find that

$$v = \frac{w - 9}{8} = \frac{-19 + \sqrt{41}}{16}$$

is the quadratic number having $(8 : 9)$ in $\mathcal{I}_{41}$ as its ideal number expression. The continued fraction expansion of v is calculated directly by the following recursive process. Let $v_0 = v$, and for $n \geq 0$, let $q_n = \lfloor v_n \rfloor$ and $v_{n+1} = \frac{1}{v_n - q_n}$. (In Appendix D, the process is demonstrated for this number v in an example, where that approach also leads to the continued fraction expansion $\langle -1, 4, \overline{1, 2, 2, 1, 5} \rangle$.)

As another observation for this example, the sequence of convergents s_i/t_i, which begins as

$$\frac{1}{1}, \quad -\frac{3}{4}, \quad -\frac{4}{5}, \quad -\frac{11}{14}, \quad -\frac{26}{33}, \quad -\frac{37}{47}, \quad -\frac{211}{268}, \quad -\frac{248}{315}, \quad -\frac{707}{898}, \quad -\frac{1662}{2111},$$

converges to v. For instance (using the final terms in the table above),

$$\frac{s_{10}}{t_{10}} = -\frac{2369}{3009} \approx -0.787304752\ldots, \quad \text{whereas } v = \frac{-19 + \sqrt{41}}{16} \approx 0.787304735\ldots,$$

providing numerical evidence for this claim.

Our main theorem for this section verifies one observation of this example in general.

Theorem 7.5.2. *Let v be a quadratic number having ideal number expression $(a : k)$ in $\mathcal{I}_\Delta$ for some positive discriminant Δ. Then the continued fraction of v is the same as the continued fraction of $(a : k)$.*

Proof. Let v have ideal number expression $(a : k)$ in $\mathcal{I}_\Delta$ for some positive discriminant Δ, and let $R : \mathcal{I}_\Delta \to \mathcal{I}_\Delta$ be the reduction map. To establish the claim of this theorem, it will suffice to show that the reduction number of $(a : k)$ is the same as $q = \lfloor v \rfloor$, and that $R(a : k)$ is an ideal number expression for the quadratic number $\frac{1}{v-q}$. Let $P(x)$ be the principal polynomial of discriminant Δ and let w be the larger root of $P(x)$. By definition, to say that $(a : k)$ is the ideal number expression of v means that $v = \frac{w-k}{a}$. Thus, $\lfloor \frac{w-k}{a} \rfloor$, the reduction number of $(a : k)$, is equal to $q = \lfloor v \rfloor$. Let $R(a : k) = (a' : k')$, so that k' is the conjugate of $aq+k$ with respect to Δ and so that a' satisfies the equation $-aa' = P(k')$. Note that

$$v - q = \frac{w - k}{a} - q = \frac{w - (aq + k)}{a} = \frac{w - \overline{k'}}{a}.$$

By Lemma 7.2.3, $(w - k')(w - \overline{k'}) = -P(k')$. So, now

$$(v - q)(w - k') = \frac{1}{a} \cdot (w - k')(w - \overline{k'}) = \frac{1}{a} \cdot -P(k') = \frac{1}{a} \cdot aa' = a'.$$

Thus, $\frac{1}{v-q} = \frac{w-k'}{a'}$, the quadratic number with ideal number expression $(a' : k')$, as we wanted to show. $\qquad\square$

We conclude this section with a second example, where we start with a quadratic number.

Example. Let v be the quadratic number $\sqrt{115}$. Applying Proposition 1.5.5 with $q = 0$, $r = 1$, $n = 1$, and $d = 115$, we find that $(1 : 0)$ is an ideal number expression for v. (We can also see that v is the major root of $P(x) = x^2 - 115$, the principal polynomial of discriminant $\Delta = 460$.) Aside from the first step, repeated application of the reduction map to $(1 : 0)$ is the same as for $(1 : -10)$, which we compiled in an example in Section 7.3. We repeat this data below, with additional rows for the convergent sequences of $(1 : 0)$:

n	0	1	2	3	4	5	6	7	8	9	10	11
a	1	15	6	11	9	10	9	11	6	15	1	15
k	0	-10	-5	-7	-4	-5	-5	-4	-7	-5	-10	-10
q	10	1	2	1	1	1	1	1	2	1	20	1
s	10	11	32	43	75	118	193	311	815	1126	23335	24461
t	1	1	3	4	7	11	18	29	76	105	2176	2281

(Note that $q_0 = \lfloor \sqrt{115} \rfloor = 10$.) Since $a_1 = a_{11}$ and $k_1 = k_{11}$, we see that the continued fraction of $v = \sqrt{115}$, or of the ideal number $(1 : 0)$ in $\mathcal{I}_{460}$, is

$$\langle 10, \overline{1, 2, 1, 1, 1, 1, 1, 2, 1, 20} \rangle.$$

One can again verify that the sequence of convergents s_i/t_i,

$$\frac{10}{1}, \quad \frac{11}{1}, \quad \frac{32}{3}, \quad \frac{43}{4}, \quad \frac{75}{7}, \quad \frac{118}{11}, \quad \frac{193}{18}, \quad \frac{311}{29}, \quad \frac{815}{76}, \quad \frac{1126}{105}, \quad \frac{23335}{2176},$$

provides increasingly accurate rational approximations of $\sqrt{115}$.

7.6 Representations of integers by indefinite ideal numbers

In this section, we apply results about continued fractions to proper representations of integers by ideal numbers of positive discriminant and their corresponding indefinite quadratic forms.

Theorem 7.6.1. *Let R be the reduction map on $\mathcal{I}_\Delta$ for some positive discriminant Δ. Let $(a : k)$ be an ideal number in $\mathcal{I}_\Delta$ and, for all $n \geq 0$, let*

$$R^n(a : k) = (a_n : k_n).$$

Let $\langle q_0, q_1, q_2, \dots \rangle$ be the continued fraction of $(a : k)$ and let s_i and t_i be the convergent sequences of $(a : k)$ as in equations (7.4) and (7.5). Let $f(x, y)$ be the quadratic form given by $(a : k)$ in ideal number notation. Then $f(s_{n-1}, t_{n-1}) = (-1)^n a_n$ for every $n \geq 0$.

Proof. Let $\langle q_0, q_1, q_2, \dots \rangle$ be the continued fraction of $(a : k)$ and consider the following matrices for $i \geq 0$:

$$U_i = \begin{bmatrix} q_i & (-1)^{i+1} \\ (-1)^i & 0 \end{bmatrix} \quad \text{and} \quad W_i = \begin{bmatrix} s_{i-1} & (-1)^i s_{i-2} \\ t_{i-1} & (-1)^i t_{i-2} \end{bmatrix}. \tag{7.6}$$

Note that W_0 is the identity matrix, and we see as follows that $W_i U_i = W_{i+1}$ for $i \geq 0$:

$$\begin{bmatrix} s_{i-1} & (-1)^i s_{i-2} \\ t_{i-1} & (-1)^i t_{i-2} \end{bmatrix} \cdot \begin{bmatrix} q_i & (-1)^{i+1} \\ (-1)^i & 0 \end{bmatrix} = \begin{bmatrix} s_{i-2} + s_{i-1} q_i & (-1)^{i+1} s_{i-1} \\ t_{i-2} + t_{i-1} q_i & (-1)^{i+1} t_{i-1} \end{bmatrix}$$

$$= \begin{bmatrix} s_i & (-1)^{i+1} s_{i-1} \\ t_i & (-1)^{i+1} t_{i-1} \end{bmatrix}.$$

Thus, each W_i is a unimodular matrix, with $W_i = U_0 \cdot U_1 \cdots U_{i-1}$ for $i \geq 1$. Finally, one can verify that

$$U_i = (-1)^i T^{(-1)^i q_i} V,$$

where T and V are the translation and involution matrices defined as in equation (4.1). By repeated application of Proposition 7.4.1, recalling that a matrix and its negative are interchangeable in the unimodular group, it follows that

$$(a : k) \cdot W_n = (a : k) \cdot U_0 \cdot U_1 \cdots U_{n-1} = \left((-1)^n a_n : k_n\right)$$

for all $n \geq 1$. As part of the implication of this equation, as in Theorem 4.1.2, if $f(x,y)$ is the quadratic form given by $(a : k)$ in ideal number notation, then $f(s_{n-1}, t_{n-1}) = (-1)^n a_n$, as claimed. $\qquad\square$

Example. The following table repeats information that we have seen in previous examples. Here, $(8 : 9)$ is an ideal number of discriminant $\Delta = 41$, for which the principal polynomial is $P(x) = x^2 + x - 10$. Since $P(9) = 80 = 8 \cdot 10$ and $P'(9) = 19$, we see that $f(x,y) = 8x^2 + 19xy + 10y^2$ is the quadratic form with ideal number expression $(8 : 9)$. In the final row of the table, we evaluate this quadratic form at each pair (s, t). As claimed in Theorem 7.6.1, in a different form, the $f(s, t)$ row contains the same entries as the a row, shifted one place to the left and alternating in sign:

i	0	1	2	3	4	5	6	7	8	9	10
a	8	1	4	2	2	4	1	4	2	2	4
k	9	−2	−3	−2	−3	−2	−3	−3	−2	−3	−2
q	−1	4	1	2	2	1	5	1	2	2	1
s	−1	−3	−4	−11	−26	−37	−211	−248	−707	−1662	−2369
t	1	4	5	14	33	47	268	315	898	2111	3009
$f(s,t)$	−1	4	−2	2	−4	1	−4	2	−2	4	−1

As we have seen, the entries in the a row repeat a pattern of (in this example) five integers indefinitely, perhaps after one or more initial terms. In the same way, the $f(s, t)$ row repeats a pattern of ten integers (the same five but alternating in sign) indefinitely. In other words, there are infinitely many integer solutions of $f(x,y) = a$ for $a = \pm 1, \pm 2, \pm 4$ in this example.

Our next result implies that when m is small enough in absolute value, the only proper representations of m occur as described in Theorem 7.6.1. Theorem 7.6.2 appears in an equivalent form in Appendix D, where its proof is based on facts about approximation of irrational real numbers by rational numbers.

Theorem 7.6.2. *Let $(a : k)$ be an ideal number in $\mathcal{I}_\Delta$, where Δ is a positive discriminant. If $P(x)$ is the principal polynomial of discriminant Δ, let $P(k) = ac$ and $P'(k) = b$, so that $f(x,y) = ax^2 + bxy + cy^2$ is the quadratic form with $(a : k)$ as its ideal number expression. Suppose that a is positive and c is negative. Let s and t be relatively prime positive integers and suppose that $f(s, t) = m$ with $|m| < \frac{\sqrt{\Delta}}{2}$. Then $\frac{s}{t}$ is a convergent in the continued fraction of $(a : k)$.*

Proof. If $f(s,t) = as^2 + bst + ct^2 = m$ with $|m| < \frac{\sqrt{\Delta}}{2}$, then

$$\left| a\left(\frac{s}{t}\right)^2 + b\left(\frac{s}{t}\right) + c \right| < \frac{\sqrt{\Delta}}{2t^2}.$$

In Theorem D.5.7, it is established that in this case, with the same restrictions on a and c as above, the fraction $\frac{s}{t}$ is a convergent in the continued fraction of the irrational number $v = \frac{-b+\sqrt{\Delta}}{2a}$. But v has ideal number expression $(a : k)$. ☐

Example. Let $\Delta = 460$, for which $P(x) = x^2 - 115$ is the principal polynomial. We have seen in previous examples that $(2 : -9)$ is a reduced ideal number in $\mathcal{I}_{460}$. Since $P(-9) = -34 = 2(-17)$ and $P'(-9) = -18$, the quadratic form with ideal number expression $(2 : -9)$ is $f(x,y) = 2x^2 - 18xy - 17y^2$. The conditions of Theorem 7.6.2 apply to this case (as they do to every reduced ideal number of positive discriminant). In the following table, we compile the data from the reduction map applied repeatedly to $(2 : -9)$, which we have seen in previous examples, together with calculation of the convergent sequences and the quadratic form evaluated at numerator and denominator pairs:

i	0	1	2	3	4	5	6
a	2	17	3	5	3	17	2
k	−9	−9	−8	−10	−10	−8	−9
q	9	1	6	4	6	1	9
s	9	10	69	286	1785	2071	20424
t	1	1	7	29	181	210	2071
$f(s,t)$	−17	3	−5	3	−17	2	−17

Here, we can see that $f(x,y)$ properly represents the integers 2, 3, −5, and −17 infinitely often, due to the periodic nature of the sequence of a values. Based on Theorem 7.6.2, we can also say 2, 3, and −5 are the only integers m with $|m| < \frac{\sqrt{460}}{2} \approx 10.7$ that are properly represented (in positive integers) by $f(x,y)$.

We consider more implications of Theorem 7.6.2 in Section 7.7.

7.7 Class group calculations

In Section 7.5, we saw some examples of computation of a class group $\mathcal{C}_\Delta$ when Δ is positive. The reduced ideal numbers of discriminant Δ give us a starting point, in that every ideal number in $\mathcal{I}_\Delta$ is equivalent to a reduced ideal number. The reduction map on reduced ideal numbers typically provides additional equivalences among those ideal numbers. In Section 7.5, we used genus equivalence to conclude that the remaining ideal numbers are not equivalent. But genus considerations are not always sufficient for drawing such a conclusion, as we illustrate with the following example.

Example. Let $\Delta = \Delta(34, 1) = 136$, for which $P(x) = x^2 - 34$. Using the method outlined in Section 7.1, we find that there are ten reduced ideal numbers in $\mathcal{I}_{136}$:

$$(1 : -5) \quad (2 : -4) \quad (3 : -5) \quad (3 : -4) \quad (5 : -3)$$
$$(5 : -2) \quad (6 : -4) \quad (6 : -2) \quad (9 : -5) \quad (9 : -4).$$

The reduction map partitions this set into two disjoint cycles, one of length four and the other of length six. Results of this algorithm appear below:

n	0	1	2	3	4	n	0	1	2	3	4	5	6
a	1	9	2	9	1	a	3	6	5	5	6	3	3
k	-5	-5	-4	-4	-5	k	-5	-4	-2	-3	-2	-4	-5
q	10	1	4	1	10	q	3	1	1	1	1	3	3

Applying Proposition 7.4.1, we have the following equivalences among reduced ideal numbers. There are at most four classes of ideal numbers in $\mathcal{I}_{136}$:

$$(1 : -5) \sim (2 : -4) \qquad (3 : -5) \sim (5 : -2) \sim (6 : -2)$$
$$(3 : -4) \sim (6 : -4) \sim (5 : -3) \qquad (9 : -5) \sim (9 : -4).$$

Can there be any other equivalences in this case? Here, with $34 = 2 \cdot 17 \equiv 2 \pmod 8$, the defined genus symbols for ideal numbers in $\mathcal{I}_{136}$ are s_2 and s_{17}. Both symbols are $+1$ for $(1 : -5)$ and $(9 : -5)$, while both are -1 for $(3 : -5)$ and $(3 : -4)$. Thus, we cannot rule out an equivalence between $(1 : -5)$ and $(9 : -5)$ by genus considerations, nor between $(3 : -5)$ and $(3 : -4)$.

In this case, however, we can apply Theorem 7.6.2 and other results to address these possibilities. We will return to this example after noting the following general statements.

Proposition 7.7.1. *Let $(a : k)$ be a reduced ideal number of positive discriminant Δ, with $a < \frac{\sqrt{\Delta}}{2}$. Let $R : \mathcal{I}_\Delta \to \mathcal{I}_\Delta$ be the reduction map, with $R^n(a : k) = (a_n : k_n)$ for all $n \geq 0$. Suppose that the continued fraction of $(a : k)$ has period ℓ and let $m = \mathrm{lcm}(\ell, 2)$. Let s_i and t_i be the numerator and denominator sequences of the convergents of this continued fraction and let*

$$W_i = \begin{bmatrix} s_{i-1} & (-1)^i s_{i-2} \\ t_{i-1} & (-1)^i t_{i-2} \end{bmatrix}$$

for $i \geq 0$. Then the group of all automorphs of $(a : k)$ is the infinite cyclic group generated by W_m.

Proof. Note that $m = \text{lcm}(\ell, 2)$ can be described as the smallest *even* positive integer for which $R^m(a : k) = (a : k)$. The matrices W_i are as defined for $i \geq 0$ in equation (7.6), and we can define the matrices U_i for $i \geq 0$ as in that equation also. The purely periodic nature of the continued fraction of a reduced ideal number shows that $U_i = U_{m+i}$ for all $i \geq 0$. Recalling that $W_n = U_0 U_1 \cdots U_n$ for all $n \geq 0$, it follows that $W_{m+i} = W_m \cdot W_i$ for all $i \geq 0$. Note that m is the smallest positive integer for which W_m is an automorph of $(a : k)$. The preceding observation shows that $W_{2m} = W_m^2$ and, more generally, $W_{jm} = W_m^j$ for all positive integers j. Since $\text{Aut}(a : k)$ is a group, all such powers of W_m are automorphs of $(a : k)$ as well, as are all negative integer powers of W_m. We will show that all automorphs of $(a : k)$ must have this form.

Let U be an automorph of $(a : k)$ and let q and r be the entries in the first column of U. Since a unimodular matrix and its negative are identified in the unimodular group, we can assume that at least one of q and r is positive. If $f(x, y) = ax^2 + bxy + cy^2$ is the quadratic form with ideal number expression $(a : k)$, then $f(q, r) = a$. Since $a < \frac{\sqrt{\Delta}}{2}$, then if q and r are both positive, we know that $\frac{q}{r}$ is a convergent in the continued fraction of $(a : k)$ by Theorem 7.6.2. Specifically, $q = s_{mi-1}$ and $r = t_{mi-1}$ for some $i \geq 1$, due to the periodicity of the continued fraction of $(a : k)$. It follows that $U = W_{mi} = W_m^i$ for some positive integer i. If q and r have opposite sign, however, then we can assume that the first column of U^{-1} is positive. But then $U^{-1} = W_m^i$ for some positive i, as above, and so $U = W_m^{-i}$. In either case, $U = (W_m)^i$ for some negative integer i. Thus, $\text{Aut}(a : k) = \langle W_m \rangle = \{W_m^i \mid i \in \mathbb{Z}\}$, as we wanted to show. $\qquad\square$

The following proposition is also a useful observation.

Proposition 7.7.2. *Let Δ be a positive discriminant, and let $(a : k)$ and $(a : \ell)$ be a pair of reduced conjugate ideal numbers in $\mathcal{I}_\Delta$. If $(a : k)$ is equivalent to a reduced ideal number $(a' : k')$, then $(a : \ell)$ is equivalent to $(a' : \ell')$, the reduced conjugate of $(a' : k')$.*

Proof. Let q and q' be the reduction numbers of $(a : k)$ and $(a' : k')$, respectively. If $(a : k) \cdot U = (a' : k')$, then we know that $(a : \overline{k}) \cdot \overline{U} = (a' : \overline{k'})$, where $\overline{U}$ is the conjugate of the unimodular matrix U. If T is the translation matrix, we then find that $(a : \ell) \cdot (T^q \overline{U} T^{-q'}) = (a' : \ell')$. We omit the details of this verification. $\qquad\square$

We illustrate the implications of these propositions by returning to our previous example.

Example. Consider again the collection of ideal numbers of discriminant $\Delta = 136$. Below we repeat the data obtained from applying the reduction map to two of the reduced ideal numbers in $\mathcal{I}_{136}$, to which we now add convergent sequence calculations:

As noted previously, these tables show that each ideal number in $\mathcal{I}_{136}$ is equivalent to at least one of the following four reduced ideal numbers: $(1 : -5)$, $(9 : -5)$, $(3 : -5)$, or $(3 : -4)$. Genus considerations imply that neither of the first two is equivalent to the last two. But we cannot eliminate the possibility that $(1 : -5)$ is equivalent to $(9 : -5)$, or

n	0	1	2	3	4
a	1	9	2	9	1
k	−5	−5	−4	−4	−5
q	10	1	4	1	10
s	10	11	54	65	704
t	1	1	5	6	65

n	0	1	2	3	4	5	6
a	3	6	5	5	6	3	3
k	−5	−4	−2	−3	−2	−4	−5
q	3	1	1	1	1	3	3
s	3	4	7	11	18	65	213
t	1	1	2	3	5	18	59

that $(3 : -5)$ is equivalent to $(3 : -4)$, via genera. We now argue as follows that neither of those options can occur.

Note that the first table shows that $W_4 = \left[\begin{smallmatrix} 65 & 54 \\ 6 & 5 \end{smallmatrix}\right]$ is an automorph of $(1 : -5)$, as is W_4^i for all integers i. Suppose that $(1 : -5)$ is equivalent to $(9 : -5)$, which we know by Proposition 7.4.1 is equivalent to $(-1 : -5)$. If $(1 : -5) \cdot U = (-1 : -5)$ for some unimodular matrix U, then $(1 : -5) \cdot (W_4^i U) = (-1 : -5)$ for every integer i. If the entries in the first column of U are q and r, for which we can assume that q is positive, we can see that it is possible to choose i sufficiently large so that both entries in the first column of $W_4^i U$ are positive. We can assume that this is true for q and r themselves. But now, if $f(x,y) = x^2 - 10xy - 9y^2$ is the quadratic form written as $(1 : -5)$ in ideal number notation, we are assuming the existence of positive integers q and r for which $f(q,r) = -1$. Then $q = s_{n-1}$ and $r = t_{n-1}$ for some positive integer n. This is impossible, since the periodicity of the continued fraction expansion of $(1 : -5)$ shows that $a_n = 1$ if and only if 4 divides n. Thus, we cannot have $f(s_{n-1}, t_{n-1}) = (-1)^n a_n = -1$.

Now suppose that $(3 : -5) \cdot U = (3 : -4)$ for some unimodular matrix U. Again using an automorph of $(3 : -5)$, which in this case is $W_6 = \left[\begin{smallmatrix} 65 & 18 \\ 18 & 5 \end{smallmatrix}\right]$, we can assume that both entries in the first column of U, say q and r, are positive. This is impossible, since $a_n = 3$ if and only if $n \equiv 5 \pmod 6$ or $n \equiv 0 \pmod 6$. In the first case, we conclude that $(3 : -5) \sim (-3 : -4)$. In this second case, we can say only that $(3 : -5)$ is equivalent to itself. Thus, $(3 : -5)$ and $(3 : -4)$ are not equivalent.

We could also make the following argument for the second observation. Since we established that $(1 : -5)$ is not equivalent to $(9 : -5)$, the principal genus in C_{134} contains two classes. Each of its cosets must also contain two classes, so the classes of $(3 : -5)$ and $(3 : -4)$ are not equal. We conclude, in any event, that C_{136} is isomorphic to $\mathbb{Z}_4$, with $[3 : -5]$ as a generator.

The following theorem summarizes observations from this example more generally and adds a condition on the number of distinct classes of ideal numbers of positive discriminant.

Theorem 7.7.3. *Let Δ be a positive discriminant and let $R : \mathcal{I}_\Delta \to \mathcal{I}_\Delta$ be the reduction map. If $(a : k)$ and $(a' : k')$ are reduced ideal numbers in $\mathcal{I}_\Delta$, then $(a : k)$ is equivalent to $(a' : k')$ if and only if*

$$R^{2n}(a : k) = (a' : k')$$

for some integer n. If R partitions the set of reduced ideal numbers into g distinct cycles, then the lengths of those cycles are either all odd or all even. In the first case, the class group C_Δ has g distinct elements. In the second case, C_Δ has 2g distinct elements.

Proof. Let Δ be a positive discriminant and let $R : \mathcal{I}_\Delta \to \mathcal{I}_\Delta$ be the reduction map. Proposition 7.4.1 shows that if $R^{2n}(a : k) = (a' : k')$ for some integer n, then $(a : k)$ is equivalent to $(a' : k')$. Conversely, suppose that $(a : k)$ is equivalent to $(a' : k')$. Let i be the smallest nonnegative integer so that $R^i(a' : k') = (a'' : k'')$ with $|a''| < \frac{\sqrt{\Delta}}{2}$. If $f(x,y)$ is the quadratic form with $(a : k)$ as its ideal number expression, then $f(q, r) = \pm a''$ for some relatively prime integers q and r. Using automorphs of $(a : k)$, we can assume that q and r are both positive. Then $q = s_{n-1}$ and $r = t_{n-1}$ for some $n \geq 0$, where s_{n-1} and t_{n-1} are convergents of $(a : k)$. Since a' is positive (as the norm of a reduced ideal number), we see that $n = 2m + i$ for some integer m. In that case, $R^{2m}(a : k) = (a' : k')$, as we wanted to show.

For the second claim, let $P(x,y)$ be the principal quadratic form of discriminant Δ, that is, the quadratic form with ideal number expression $(1 : 0)$. Let $f(x,y) = ax^2 + bxy + cy^2$ be a quadratic form of discriminant Δ with ideal number expression $(a : k)$. Then Proposition 1.4.3 implies that $P(q, r) = -1$ for some integers q and r if and only if $f(q - kr, ar) = -a$, with $\gcd(q - kr, ar) = 1$. But now there are integers s and t so that $U = \left[\begin{smallmatrix} q-kr & s \\ ar & t \end{smallmatrix}\right]$ is unimodular, and we find that $(a : k) \cdot U$ is equivalent to $(-a : k)$. (The norm of $(a : k) \cdot U$ is $f(q - kr, ar) = -a$, while the character is $k + a(q - kr)s + b(ar)s + c(ar)t \equiv k$ (mod a). A translation then shows that $(a : k) \sim (-a : k)$.) In each cycle of reduced ideal numbers under the reduction map R, we can select $(a : k)$ of minimal norm, in which case $a < \frac{\sqrt{\Delta}}{2}$. But this occurs if and only if ℓ is odd so that $f(s_{\ell-1}, t_{\ell-1}) = (-1)^\ell a = -a$. In a cycle of odd length, each reduced ideal number is equivalent to all others. Thus, if there are g distinct cycles, all of odd length, then there are g distinct classes of ideal numbers. On the other hand, in a cycle of even length, $R^m(a : k) \sim R^n(a : k)$ only when m and n have the same parity. In that case, each cycle has two distinct classes of ideal numbers and there are $2g$ classes in total. □

Example. Let $\Delta = 401$, with principal polynomial $P(x) = x^2 + x - 100$. Using the following table of negative values of $P(x)$, one can establish that there are precisely nineteen reduced ideal numbers in $\mathcal{I}_{401}$:

x	−10	−9	−8	−7	−6	−5	−4	−3	−2	−1
$P(x)$	−10	−28	−44	−58	−70	−80	−88	−94	−98	−100

These ideal numbers appear in the following tables, where we find that the reduction map partitions this collection into five cycles, three of them with length three and two with length five:

n	0	1	2	3
a	1	10	10	1
k	−10	−10	−1	−10
q	19	1	1	19

n	0	1	2	3
a	2	14	5	2
k	−10	−9	−6	−10
q	9	1	3	9

n	0	1	2	3
a	2	5	14	2
k	−9	−10	−6	−9
q	9	3	1	9

n	0	1	2	3	4	5
a	4	11	8	10	7	4
k	−9	−8	−4	−5	−6	−9
q	4	1	1	1	2	4

n	0	1	2	3	4	5
a	4	7	10	8	11	4
k	−8	−9	−6	−5	−4	−8
q	4	2	1	1	1	4

The period of each reduced ideal number is odd, thus each of them is equivalent to all other reduced ideal numbers in the same cycle. For instance, applying R^2 repeatedly to $(4 : -9)$ shows that

$$(4 : -9) \sim (8 : -4) \sim (7 : -6) \sim (11 : -8) \sim (10 : -5).$$

We conclude that $\mathcal{C}_{401}$ has five elements, so is isomorphic to $\mathbb{Z}_5$, with $[2 : -1]$ (or any other nonidentity element) as a generator. Since 401 is prime, the only defined genus symbol for $\mathcal{I}_{401}$ is s_{401}, which is +1 for each ideal number.

A Appendix: Review of number theory

In Appendix A, we review some definitions and results from elementary number theory that we will require in the text. We will emphasize computational methods, mostly as applications of the Euclidean algorithm, for solving linear equations and congruences, including systems of two or more linear congruences. In Sections A.7 and A.8, we review standard results about quadratic congruences, proving the quadratic reciprocity theorem and other facts needed to determine the number of solutions of a quadratic congruence modulo a prime number. (Techniques for solving quadratic congruences modulo composite integers are developed in the text in the context of ideal number composition.)

A.1 Divisibility and congruence

We assume the familiar algebraic properties of addition and multiplication in the set of integers, and the properties of the usual order relation on integers, including the following *well-ordering principle*: Every nonempty set of positive integers has a smallest element.

Definition. If m and n are integers, we say that m *divides* n, and write $m \mid n$, if there is an integer q so that $n = mq$.

If m divides n with $m \neq 0$, then the integer q is unique, and we may write $q = \frac{n}{m}$. The following theorem gives us a method of testing whether m divides n when m is positive.

Theorem A.1.1 (Division algorithm). *If n and m are integers with $m > 0$, then there are unique integers q and r such that $n = mq + r$ and $0 \leq r < m$.*

Proof. Let T be the set $\{n - mx \mid x \in \mathbb{Z}\}$. The subset S of nonnegative elements of T is nonempty—if $n \geq 0$, then $n = n - m(0)$ is in S; if $n < 0$, then $n - m(n) = -n(m - 1)$ is in S. Thus, there is a smallest element, say r, in S. By definition, $r = n - m(q)$ for some integer q, so that $n = mq + r$ with $r \geq 0$. We also have $r < m$, since otherwise $r - m = n - m(q + 1) < r$ is an element of S, contradicting the definition of r.

To show that q and r are unique, suppose that we also have $n = ms + t$ with $0 \leq t < m$. This implies that $m(q - s) = t - r$. But with $0 \leq r, t < m$, we find that $-m < t - r < m$. The only multiple of m strictly between $-m$ and m is $0 = m(0)$. We conclude that $q - s = 0$, so that $q = s$ and $r = t$. $\square$

Definition. If $n = mq + r$ with $0 \leq r < m$ as under the division algorithm, we write the *quotient* q as n div m and the *remainder* r as n mod m.

If m is positive, then m divides n if and only if n mod $m = 0$, and $-m$ divides n if and only if m divides n. If $m = 0$, then m divides n if and only if $n = 0$. Thus, we can use the division algorithm to test divisibility of one integer by another in every case.

https://doi.org/10.1515/9783111319360-008

Definition. If a and b are integers and m is a positive integer, we say that a is *congruent to* b *modulo* m, and write $a \equiv b \pmod{m}$, if m divides $a - b$.

The following proposition lists some important properties of the divisibility and congruence relations on the set of integers.

Proposition A.1.2. *Let a, b, c, d, and m be integers, with $m > 0$. Then the following statements are true:*

1. $a \mid a$.
2. *If $a \mid b$ and $b \mid a$, then either $a = b$ or $a = -b$.*
3. *If $a \mid b$ and $b \mid c$, then $a \mid c$.*
4. $a \equiv a \pmod{m}$.
5. *If $a \equiv b \pmod{m}$, then $b \equiv a \pmod{m}$.*
6. *If $a \equiv b \pmod{m}$ and $b \equiv c \pmod{m}$, then $a \equiv c \pmod{m}$.*
7. *If $a \equiv c \pmod{m}$ and $b \equiv d \pmod{m}$, then $a + b \equiv c + d \pmod{m}$.*
8. *If $a \equiv c \pmod{m}$ and $b \equiv d \pmod{m}$, then $ab \equiv cd \pmod{m}$.*
9. *If $a \equiv b \pmod{m}$ and d is a positive divisor of m, then $a \equiv b \pmod{d}$.*
10. $a \equiv b \pmod{m}$ *if and only if $a \bmod m = b \bmod m$, that is, if and only if a and b have the same remainder, under the division algorithm, on division by m.*

Proof. We prove statements (6) and (8), leaving the remaining proofs as exercises. For (6), suppose that $a \equiv b \pmod{m}$ and $b \equiv c \pmod{m}$. Then m divides both $a - b$ and $b - c$, say with $a - b = mq$ and $b - c = mr$ for some integers q and r. But now

$$a - c = (a - b) + (b - c) = mq + mr = m(q + r).$$

Since $q + r$ is an integer, then m divides $a - c$, so that $a \equiv c \pmod{m}$.

For (8), suppose that $a \equiv c \pmod{m}$ and $b \equiv d \pmod{m}$, so that $a - c = mq$ and $b - d = mr$ for some integers q and r. Then

$$ab - cd = ab - cb + cb - cd = (a - c)b + c(b - d) = mqb + cmr = m(qb + cr),$$

and so $ab \equiv cd \pmod{m}$, since $qb + cr$ is an integer. □

Parts (4)–(6) of Proposition A.1.2 show that congruence modulo m is an equivalence relation on the set of integers. We write the set of all equivalence classes under this relation (which we also call *congruence classes* modulo m) as $\mathbb{Z}_m$. By definition, the congruence class of a modulo m is

$$[a] = \{x \in \mathbb{Z} \mid x \equiv a \pmod{m}\} = \{x \in \mathbb{Z} \mid x = a + mq \text{ for some integer } q\}.$$

Part (10) of Proposition A.1.2 shows that $\mathbb{Z}_m = \{[0], [1], [2], \ldots, [m-1]\}$, and parts (8) and (9) imply that the following operations of addition and multiplication are well-defined on $\mathbb{Z}_m$:

$$[a] + [b] = [a + b] \quad \text{and} \quad [a] \cdot [b] = [ab].$$

We usually write $\mathbb{Z}_m$ as $\{0, 1, 2, \ldots, m-1\}$ when it is clear that we are working modulo m.

If $f(x)$ is a polynomial with integer coefficients, and m is a positive integer, a general problem is to find all integers x for which m divides $f(x)$, or equivalently, for which $f(x) \equiv 0 \pmod{m}$. Repeated application of results from Proposition A.1.2 shows that if $x \equiv y \pmod{m}$, then $f(x) \equiv f(y) \pmod{m}$. Thus, it suffices to find all solutions of the *polynomial congruence* $f(x) \equiv 0 \pmod{m}$ in the set $\mathbb{Z}_m$. We will establish a method of solving $f(x) \equiv 0 \pmod{m}$ when $f(x)$ is a *linear* polynomial in Section A.5, and describe results concerning $f(x) \equiv 0 \pmod{m}$ when $f(x)$ is *quadratic* in Section A.7.

A.2 Combinations

The following definition leads to several important results about divisibility.

Definition. If a and b are integers, let

$$\langle a \rangle = \{aq \mid q \in \mathbb{Z}\} \quad \text{and} \quad \langle a, b \rangle = \{as + bt \mid s, t \in \mathbb{Z}\}.$$

We say that n is a *multiple* of a if n is in $\langle a \rangle$, and that n is a *combination* of a and b if n is in $\langle a, b \rangle$.

Proposition A.2.1. *If a and b are integers, then $\langle a \rangle$ is a subset of $\langle b \rangle$ if and only if b divides a.*

Proof. If $\langle a \rangle \subseteq \langle b \rangle$, then $a = a(1)$ is in $\langle b \rangle$, so that $a = bq$ for some integer q. But then b divides a. Conversely, suppose that b divides a, say with $a = bq$ for some integer q. If x is an element of $\langle a \rangle$, then $x = ar = (bq)r = b(qr)$ for some integer r. But then x is an element of $\langle b \rangle$, and it follows that $\langle a \rangle$ is a subset of $\langle b \rangle$. □

It follows that $\langle a \rangle = \langle b \rangle$ if and only if $a = b$ or $a = -b$, using statement (2) from Proposition A.1.2.

Theorem A.2.2. *Let a and b be integers. Then there is an integer d such that $\langle a, b \rangle = \langle d \rangle$. In this case, d has the following properties:*
1. *d is a common divisor of a and b, that is, d divides a and d divides b.*
2. *If c is a common divisor of a and b, then c divides d.*

Proof. If $a = 0 = b$, then $\langle a, b \rangle = \{0\} = \langle 0 \rangle$. If a and b are not both zero, then $\langle a, b \rangle$ must contain a positive integer, since a, b, $-a$, and $-b$ are elements of $\langle a, b \rangle$. So, the set of positive elements in $\langle a, b \rangle$ is nonempty, and must contain a smallest element d. We show that $\langle a, b \rangle = \langle d \rangle$ in this case.

By definition, d in $\langle a, b \rangle$ implies that $d = as + bt$ for some integers s and t. Now for any q we see that $dq = a(sq) + b(tq)$ is an element of $\langle a, b \rangle$, and so $\langle d \rangle$ is a subset of $\langle a, b \rangle$. To show the reverse inclusion, let n be an element of $\langle a, b \rangle$, say $n = ax + by$ for some integers x and y. Since d is positive, we can apply the division algorithm to write $n = dq + r$ for some integers q and r with $0 \le r < d$. Note that $r = n - dq = a(x - sq) + b(y - tq)$ is an element of $\langle a, b \rangle$. If $r > 0$, this contradicts the definition of d as the smallest positive integer in $\langle a, b \rangle$, so we must conclude that $r = 0$, and then $n = dq$ is in $\langle d \rangle$. Thus, $\langle a, b \rangle$ is a subset of $\langle d \rangle$, and we conclude that $\langle a, b \rangle = \langle d \rangle$.

As noted, a and b are elements of $\langle a, b \rangle$, so if $\langle a, b \rangle = \langle d \rangle$, then d divides both a and b by definition. If c divides both a and b, say with $a = cq$ and $b = cr$, then $d = as + bt = c(qs + rt)$, so that c divides d. □

Definition. Let a and b be integers. We say that $d \ge 0$ is the *greatest common divisor* of a and b, and write $d = \gcd(a, b)$, if d is a common divisor of a and b that is a multiple of every other common divisor of a and b. If $\gcd(a, b) = 1$, we say that a is *relatively prime* to b, or that a and b are *coprime*.

Theorem A.2.2 shows that n is a combination of a and b if and only if $d = \gcd(a, b)$ divides n. In particular, d itself is a combination of a and b. This fact has several useful consequences for divisibility in the set of integers, which we compile in the remainder of this section.

Corollary A.2.3. *If a and b are integers, then a is relatively prime to b if and only if there are integers s and t so that $as + bt = 1$.*

Proof. Let $d = \gcd(a, b)$, so that $\langle a, b \rangle = \langle d \rangle$. If $d = 1$, then 1 is a combination of a and b. Conversely, if $as + bt = 1$ for some s and t, then 1 is in $\langle d \rangle$. It follows that d divides 1, and so $d = 1$ since $d \ge 0$. □

Corollary A.2.4. *If a, b, and c are integers with $\gcd(a, b) = d$, and a divides bc, then a divides cd. In particular, if a divides bc with a and b coprime, then a divides c.*

Proof. If a divides bc, then $bc = aq$ for some integer q. Let $\gcd(a, b) = d$, so that there are integers s and t with $d = as + bt$. Then

$$cd = c(as + bt) = acs + (bc)t = acs + (aq)t = a(cs + qt).$$

So, a divides cd, since $cs + qt$ is an integer. □

Corollary A.2.5 (Congruence cancellation property). *Let a, b, and c be integers, let m be a positive integer, and let $\gcd(a, m) = d$. Then $ab \equiv ac \pmod{m}$ if and only if $b \equiv c \pmod{\frac{m}{d}}$.*

Proof. Suppose that $ab \equiv ac \pmod{m}$, so that m divides $ab - ac = a(b - c)$. If $d = \gcd(a, m)$, then Corollary A.2.4 implies that $\frac{m}{d}$ divides $b - c$. Conversely, suppose that $b \equiv c \pmod{\frac{m}{d}}$, where $d = \gcd(a, m)$ for some integer a. Then $b - c = \frac{m}{d}q$ for some integer q, so that $ab - ac = a \cdot \frac{m}{d}q = m \cdot \frac{a}{d}q$. Since $\frac{a}{d}$ is an integer, we can conclude that $ab \equiv ac \pmod{m}$. □

Corollary A.2.6. *If a divides n and b divides n, and* $\gcd(a, b) = d$, *then ab divides nd. In particular, if a and b are coprime, then ab divides n.*

Proof. We have that $n = bq$ for some integer q. So, a divides bq and by Corollary A.2.4, then a divides qd, say that $qd = ar$ for some integer r. But now

$$nd = (bq)d = b(qd) = b(ar) = (ab)r,$$

and so ab divides nd. □

Definition. If a and b are integers, we say that an integer $m \geq 0$ is the *least common multiple* of a and b, and write $m = \mathrm{lcm}(a, b)$, if:
1. a divides m and b divides m, and
2. if a and b both divide n, then m divides n.

Corollary A.2.7. *If a and b are integers, then* $|ab| = \gcd(a, b) \cdot \mathrm{lcm}(a, b)$.

Proof. Assume for simplicity that a and b are both nonnegative. If $a = 0 = b$, then $\mathrm{lcm}(a, b) = 0$, since 0 is the only common multiple of a and b. In all other cases, $d = \gcd(a, b)$ is positive, and we can consider the integer $m = \frac{ab}{d}$. Note that $m = a \cdot \frac{b}{d} = b \cdot \frac{a}{d}$ is a common multiple of a and b, since $\frac{b}{d}$ and $\frac{a}{d}$ are integers. If a and b both divide n, then Corollary A.2.6 shows that m divides n. Thus, $m = \mathrm{lcm}(a, b)$ by definition. □

A.3 Prime factorization

An integer $p > 1$ is called *prime* if its only positive divisors are 1 and p. If $n > 1$ is not prime, it is called *composite*. Note that 1 is neither prime nor composite by definition. Using results from Section A.2, we establish the existence and uniqueness of prime factorization in the set of integers.

Theorem A.3.1 (Euclid's lemma). *Let p be a prime number, and suppose that p divides ab for some integers a and b. Then either p divides a or p divides b.*

Proof. Let $d = \gcd(a, p)$. Since d is a positive divisor of p, either $d = p$ or $d = 1$. If $d = p$, then p divides a by definition. If $d = 1$ and p divides ab, then p divides b by Corollary A.2.4. □

When n is composite, then it is always possible to find integers a and b such that n divides the product ab without dividing either term in the product. In fact, if we let a be any positive divisor of n other than 1 and n, so that $n = ab$ for some integer b, then we have that n divides ab, but does not divide either a or b.

Theorem A.3.2 (Fundamental theorem of arithmetic). *Every integer n > 1 can be written as a product of primes. This product is unique, aside from the order of the factors.*

Proof. We first establish the existence of such a factorization. If not every integer $n > 1$ can be written as a product of primes, we can assume that n is the smallest integer that cannot be so written. Then n must be composite, since we regard a prime number as a product of primes with just one factor. So, n can be written as $n = ab$ with $1 < a, b < n$. But then, by assumption, a and b can be written as product of primes, and thus $n = ab$ has that property as well, contradicting our assumption.

Now suppose that there is an integer n larger than 1 that can be written in two ways as a product of primes, say that $n = p_1 p_2 \cdots p_k$ and $n = q_1 q_2 \cdots q_\ell$ with each p_i and q_j prime. We may again take n to be as small as possible with this property. In particular, we can assume that $p_i \neq q_j$ for $1 \leq i \leq k$ and $1 \leq j \leq \ell$, since otherwise we could cancel that common term from both products and start over with the smaller resulting integer. Here, p_1 divides n since $p_2 p_3 \cdots p_k$ is an integer, and so p_1 divides the product $q_1 q_2 \cdots q_\ell$. Applying Euclid's lemma repeatedly, we can say that p_1 divides at least one term in that product. Rearranging those terms if necessary, we can assume that p_1 divides q_1. But q_1 is prime, so it has no positive divisors other than itself and 1. Since $p_1 > 1$, we must conclude that $p_1 = q_1$. This contradicts the assumption that no p_i is the same as any q_j. So, in fact, the two expressions for n are the same, aside from rearrangement of the terms. □

The following notation and facts will occasionally be useful in reference to prime factorization.

Definition. Let n be a positive integer and let p be a prime number. Then we write $e_p(n) = t$ if p^t divides n but p^{t+1} does not divide n. We refer to $e_p(n)$ as the *exponent* of p in n.

We can define a positive integer n by specifying $e_p(n) \geq 0$ for all primes p, as long as $e_p(n) = 0$ for all but finitely many primes. The uniqueness of prime factorization implies that then

$$n = \prod_p p^{e_p(n)},$$

where the product is taken over all primes. (All but finitely many terms of this product equal 1, so this infinite product is actually finite in practice.)

Proposition A.3.3. *If m and n are positive integers, then*

$$e_p(mn) = e_p(m) + e_p(n)$$

for every prime number p.

Proof. Let $e_p(m) = s$ and $e_p(n) = t$, so that $m = p^s a$ and $n = p^t b$ for some positive integers a and b. Then $mn = p^s a \cdot p^t b = p^{s+t} ab$. If p divides ab, then p divides either a or b by Euclid's lemma. But that is impossible since p^{s+1} does not divide m and p^{t+1} does not divide n. So, we see that p^{s+t} divides mn but p^{s+t+1} does not divide mn, and conclude that $e_p(mn) = s + t = e_p(m) + e_p(n)$, as we wanted to show. □

Corollary A.3.4. *Let m and n be positive integers. Then the following statements are true:*
1. *m divides n if and only if $e_p(m) \le e_p(n)$ for every prime p.*
2. *$e_p(\gcd(m, n)) = \min(e_p(m), e_p(n))$ for every prime p.*
3. *$e_p(\text{lcm}(m, n)) = \max(e_p(m), e_p(n))$ for every prime p.*

Proof. Let m and n be positive integers.

(1) Suppose that m divides n, say that $n = mq$ for some integer q, which is clearly positive. Then $e_p(n) = e_p(mq) = e_p(m) + e_p(q)$ for every prime p by Proposition A.3.3. Since $e_p(q) \ge 0$ for all p, it follows that $e_p(n) \ge e_p(m)$ for all p. Conversely, if $e_p(m) \le e_p(n)$ for all p, then $e_p(n) - e_p(m) \ge 0$, with equality holding for all but finitely many primes p since $e_p(n) = 0$ in all but finitely many cases. So, we can define a positive integer q by stating that $e_p(q) = e_p(n) - e_p(m)$ for all p, and then $n = mq$ by Proposition A.3.3. Thus, m divides n.

(2) Let $d = \gcd(a, b)$. Since d divides a and d divides b, it follows that $e_p(d) \le e_p(a)$ and $e_p(d) \le e_p(b)$, that is, $e_p(d) \le \min(e_p(a), e_p(b))$ for all primes p. If c is defined so that $e_p(c) = \min(e_p(a), e_p(b))$ for all p, then we find that c is a common divisor of a and b, so must divide d. But then $e_p(d) \ge \min(e_p(a), e_p(b))$ for all primes p, and our result follows. The proof of statement (3) is similar, and is omitted. □

A.4 The Euclidean algorithm

As we saw in Section A.2, the greatest common divisor of a pair of integers a and b is also a combination of a and b. We now describe an efficient algorithm for calculating the greatest common divisor d of a pair of integers a and b, and for expressing d as a combination of a and b. This method is based on repeated application of the division algorithm, together with the following observations.

Proposition A.4.1. *Let a and b be integers with $a \ge b \ge 0$.*
1. *If $b > 0$ and $a = bq + r$ for some integers q and r, then $\gcd(a, b) = \gcd(b, r)$.*
2. *If $b = 0$, then $\gcd(a, b) = a$.*

Proof. For statement (1), we find that if $a = bq + r$, then

$$as + bt = (bq + r)s + bt = b(qs + t) + rt$$

and

$$bs + rt = bs + (a - bq)t = at + b(s - qt).$$

It follows that $\langle a, b \rangle = \langle b, r \rangle$, and so $\gcd(a, b) = \gcd(b, r)$. For statement (2), note that every integer divides 0, so that the common divisors of a and 0 are precisely the divisors of a. □

We will illustrate the procedure of the Euclidean algorithm with an example before describing it more formally.

Example. Let $a = 455$ and $b = 143$. If we divide a by b, we obtain the equation $455 = 143 \cdot 3 + 26$. Now part (1) of Proposition A.4.1 implies that $\gcd(455, 143) = \gcd(143, 26)$, and we can similarly divide 143 by 26 to obtain a still smaller pair of integers with the same greatest common divisor. We continue the process in the left-hand list of equations below, eventually obtaining 0 as a remainder:

$$
\begin{aligned}
455 &= 143 \cdot 3 + 26, & 26 &= 455 - 143 \cdot 3 = a - 3b, \\
143 &= 26 \cdot 5 + 13, & 13 &= 143 - 26 \cdot 5 = -5a + 16b, \\
26 &= 13 \cdot 2 + 0.
\end{aligned}
$$

Proposition A.4.1 implies in turn that

$$\gcd(455, 143) = \gcd(143, 26) = \gcd(26, 13) = \gcd(13, 0) = 13,$$

with $\gcd(a, b)$ the last nonzero remainder in the left-hand column. Now we solve each equation for its remainder in the right-hand list. By substitution, we see that each remainder can be written as a combination of a and b. (For instance, $13 = 143 - 26 \cdot 5 = b - 5(a - 3b) = -5a + 16b$.) In particular, the last nonzero remainder, $\gcd(a, b)$, has that form. In this example, we conclude that

$$\gcd(455, 143) = 13 = 455(-5) + 143(16).$$

We now describe the Euclidean algorithm in a "pseudocode" form that can be easily programmed into any computer language. For simplicity, we will assume that the inputs a and b satisfy $a \geq b \geq 0$.

Given integers $a \geq b \geq 0$:
1. Initialize sequences r_i, s_i, and t_i as follows:
   - Let $r_0 = a$ and $r_1 = b$
   - Let $s_0 = 1$ and $s_1 = 0$
   - Let $t_0 = 0$ and $t_1 = 1$
2. Let $k = 0$ and, while $r_{k+1} > 0$, do the following:
   - Let $q = r_k$ div r_{k+1}
   - Let $r_{k+2} = r_k$ mod r_{k+1}
   - Let $s_{k+2} = s_k - s_{k+1} \cdot q$
   - Let $t_{k+2} = t_k - t_{k+1} \cdot q$
   - Replace k by $k + 1$

3.   At the conclusion of this **while** loop, we have

$$\gcd(a,b) = r_k = a \cdot s_k + b \cdot t_k$$

We will not prove that this process produces the values claimed, but will illustrate the steps of the algorithm with a table in the following example.

Example. Let $a = 263$ and $b = 49$. The terms of the sequences r_i, s_i, and t_i appear below, along with the values of q that arise in the algorithm:

i	0	1	2	3	4	5	6	7	8
q			5	2	1	2	1	1	2
r	263	49	18	13	5	3	2	1	0
s	1	0	1	−2	3	−8	11	−19	49
t	0	1	−5	11	−16	43	−59	102	−263

The initial terms of the sequences appear in the columns for $i = 0$ and $i = 1$. When $k = 0$, we fill in the terms of column $k+2 = 2$ with $q = 263$ div $49 = 5$, $r = 263$ mod $49 = 18$, $s = 1 - 0 \cdot 5 = 1$, and $t = 0 - 1 \cdot 5 = -5$. We then replace k by $k + 1 = 1$ and, since $r_2 > 0$, move on to the entries in column $k + 2 = 3$. The **while** loop must terminate, since r_i is a strictly decreasing sequence of nonnegative integers, Note that with the final iteration of k, the data of interest appears in the next-to-last column:

$$\gcd(263, 49) = 1 = 263(-19) + 49(102).$$

As s_i and t_i are initialized and calculated, it is always the case that $r_i = a \cdot s_i + b \cdot t_i$. In particular, if $b = 0$, then the **while** loop does not run, and we conclude, since $k = 0$, that $\gcd(a,b) = a = a(1) + b(0)$. If $b > 0$ and b divides a, then the **while** loop runs only once, leaving $k = 1$. In that case, $\gcd(a,b) = b = a(0) + b(1)$.

In the remainder of Appendix A, we will write

$$\textbf{euclid}(a, b) = (d, s, t)$$

to mean that $\gcd(a, b) = d = as + bt$, with d, s, and t calculated using the algorithm outlined in this section.

A.5 Linear equations and congruences

The Euclidean algorithm allows us to write $d = \gcd(a, b)$ as a combination of a and b in practice. Any multiple of d can then be similarly expressed. For example, since $13 = 455(-5) + 143(16)$ from an example in Section A.4, then

$$n = 585 = 45 \cdot 13 = 45\big(455(-5) + 143(16)\big)$$
$$= 455(45 \cdot -5) + 143(45 \cdot 16) = 455(-225) + 143(720).$$

This expression is not unique, but produces all other integer solutions of the equation $ax + by = n$, as we see in the following theorem.

Theorem A.5.1. *Let a and b be integers (not both zero), and let s and t be a pair of integers for which $\gcd(a, b) = d = as + bt$. Let n be a multiple of d, say with $n = dr$ for some r in $\mathbb{Z}$. Then all integer solutions of $ax + by = n$ are given by*

$$(x, y) = \left(sr + \frac{b}{d} \cdot q, \ tr - \frac{a}{d} \cdot q \right), \tag{A.1}$$

where q is an arbitrary integer.

Note that $\frac{a}{d}$ and $\frac{b}{d}$ are integers since d is a positive common divisor of a and b.

Proof. Any pair given by equation (A.1) satisfies $ax + by = n$:

$$a\left(sr + \frac{b}{d} \cdot q \right) + b\left(tr - \frac{a}{d} \cdot q \right) = asr + \frac{abq}{d} + btr - \frac{abq}{d} = (as + bt)r = dr = n.$$

Conversely, suppose that (x, y) is an integer pair for which $ax + by = n$. Then $ax + by = n = asr + btr$ so that $a(x - sr) = b(tr - y)$. Now b divides $a(x - sr)$, so that $\frac{b}{d}$ divides $x - sr$ by Corollary A.2.4. Thus, there is an integer q so that $x - sr = \frac{b}{d} \cdot q$. Substituting this expression into $a(x - sr) = b(tr - y)$, we then see that $tr - y = \frac{a}{d} \cdot q$. Solving these equations for x and y gives us the formula of (A.1). $\qquad\square$

Example. All integer solutions of $455x + 143y = 585$ are given by

$$(x, y) = \left(-225 + \frac{143}{13} \cdot q, \ 720 - \frac{455}{13} \cdot q \right) = (-225 + 11q, 720 - 35q),$$

with q an integer. Note that this expression for solutions is not unique. For instance, if $q = 21$, we find that $(x, y) = (6, -15)$ is also a solution. We could then say that all solutions of $455x + 143y = 585$ are given by $(x, y) = (6 + 11q, -15 - 35q)$. (Now we obtain $(x, y) = (-225, 720)$ when $q = -21$, for example.)

The method used to solve a linear equation also applies to the typical *linear congruence*, traditionally written in the form $ax \equiv b \pmod{m}$. (This has the form $f(x) \equiv 0 \pmod{m}$ where $f(x) = ax - b$.)

Theorem A.5.2. *Let a, b, and m be integers, with m positive, and let $d = \gcd(a, m)$. If d divides b, then $ax \equiv b \pmod{m}$ has d distinct solutions in $\mathbb{Z}_m$, and any two solutions of $ax \equiv b \pmod{m}$ are congruent modulo $\frac{m}{d}$. If d does not divide b, then $ax \equiv b \pmod{m}$ has no solutions.*

Proof. Notice that x satisfies $ax \equiv b$ (mod m) if and only if there is some y so that (x, y) is a solution of $ax + my = b$. This equation has solutions if and only if $d = \gcd(a, m)$ divides b. If d divides b and (u, v) is a solution of $ax + my = b$ with $u \geq 0$ as small as possible, then Theorem A.5.1 implies that all solutions of $ax + my = b$ have the form $(x, y) = (u + \frac{m}{d} \cdot q, v - \frac{a}{d} \cdot q)$ where q is an integer. There are d distinct solutions for x in $\mathbb{Z}_m$, namely $x = u + \frac{m}{d} \cdot q$ for $0 \leq q < d$. $\qquad\square$

Example. Solutions of the congruence $266x \equiv 301$ (mod 413) are the same as the x-coordinates in integer solutions of $266x + 413y = 301$. If

$$\textbf{euclid}(266, 413) = (7, 14, -9)$$

from the Euclidean algorithm of Section A.4, and we note that $301 = 7 \cdot 43$, then all solutions of $266x + 413y = 301$ are given by

$$\left(43(14) + \frac{413}{7} \cdot q \,,\, 43(-9) - \frac{266}{7} \cdot q \right) = (602 + 59q \,,\, -387 - 38q),$$

with q an arbitrary integer. If we observe that $602 + 59(-10) = 12$, then all solutions of $266x \equiv 301$ (mod 413) have the form $x = 12 + 59q$. There are seven *distinct* solutions in $\mathbb{Z}_{413}$, namely 12, 71, 130, 189, 248, 307, and 366.

The preceding example illustrates that we can solve an arbitrary linear congruence systematically, using the Euclidean algorithm. Often, particularly when solving a congruence by hand, we can also adopt a trial-and-error approach, as illustrated in the following example. The congruence cancellation property (Corollary A.2.5) is our main tool here.

Example. Consider the congruence $60x \equiv 66$ (mod 111), where $a = 60$ and $b = 66$ have a common divisor of 6. We can cancel this common factor, but we must adjust the modulus $m = 111$ according to Corollary A.2.5. Since $\gcd(6, 111) = 3$, we find that $10x \equiv 11$ (mod 37). Now $\gcd(10, 11) = 1$, but we might cancel other factors of the linear coefficient by replacing 11 by an integer to which it is congruent modulo 37. For instance, $11 \equiv 48$ (mod 37), so that $2 \cdot 5x \equiv 2 \cdot 24$ (mod 37). This time, since $\gcd(2, 37) = 1$, we can cancel the common factor of 2 without affecting the modulus. So, now $5x \equiv 24$ (mod 37). With some additional searching, $24 \equiv 61 \equiv 98 \equiv 135$ (mod 37), we can replace this congruence by $5x \equiv 5 \cdot 27$ (mod 37). We conclude that $x \equiv 27$ (mod 37) since $\gcd(5, 37) = 1$. So, the solutions of our original congruence are of the form $x = 27 + 37q$ with q an integer. Note that there are three distinct solutions in $\mathbb{Z}_{111}$, as predicted by Theorem A.5.2, namely 27, 64, and 101.

If $\gcd(a, b) = d$ divides n, then $ax + by = n$ has the same solutions as $\frac{a}{d}x + \frac{b}{d}y = \frac{n}{d}$. Similarly, if $\gcd(a, m) = d$ divides b, then the congruences

$$ax \equiv b \pmod{m} \quad \text{and} \quad \frac{a}{d}x \equiv \frac{b}{d} \left(\bmod\ \frac{m}{d}\right)$$

have the same integer solutions (although we may need to present these solutions as elements of $\mathbb{Z}_m$ or of $\mathbb{Z}_{m/d}$, depending on how the question is stated). In the next section, we will restrict our attention to linear equations $ax + by = n$ with $\gcd(a, b) = 1$, or to linear congruences $ax \equiv b \pmod{m}$ with $\gcd(a, m) = 1$.

A.6 Systems of linear congruences

If $\gcd(a, m) = 1$, then $ax \equiv 1 \pmod{m}$ has a unique solution in $\mathbb{Z}_m$, which we call the *inverse* of a modulo m, and write as a^{-1} if m is apparent from context. In this case, every congruence $ax \equiv b \pmod{m}$ has a unique solution in $\mathbb{Z}_m$, namely $x = a^{-1}b$ modulo m. Note that if

$$\mathbf{euclid}(a, b) = (1, s, t),$$

that is, if $\gcd(a, b) = 1 = as + bt$, then $as \equiv 1 \pmod{m}$, so that $a^{-1} \equiv s \pmod{m}$.

In this section, we demonstrate that we can solve $ax \equiv b \pmod{m}$, when $\gcd(a, m) = 1$, by solving $ax \equiv b \pmod{d}$ for certain divisors d of m. While this method is typically not more efficient than the Euclidean algorithm approach, it suggests general results that we will see are helpful in solving quadratic congruences in the text.

Theorem A.6.1 (Chinese remainder theorem). *Let $m_1, m_2, \ldots, m_k$ be pairwise relatively prime positive integers, that is, with $\gcd(m_i, m_j) = 1$ when $i \neq j$, and let $a_1, a_2, \ldots, a_k$ be a collection of integers, not necessarily distinct. Then there is a unique integer x modulo $m = m_1 m_2 \cdots m_k$ that simultaneously satisfies*

$$
\begin{aligned}
x &\equiv a_1 \pmod{m_1} \\
x &\equiv a_2 \pmod{m_2} \\
&\ \ \vdots \\
x &\equiv a_k \pmod{m_k}.
\end{aligned}
\tag{A.2}
$$

Proof. If x and y both satisfy all congruences in (A.2), then $x - y$ is divisible by m_1, $m_2, \ldots, m_k$. It follows that $x - y$ must be divisible by the least common multiple of these moduli. With those values *pairwise* relatively prime, it is not difficult to see that this least common multiple is $m = m_1 m_2 \cdots m_k$. Hence, if a solution exists, it is unique modulo m. We show that a solution exists by constructing an example.

For $i = 1, 2, \ldots, k$, let $n_i = \frac{m}{m_i} = m_1 \cdots m_{i-1} m_{i+1} \cdots m_k$, the product of each of the moduli excluding m_i. Since m_j is relatively prime to m_i if $i \neq j$, we see that n_i is relatively prime to m_i. Thus, n_i has an inverse modulo m_i, which we denote by n_i^{-1}. Now consider the integer

$$x = a_1 n_1 n_1^{-1} + a_2 n_2 n_2^{-1} + \cdots + a_k n_k n_k^{-1}. \tag{A.3}$$

Here, $n_i n_i^{-1} \equiv 1 \pmod{m_i}$ for all i, but $n_i n_i^{-1} \equiv 0 \pmod{m_j}$ if $i \neq j$ since in that case m_j is a factor of n_i. Thus, we find that

$$x \equiv a_1 \cdot 1 + a_2 \cdot 0 + \cdots + a_k \cdot 0 \equiv a_1 \pmod{m_1},$$

and in a similar way, that x satisfies each congruence in (A.2). □

Example. Consider the linear congruence $ax \equiv b \pmod{m}$ where $a = 331$, $b = 857$, and $m = 1071 = 3^2 \cdot 7 \cdot 17$. Any solution of this congruence must also satisfy $ax \equiv b \pmod{d}$ if d divides m. Consider $ax \equiv b \pmod{m_i}$ for $m_1 = 9$, $m_2 = 7$, and $m_3 = 17$. (We select these moduli because they are pairwise relatively prime and their product equals m.) Each resulting congruence simplifies greatly and can be solved relatively easily.
1. $ax \equiv b \pmod 9$ simplifies to $7x \equiv 2 \pmod 9$, with solution $x = 8$.
2. $ax \equiv b \pmod 7$ simplifies to $2x \equiv 3 \pmod 7$, with solution $x = 5$.
3. $ax \equiv b \pmod{17}$ simplifies to $8x \equiv 7 \pmod{17}$, with solution $x = 3$.

The solution of $ax \equiv b \pmod{1071}$ must simultaneously satisfy

$$\begin{aligned} x &\equiv r \pmod 9 \\ x &\equiv s \pmod 7 \\ x &\equiv t \pmod{17} \end{aligned} \tag{A.4}$$

with $r = 8$, $s = 5$, and $t = 3$. The Chinese remainder theorem shows that such an x must exist, and is unique modulo $m = 9 \cdot 7 \cdot 17 = 1071$. To illustrate the formula in the proof of Theorem A.6.1, we find a general solution of system (A.4) in terms of r, s, and t. If $n_1 = 7 \cdot 17 = 119$, then n_1^{-1} is the solution of $119x \equiv 1 \pmod 9$, which we find to be $n_1^{-1} = 5$. Likewise, for $n_2 = 9 \cdot 17 = 153$, we find that $153x \equiv 1 \pmod 7$ has solution $n_2^{-1} = 6$. For $n_3 = 9 \cdot 7 = 63$, we find that $63x \equiv 1 \pmod{17}$ has solution $n_3^{-1} = 10$. Thus, the general solution of system (A.4) is

$$x \equiv r \cdot 119 \cdot 5 + s \cdot 153 \cdot 6 + t \cdot 63 \cdot 10 \equiv 595r + 918s + 630t \pmod{1071}.$$

When $r = 8$, $s = 5$, and $t = 3$, we obtain $x \equiv 11240 \equiv 530 \pmod{1071}$. So, $x = 530$ is the unique solution of $331x \equiv 857 \pmod{1071}$.

We note a special case of the Chinese remainder theorem, in which the formula in equation (A.3) arises from the Euclidean algorithm. We can often invoke this approach, even for systems of more than two congruences, by solving two of them at a time.

Corollary A.6.2. *Let m and n be positive integers with $\gcd(m, n) = 1$, and suppose that $1 = ms + nt$ for some integers s and t. Then for all integers a and b, the unique solution of the pair of congruences*

$$x \equiv a \pmod{m} \quad and \quad x \equiv b \pmod{n}$$

is given by

$$x \equiv ant + bms \pmod{mn}.$$

Proof. The system has a unique solution modulo mn by Theorem A.6.1. The equation $1 = ms + nt$ implies that $ms \equiv 1 \pmod{n}$ and $nt \equiv 1 \pmod{m}$. But then we have

$$ant + bms \equiv a(1) + (0)bs \equiv a \pmod{m}$$

and

$$ant + bms \equiv (0)at + b(1) \equiv b \pmod{n}.$$

Thus, $x = ant + bms$ satisfies both congruences in the system and must be the unique solution modulo mn. □

For later use, we will write

$$\mathbf{CRT}(m, n, a, b) = x$$

for the unique solution of $x \equiv a \pmod{m}$ and $x \equiv b \pmod{n}$ modulo mn when m and n are relatively prime. We compute this value as follows:
1. Calculate **euclid**$(m, n) = (d, s, t)$ and verify that $d = 1$
2. Return $x = (ant + bms) \bmod mn$

When $\gcd(m, n) > 1$, the system $x \equiv a \pmod{m}$ and $x \equiv b \pmod{n}$ may or may not have solutions. We will not need to pursue that case in this text.

A.7 Quadratic congruences modulo primes

We now turn our attention to the next general case of polynomial congruences, $f(x) \equiv 0 \pmod{m}$ where $f(x)$ is a quadratic polynomial. In this appendix, we will restrict our attention to the case where $m = p$ is a prime number, giving a complete description of the *number* of solutions of a quadratic congruence modulo a prime number.

Let $f(x) = ax^2 + bx + c$ with a, b, and c integers, and define the *discriminant* of $f(x)$ to be

$$\Delta = \Delta(f) = b^2 - 4ac.$$

Let p be a prime number and consider the congruence $ax^2 + bx + c \equiv 0 \pmod{p}$. We will assume that p does not divide a, since otherwise the quadratic congruence reduces to a linear congruence. It is also convenient to treat $p = 2$ separately, in Proposition A.7.1, whose proof is left as an exercise.

Proposition A.7.1. *Let $f(x) = ax^2 + bx + c$ with a odd.*
1. *If b is odd and c is even, then $f(x) \equiv 0 \pmod{2}$ has two solutions in $\mathbb{Z}_2$, namely $x = 0$ and $x = 1$.*
2. *If b and c are odd, then $f(x) \equiv 0 \pmod{2}$ has no solutions in $\mathbb{Z}_2$.*
3. *If b is even, then $f(x) \equiv 0 \pmod{2}$ has one solution in $\mathbb{Z}_2$, namely $x = 0$ if c is even, and $x = 1$ if c is odd.*

If $f(x) = ax^2 + bx + c$ with a odd, then we can rephrase Proposition A.7.1 as follows:
1. If $\Delta \equiv 1 \pmod{8}$, then $f(x) \equiv 0 \pmod{2}$ has two solutions in $\mathbb{Z}_2$.
2. If $\Delta \equiv 5 \pmod{8}$, then $f(x) \equiv 0 \pmod{2}$ has no solutions in $\mathbb{Z}_2$.
3. If $\Delta \equiv 0 \pmod{4}$, then $f(x) \equiv 0 \pmod{2}$ has exactly one solution in $\mathbb{Z}_2$.

When p is an odd prime, the discriminant of $f(x)$ likewise determines the number of solutions of $f(x) \equiv 0 \pmod{p}$. In that case, a quadratic congruence $ax^2 + bx + c \equiv 0 \pmod{p}$ can be transformed into a specialized form by *completing the square*. Multiplying both sides of the congruence by $4a$ and adding $b^2 - 4ac$ to both sides yields

$$4a^2x^2 + 4abx + b^2 = (2ax + b)^2 \equiv b^2 - 4ac \pmod{p}. \tag{A.5}$$

Notice that $\gcd(4a, p) = 1$ if p is odd and does not divide a. Thus, we can likewise cancel $4a$ from both sides of the resulting congruence by the congruence cancellation property, and no unwanted solutions of the original congruence are introduced. The situation is now analogous to that of solving a quadratic equation over the real numbers. In that case, the question of whether solutions exist is determined purely by whether the discriminant $\Delta(f) = b^2 - 4ac$ is greater than or equal to zero. If so, we can take square roots of both sides and solve for x, obtaining the familiar *quadratic formula*:

$$x = \frac{-b \pm \sqrt{b^2 - 4ac}}{2a}.$$

In our situation, we must determine whether Δ has a "square root modulo p" and then solve a linear congruence for x. We illustrate with an example.

Example. Find all solutions of $x^2 - 3x + 3 \equiv 0 \pmod{13}$.

Here, $a = 1$, $b = -3$, and $c = 3$, so that $\Delta = b^2 - 4ac = -3$, and our congruence can be replaced by $(2x - 3)^2 \equiv -3 \pmod{13}$, as in (A.5). There is no integer whose square is -3, but we can replace -3 by any integer to which it is congruent modulo 13. By trial-and-error, we find that $-3 \equiv 10 \equiv 23 \equiv 36 \pmod{13}$, and recognize that $36 = (\pm 6)^2$. So, two solutions of the original congruence are given by solving $2x - 3 \equiv 6 \pmod{13}$ and $2x - 3 \equiv -6 \pmod{13}$ for $x = 11$ and $x = 5$, respectively. (We will see in Proposition A.7.2 that these must be the only possibilities.)

This example illustrates that the process of solving a quadratic congruence modulo an odd prime is fairly mechanical except for the step of determining whether $b^2 - 4ac$ is congruent to a square. So now, changing our notation, we will concentrate on quadratic congruences of the form $x^2 \equiv a$ (mod p) where p is an odd prime number and a is an integer.

For a fixed prime p, we can solve a congruence of the form $x^2 \equiv a$ (mod p) by calculating x^2 mod p for $0 \le x < p$, as in the table below for $p = 13$:

x	0	1	2	3	4	5	6	7	8	9	10	11	12
x^2	0	1	4	9	16	25	36	49	64	81	100	121	144
x^2 mod 13	0	1	4	9	3	12	10	10	12	3	9	4	1

We see, for instance, that $x^2 \equiv 10$ (mod 13) has two solutions, $x = 6$ and $x = 7$. (Any number congruent to one of these modulo 13, such as 19, is also a solution, but as before we take the number of solutions of $x^2 \equiv 10$ (mod 13) to mean the number of solutions in $\mathbb{Z}_{13}$.) We also have that $x^2 \equiv -3$ (mod 13) has the same two solutions, since $-3 \equiv 10$ (mod 13). On the other hand, $x^2 \equiv 11$ (mod 13) has no solutions since 11 does not appear in the last row of the table.

Proposition A.7.2. *Let p be an odd prime and let a be an integer. If p does not divide a, then $x^2 \equiv a$ (mod p) has either two solutions or no solutions. If p divides a, then $x^2 \equiv a$ (mod p) has exactly one solution, namely $x = 0$.*

Proof. We have seen by example that $x^2 \equiv a$ (mod p) might have no solutions. So, suppose that $x^2 \equiv a$ (mod p) has a solution, say $x = b$. Then $-b$ is also a solution since $(-b)^2 = b^2$. If c is yet another solution, then $c^2 \equiv a \equiv b^2$ (mod p), implying that p divides $c^2 - b^2 = (c - b)(c + b)$. Since p is prime, it follows that p divides $c - b$ or p divides $c + b$, that is, either $c \equiv b$ (mod p) or $c \equiv -b$ (mod p). So, c must be the same solution as b or $-b$ and, therefore, $x^2 \equiv a$ (mod p) has a maximum of two distinct solutions.

It is possible, in the same way, that $-b$ is simply another name for the solution b, but that would imply that $b \equiv -b$ (mod p), so that p divides $b - (-b) = 2b$. Since $p \ne 2$ is prime, this occurs if and only if p divides b, so that $b \equiv 0$ (mod p). But then $b^2 \equiv a \equiv 0$ (mod p). So, $x^2 \equiv a$ (mod p) has exactly one solution if $a \equiv 0$ (mod p), but must have either two distinct solutions or no solutions if $a \ne 0$ (mod p). □

The following notation summarizes these possibilities.

Definition. Let p be an odd prime and a any integer. Then the *Legendre symbol*, written as $(\frac{a}{p})$, is defined as

$$
\left(\frac{a}{p}\right) =
\begin{cases}
1, & \text{if } x^2 \equiv a \pmod p \text{ has two solutions,} \\
0, & \text{if } x^2 \equiv a \pmod p \text{ has only one solution,} \\
-1, & \text{if } x^2 \equiv a \pmod p \text{ has no solutions.}
\end{cases}
$$

If $\left(\frac{a}{p}\right) = 1$, then we say that a is a *quadratic residue* modulo p.

Example. Referring to the table above, we have for instance that $\left(\frac{10}{13}\right) = 1$ and $\left(\frac{11}{13}\right) = -1$.

We conclude this section with a list of rules that allow us to calculate a Legendre symbol $\left(\frac{a}{p}\right)$ without directly solving the quadratic congruence, $x^2 \equiv a \pmod p$, to which it refers. Doing so does not help solve $x^2 \equiv a \pmod p$, when solutions exist. In practice, it may be necessary to test whether $x^2 \equiv a \pmod p$ for $0 \le x \le \frac{p-1}{2}$, as in the proof of Proposition A.7.2. But we can know beforehand whether a search for solutions will be successful. Aside from the first three statements, whose explanations are straightforward, we will postpone the proofs of these claims until Section A.8.

Facts. Let p be an odd prime and let a and b be integers.
1. $\left(\frac{a}{p}\right) = 0$ if and only if p divides a. (This is part of Proposition A.7.2.)
2. $\left(\frac{1}{p}\right) = 1$. (The congruence $x^2 \equiv 1 \pmod p$ has solutions 1 and –1.)
3. If $a \equiv b \pmod p$, then $\left(\frac{a}{p}\right) = \left(\frac{b}{p}\right)$. (In that case, $x^2 \equiv a \pmod p$ and $x^2 \equiv b \pmod p$ then have the same solutions, and so the same number of solutions.)
4. $a^{\frac{p-1}{2}} \equiv \left(\frac{a}{p}\right) \pmod p$. (This is called *Euler's criterion.*)
5. $\left(\frac{ab}{p}\right) = \left(\frac{a}{p}\right)\left(\frac{b}{p}\right)$.
6.

$$
\left(\frac{-1}{p}\right) = (-1)^{\frac{p-1}{2}} =
\begin{cases}
1, & \text{if } p \equiv 1 \pmod 4 \\
-1, & \text{if } p \equiv 3 \pmod 4.
\end{cases}
$$

7.

$$
\left(\frac{2}{p}\right) = (-1)^{\frac{p^2-1}{8}} =
\begin{cases}
1, & \text{if } p \equiv 1 \pmod 8 \text{ or } p \equiv 7 \pmod 8 \\
-1, & \text{if } p \equiv 3 \pmod 8 \text{ or } p \equiv 5 \pmod 8.
\end{cases}
$$

8. If q is an odd prime and $q \ne p$, then

$$
\left(\frac{q}{p}\right) =
\begin{cases}
\left(\frac{p}{q}\right), & \text{if } p \equiv 1 \pmod 4 \text{ or } q \equiv 1 \pmod 4 \\
-\left(\frac{p}{q}\right), & \text{if } p \equiv 3 \pmod 4 \text{ and } q \equiv 3 \pmod 4.
\end{cases}
$$

The final claim is known as the *quadratic reciprocity theorem.*

Example. Consider the congruence $x^2 \equiv 30 \pmod{59}$. By property (5), we see that

$$
\left(\frac{30}{59}\right) = \left(\frac{2}{59}\right)\left(\frac{3}{59}\right)\left(\frac{5}{59}\right).
$$

So, we look at these symbols separately.

Since $59 \equiv 3 \pmod 8$, property (7) implies that $(\frac{2}{59}) = -1$.

Both 3 and 59 are congruent to 3 modulo 4, so property (8) implies that $(\frac{3}{59}) = -(\frac{59}{3})$. But now we can simplify the resulting symbol, working modulo 3. Since $59 \equiv 2 \pmod 3$, we have that $-(\frac{59}{3}) = -(\frac{2}{3})$. By property (7), $-(\frac{2}{3}) = -(-1) = 1$. So $(\frac{3}{59}) = 1$.

Finally, since $5 \equiv 1 \pmod 4$, property (8) implies that $(\frac{5}{59}) = (\frac{59}{5}) = (\frac{4}{5})$. Now 4 is clearly a square modulo 5, or we could note that $(\frac{4}{5}) = (\frac{2 \cdot 2}{5}) = (\frac{2}{5})^2$ by property (5). Whether $(\frac{2}{5})$ is 1 or −1 (it is −1 by property (7)), its square is 1. So, $(\frac{5}{59}) = 1$.

Therefore,

$$\left(\frac{30}{59}\right) = \left(\frac{2}{59}\right)\left(\frac{3}{59}\right)\left(\frac{5}{59}\right) = -1 \cdot 1 \cdot 1 = -1.$$

By definition, this means that $x^2 \equiv 30 \pmod{59}$ has no solutions.

This example illustrates how we can use a process of repeated simplification to calculate a particular Legendre symbol. The quadratic reciprocity theorem also allows us to answer more general questions about a particular congruence modulo *all* primes, as in the following example.

Example. For what primes p does the congruence $x^2 \equiv 3 \pmod p$ have a solution?

Direct calculation shows that a solution exists when $p = 2$ or $p = 3$, so assume that $p > 3$. By definition, $x^2 \equiv 3 \pmod p$ has a solution if and only if $(\frac{3}{p}) = 1$. Now p is congruent to 1 or 3 modulo 4, and since $3 \equiv 3 \pmod 4$, quadratic reciprocity allows us to say that

$$\left(\frac{3}{p}\right) = \begin{cases} (\frac{p}{3}), & \text{if } p \equiv 1 \pmod 4 \\ -(\frac{p}{3}), & \text{if } p \equiv 3 \pmod 4. \end{cases}$$

For $p \neq 3$, either $p \equiv 1 \pmod 3$ or $p \equiv 2 \pmod 3$, and we find that $(\frac{1}{3}) = 1$ and $(\frac{2}{3}) = -1$. (We simply need to note that $1^2 \equiv 1 \pmod 3$ and $2^2 \equiv 1 \pmod 3$, or we could use property (7).) Combining this calculation with the previous formula for $(\frac{3}{p})$, we conclude that

$$\left(\frac{3}{p}\right) = \begin{cases} 1, & \text{if } p \equiv 1 \pmod{12} \text{ or } p \equiv 11 \pmod{12} \\ -1, & \text{if } p \equiv 5 \pmod{12} \text{ or } p \equiv 7 \pmod{12}. \end{cases}$$

Thus, $x^2 \equiv 3 \pmod p$ has a solution if and only if $p = 2$, $p = 3$, $p \equiv 1 \pmod{12}$, or $p \equiv 11 \pmod{12}$.

A.8 Quadratic reciprocity theorem

In this section, we prove the quadratic reciprocity theorem and other facts about Legendre symbols listed in Section A.7. The proof presented requires Euler's criterion and Wilson's theorem, which we establish first. (Our proof of Euler's criterion, with Wilson's theorem as a corollary, is due to Dirichlet [4] and the proof of quadratic reciprocity is adapted from Sey Y. Kim [5]. Both proofs are based on calculating products of elements in a particular set in different ways.)

Theorem A.8.1 (Euler's criterion). *If p is an odd prime number and a is an integer, then*

$$a^{\frac{p-1}{2}} \equiv \left(\frac{a}{p}\right) \pmod{p}.$$

Proof. If p divides a, then $a^{\frac{p-1}{2}} \equiv 0^{\frac{p-1}{2}} \equiv 0 \pmod{p}$ and $\left(\frac{a}{p}\right) = 0$, so assume instead that p does not divide a. Consider the set $S = \{1, 2, \ldots, p-1\}$. For each q in S, the linear congruence $qx \equiv a \pmod{p}$ has a unique solution r in S. Note that $r = q$ if and only if $q^2 \equiv a \pmod{p}$. We consider two cases for a.

Suppose that $\left(\frac{a}{p}\right) = -1$, which by definition means that $x^2 \equiv a \pmod{p}$ has no solutions. In this case, the set S consists of $\frac{p-1}{2}$ pairs of *distinct* elements q and r for which $qr \equiv a \pmod{p}$. Grouping these pairs of elements together in the product of the elements of S, it follows that

$$(p-1)! \equiv a^{\frac{p-1}{2}} \pmod{p}, \quad \text{if } \left(\frac{a}{p}\right) = -1. \tag{A.6}$$

Suppose now that $\left(\frac{a}{p}\right) = 1$, so that $x^2 \equiv a \pmod{p}$ has two solutions. If b is one of these solutions, then the second is $p - b \equiv -b \pmod{p}$. Now S consists of b, $p - b$, and $\frac{p-3}{2}$ pairs of distinct elements q and r for which $qr \equiv a \pmod{p}$. Note that $b(p-b) \equiv b(-b) \equiv -b^2 \equiv -a \pmod{p}$, since $b^2 \equiv a \pmod{p}$. Thus, we find that

$$(p-1)! \equiv -a \cdot a^{\frac{p-3}{2}} \equiv -a^{\frac{p-1}{2}} \pmod{p}, \quad \text{if } \left(\frac{a}{p}\right) = 1. \tag{A.7}$$

Congruence (A.7) applies when $a = 1$, and it follows that

$$(p-1)! \equiv -\left(1^{\frac{p-1}{2}}\right) \equiv -1 \pmod{p}, \quad \text{for odd primes } p. \tag{A.8}$$

Now congruences (A.6) and (A.7) can be combined as

$$a^{\frac{p-1}{2}} \equiv -\left(\frac{a}{p}\right) \cdot (p-1)! \equiv \left(\frac{a}{p}\right) \pmod{p},$$

completing the proof of Euler's criterion. $\qquad\square$

We will refer to the congruence established in (A.8) as *Wilson's theorem*. (This theorem states more precisely that $(p - 1)! \equiv -1 \pmod{p}$ if and only if p is prime.) We can now prove the following corollaries of Euler's criterion, assumed in Section A.7.

Corollary A.8.2. *If p is an odd prime and a and b are integers, then*

$$\left(\frac{ab}{p}\right) = \left(\frac{a}{p}\right)\left(\frac{b}{p}\right).$$

Proof. Euler's criterion and properties of exponents show that

$$\left(\frac{ab}{p}\right) \equiv (ab)^{\frac{p-1}{2}} \equiv a^{\frac{p-1}{2}} \cdot b^{\frac{p-1}{2}} \equiv \left(\frac{a}{p}\right)\left(\frac{b}{p}\right) \pmod{p}.$$

Since Legendre symbols and their products are equal to 1, −1, or 0, and since $p > 2$, the congruence $\left(\frac{ab}{b}\right) \equiv \left(\frac{a}{p}\right)\left(\frac{b}{p}\right) \pmod{p}$ implies the equation $\left(\frac{ab}{p}\right) = \left(\frac{a}{p}\right)\left(\frac{b}{p}\right)$. □

Corollary A.8.3. *If p is an odd prime, then*

$$\left(\frac{-1}{p}\right) = (-1)^{\frac{p-1}{2}} = \begin{cases} 1, & \text{if } p \equiv 1 \pmod{4}, \\ -1, & \text{if } p \equiv 3 \pmod{4}. \end{cases}$$

Proof. Euler's criterion implies that $\left(\frac{-1}{p}\right) \equiv (-1)^{\frac{p-1}{2}} \pmod{p}$. We can again replace this congruence by an equation since both sides are 1 or −1. The second equation is simply the observation that $\frac{p-1}{2}$ is even if $p \equiv 1 \pmod{4}$ and is odd if $p \equiv 3 \pmod{4}$. □

Euler's criterion provides a similar formula for $\left(\frac{2}{p}\right)$, although here the outcome is less immediate.

Corollary A.8.4. *If p is an odd prime, then*

$$\left(\frac{2}{p}\right) = (-1)^{\frac{p^2-1}{8}} = \begin{cases} 1, & \text{if } p \equiv 1 \pmod{8} \text{ or } p \equiv 7 \pmod{8}, \\ -1, & \text{if } p \equiv 3 \pmod{4} \text{ or } p \equiv 5 \pmod{8}. \end{cases}$$

Proof. Let p be an odd prime and consider the following two sets:

$$S = \left\{2n \mid 1 \leq n \leq \frac{p-1}{2}\right\} = \{2, 4, \ldots, p - 1\},$$

containing all even positive integers less than p, and

$$T = \left\{(-1)^n n \mid 1 \leq n \leq \frac{p-1}{2}\right\} = \left\{-1, 2, -3, \ldots, \pm\frac{p-1}{2}\right\},$$

in which terms alternative between negative odd integers and positive even integers. Both sets contain $\frac{p-1}{2}$ elements, and we see as follows that the elements of S are congruent modulo p to the elements of T in some order. If $0 < 2n < \frac{p}{2}$, then $2n$ is in both S and

T, whereas if $\frac{p}{2} < 2n < p$, so that $-\frac{p}{2} < 2n - p < 0$, then $2n$ is in S and $2n - p$ is in T, with $2n \equiv 2n - p \pmod{p}$.

It follows that the products of the elements of S and of T are congruent to each other modulo p. The product of the elements of S is

$$(2 \cdot 1) \cdot (2 \cdot 2) \cdots \left(2 \cdot \frac{p-1}{2}\right) = 2^{\frac{p-1}{2}} \cdot \left(\frac{p-1}{2}\right)! \equiv \left(\frac{2}{p}\right)\left(\frac{p-1}{2}\right)! \pmod{p},$$

with the final congruence due to Euler's criterion. On the other hand, if m is the number of negative elements in T, then the product of the elements of T is

$$(-1)(2)(-3) \cdots \left(\pm \frac{p-1}{2}\right) = (-1)^m \left(1 \cdot 2 \cdot 3 \cdots \frac{p-1}{2}\right) = (-1)^m \left(\frac{p-1}{2}\right)!.$$

Thus, $\left(\frac{2}{p}\right)\left(\frac{p-1}{2}\right)! \equiv (-1)^m \left(\frac{p-1}{2}\right)! \pmod{p}$, implying that $\left(\frac{2}{p}\right) = (-1)^m$ since $\left(\frac{p-1}{2}\right)!$ is relatively prime to p.

The number of negative elements in T is the number of odd integers, $2k-1$, satisfying $1 \le 2k-1 \le \frac{p-1}{2}$. This is the number of integers k with $1 \le k \le \frac{p+1}{4}$, which we can write as $m = \lfloor \frac{p+1}{4} \rfloor$, the largest integer less than or equal to $\frac{p+1}{4}$. If we write $p = 8q + r$ with $r = 1$, 3, 5, or 7, then $m = \lfloor \frac{p+1}{4} \rfloor = 2q + \lfloor \frac{r+1}{4} \rfloor$. We then see that if $r = 1$, then $m = 2q$ is even; if $r = 3$ or $r = 5$, then $m = 2q + 1$ is odd; and if $r = 7$, then $m = 2q + 2$ is even. The formula for $\left(\frac{2}{p}\right)$ follows. $\square$

We now turn to the proof of quadratic reciprocity theorem. Throughout the remainder of this section, let p and q be distinct odd primes. The Chinese remainder theorem shows that $x^2 \equiv 1 \pmod{pq}$ has four solutions, obtained by solving all possible pairs of congruences $x \equiv \pm 1 \pmod{p}$ and $x \equiv \pm 1 \pmod{q}$. We can write these solutions as $1, -1$, b, and $-b$ for some b. On the other hand, using Corollary A.8.3, $x^2 \equiv -1 \pmod{pq}$ has no solutions if $p \equiv 3 \pmod{4}$ or $q \equiv 3 \pmod{4}$, but has four solutions, say $c, d, -c$, and $-d$, if $p \equiv 1 \pmod{4}$ and $q \equiv 1 \pmod{4}$. Here, $(cd)^2 = c^2 d^2 \equiv (-1)(-1) \equiv 1 \pmod{pq}$, so that cd is congruent to one of the four solutions of $x^2 \equiv 1 \pmod{pq}$. In fact, $cd \equiv \pm b \pmod{pq}$, since cd is not congruent to either $c(c) \equiv -1 \pmod{pq}$ or $c(-c) \equiv 1 \pmod{pq}$. Notice that then $bcd \equiv b(\pm b) \equiv \pm b^2 \equiv \pm 1 \pmod{pq}$.

Now consider the following subsets of $U = \{1, 2, \ldots, \frac{pq-1}{2}\}$:

$$P = \{a \in U \mid \gcd(a, pq) = p\}, \quad Q = \{a \in U \mid \gcd(a, pq) = q\},$$

and

$$R = \{a \in U \mid \gcd(a, pq) = 1\}.$$

Since $\gcd(a, pq)$ is a divisor of pq, but there is no multiple of pq in U, then every element of U is in one and only one of the sets P, Q, or R. From the fact that

$$\frac{pq-1}{2} = \frac{q-1}{2} \cdot p + \frac{p-1}{2} = \frac{p-1}{2} \cdot q + \frac{q-1}{2},$$

we find that P and Q consist of the first $\frac{q-1}{2}$ positive multiples of p and the first $\frac{p-1}{2}$ positive multiples of q, respectively. For every a with $\gcd(a, pq) = 1$, either a or $-a$, but not both, is congruent to an element of R modulo pq. We can assume, in particular, that R contains b and, when $p \equiv 1 \pmod 4$ and $q \equiv 1 \pmod 4$, the elements c and d as defined in the preceding paragraph. To establish the quadratic reciprocity theorem, we calculate the product of the elements of R modulo p, modulo q, and modulo pq via the following lemma.

Lemma A.8.5. *Let p and q be distinct odd primes and let r be the product of the elements in $R = \{a \in U \mid \gcd(a, pq) = 1\}$. Then the following statements are true:*
1. $r \equiv (\frac{-1}{q})(\frac{q}{p}) \pmod p$,
2. $r \equiv (\frac{-1}{p})(\frac{p}{q}) \pmod q$,
3. $r \equiv 1$ *or* $-1 \pmod{pq}$ *if and only if* $p \equiv 1 \pmod 4$ *and* $q \equiv 1 \pmod 4$.

Proof. For statement (1), let s be the product of the elements in the set $S = Q \cup R$. Since R and $Q = \{q, 2q, \ldots, \frac{p-1}{2}q\}$ are disjoint, then

$$s = r \cdot q^{\frac{p-1}{2}}\left(\frac{p-1}{2}\right)! \equiv r \cdot \left(\frac{q}{p}\right)\left(\frac{p-1}{2}\right)! \pmod p, \qquad (A.9)$$

using Euler's criterion for the congruence. On the other hand, since $S = U - P$, we find that S contains $\frac{q-1}{2}$ copies of the set $\{1, 2, \ldots, p-1\}$ and an additional copy of $\{1, 2, \ldots, \frac{p-1}{2}\}$ when its elements are reduced modulo p. Thus,

$$s \equiv ((p-1)!)^{\frac{q-1}{2}} \cdot \left(\frac{p-1}{2}\right)! \equiv (-1)^{\frac{q-1}{2}} \cdot \left(\frac{p-1}{2}\right)! \equiv \left(\frac{-1}{q}\right)\left(\frac{p-1}{2}\right)! \pmod p, \qquad (A.10)$$

here using Wilson's theorem and Euler's criterion. Combining (A.9) and (A.10) and using the fact that $(\frac{p-1}{2})!$ is relatively prime to p, then $r \equiv (\frac{-1}{q})(\frac{q}{p}) \pmod p$.

For statement (2), let t be the product of the elements of $T = P \cup R = U - Q$. The argument for statement (1) applies in precisely the same way with p and q interchanged. Thus, we likewise conclude that $r \equiv (\frac{-1}{p})(\frac{p}{q}) \pmod q$.

If a is in R, then there is a unique a' in R so that either $aa' \equiv 1 \pmod{pq}$ or $aa' \equiv -1 \pmod{pq}$. (If a' satisfies the first congruence, then $-a'$ satisfies the second, and vice versa. But exactly one of a' and $-a'$ is congruent modulo pq to an element of R.) For all p and q, we find that $a' = a$ for 1 and b, two of the solutions of $x^2 \equiv 1 \pmod{pq}$. When $p \equiv 1 \pmod 4$ and $q \equiv 1 \pmod 4$, we also have $a' = a$ for c and d, two of the solutions of $x^2 \equiv -1 \pmod{pq}$. Thus, if either $p \equiv 3 \pmod 4$ or $q \equiv 3 \pmod 4$, then R consists of 1, b, and pairs of distinct elements whose product modulo pq is ± 1. In that case, $r \equiv 1 \cdot b \cdot (\pm 1) \equiv \pm b \pmod{pq}$. However, if $p \equiv 1 \pmod 4$ and $q \equiv 1 \pmod 4$, then R contains 1, b, c, d, and pairs of distinct elements whose product is ± 1. In this case,

$r \equiv 1 \cdot b \cdot c \cdot d \cdot (\pm 1) \equiv \pm 1 \pmod{pq}$, using the previous observation that $bcd \equiv \pm 1 \pmod{pq}$. This establishes statement (3). □

We can now prove our main result in the following form.

Theorem A.8.6 (Quadratic reciprocity theorem). *If p and q are distinct odd primes, then*

$$\left(\frac{p}{q}\right)\left(\frac{q}{p}\right) = \begin{cases} 1, & \text{if } p \equiv 1 \pmod{4} \text{ or } q \equiv 1 \pmod{4}, \\ -1, & \text{if } p \equiv 3 \pmod{4} \text{ and } q \equiv 3 \pmod{4}. \end{cases}$$

Proof. Lemma A.8.5 implies that $(\frac{-1}{q})(\frac{q}{p})$ and $(\frac{-1}{p})(\frac{p}{q})$ are equal, so that $r \equiv \pm 1 \pmod{pq}$, if and only if $p \equiv 1 \pmod 4$ and $q \equiv 1 \pmod 4$. Put another way,

$$\left(\frac{p}{q}\right)\left(\frac{q}{p}\right) = \begin{cases} (\frac{-1}{p})(\frac{-1}{q}), & \text{if } p \equiv 1 \pmod 4 \text{ and } q \equiv 1 \pmod 4, \\ -(\frac{-1}{p})(\frac{-1}{q}), & \text{if } p \equiv 3 \pmod 4 \text{ or } q \equiv 3 \pmod 4. \end{cases}$$

Therefore, by Corollary A.8.3,

$$\left(\frac{p}{q}\right)\left(\frac{q}{p}\right) = \begin{cases} 1 \cdot 1 = 1, & \text{if } p \equiv 1 \pmod 4 \text{ and } q \equiv 1 \pmod 4, \\ -(1 \cdot -1) = 1, & \text{if } p \equiv 1 \pmod 4 \text{ and } q \equiv 3 \pmod 4, \\ -(-1 \cdot 1) = 1, & \text{if } p \equiv 3 \pmod 4 \text{ and } q \equiv 1 \pmod 4, \\ -(-1 \cdot -1) = -1, & \text{if } p \equiv 3 \pmod 4 \text{ and } q \equiv 3 \pmod 4, \end{cases}$$

which is what we wanted to show. □

B Appendix: Review of groups

Groups arise in various ways in the text, including groups of unimodular matrices and groups of classes or congruence classes of ideal numbers. In this appendix, we review some terminology and results concerning groups, with the goal of proving the fundamental theorem of finite Abelian groups, which we use in the text to describe class groups.

B.1 Groups and subgroups

A binary operation $*$ on a set G is a function from $G \times G$ to G, that is, a rule that assigns to each ordered pair of elements a and b in G one and only one element, written as $a * b$, in the set G. (To emphasize that $a * b$ must be an element of G, we say that G is *closed* under the $*$ operation or that $*$ has the *closure property*.) Operations that satisfy certain algebraic properties are of particular importance.

Definition. Let G be a set on which an operation $*$ is defined. We say that G is a *group* under $*$ if $*$ has the following properties:
1. *Associative property*: $a * (b * c) = (a * b) * c$ for every a, b, c in G.
2. *Identity property*: There is an element e in G so that $a * e = a = e * a$ for every a in G.
3. *Inverse property*: For every a in G, there is some b in G so that $a * b = e = b * a$.

We say that G is an *Abelian group* under $*$ if it also has the *commutative property*: $a * b = b * a$ for all a and b in G.

The following proposition lists some additional properties that hold in every group. The first pair of these claims are known as the *cancellation properties* for a group. We leave the proof of these claims as an exercise.

Proposition B.1.1. *If G is a group under $*$, then the following are true:*
1. *If $a * b = a * c$ or if $b * a = c * a$ for some a, b, and c in G, then $b = c$.*
2. *The identity element of G is unique.*
3. *For all a in G, the inverse of a is unique.*

When making general statements about groups, we will typically write the operation of G as multiplication, (i. e., $a \cdot b$, or simply ab, rather than $a * b$), the identity element as 1, and the inverse of an element a as a^{-1}. This notation must, of course, be interpreted correctly in individual examples of groups.

Definition. A subset H of a group G is called a *subgroup* of G if the following are true:
1. H contains the identity element, 1, of G.
2. For every a in H, the inverse, a^{-1}, of a, as defined in G, is also in H.
3. For every a and b in H, the product ab is an element of H.

https://doi.org/10.1515/9783111319360-009

We can also say that a subset H of G is a subgroup of G if and only if H is a group under the operation of G.

To give a general example of a subgroup, we introduce the following *exponent notation* in an arbitrary group, written multiplicatively. For each a in a group G, we define a^n for integers $n \geq 0$ recursively by saying that $a^0 = 1$ and $a^n = a^{n-1} \cdot a$ for $n > 0$. If $n = -m$ is negative, we let $a^n = (a^{-1})^m$, where a^{-1} is the inverse of a. Using associativity of the operation in G, we can show that $a^m a^n = a^{m+n}$ and $(a^m)^n = a^{mn}$ for all integers m and n. If a and b are elements in an *Abelian* group, then $(ab)^n = a^n b^n$ for all integers n. (The last claim is not usually true in a non-Abelian group, as $(ab)^2 = (ab)(ab)$ is not necessarily the same as $a^2 b^2 = (aa)(bb)$. The following proposition holds in every group though.)

Proposition B.1.2. *If a and b are elements of a group G, then $(ab)^{-1} = b^{-1}a^{-1}$.*

Proof. Since $(ab)(b^{-1}a^{-1}) = a(bb^{-1})a^{-1} = a \cdot 1 \cdot a^{-1} = aa^{-1} = 1$, and $(b^{-1}a^{-1})(ab) = 1$ by similar calculations, then $b^{-1}a^{-1}$ has the property of the unique inverse of ab in G. (Note that we used only the associative property and did not assume the commutative property in these calculations.) □

Proposition B.1.3. *Let G be a group and let a be an element of G. Then $\langle a \rangle = \{a^n \mid n \in \mathbb{Z}\}$ is a subgroup of G.*

Proof. Assuming the exponent rules noted above, we can say the following:
1. The identity element $1 = a^0$ is in $\langle a \rangle$.
2. For every a^n in $\langle a \rangle$, its inverse $(a^n)^{-1} = a^{-n}$ is in $\langle a \rangle$.
3. If a^m and a^n are in $\langle a \rangle$, then $a^m \cdot a^n = a^{m+n}$ is in $\langle a \rangle$.

Thus, $\langle a \rangle$ is a subgroup of G. □

We refer to $\langle a \rangle$ as the *cyclic subgroup* of G generated by a. The following proposition and terminology help us classify these subgroups, as we will see in Corollary B.1.5.

Proposition B.1.4. *Let G be a group with finitely many elements. Then for all a in G, there is an integer $t > 0$ so that $a^t = 1$. If t is the smallest positive integer with this property, then $a^n = 1$ if and only if t divides n, and $a^r = a^s$ if and only if $r \equiv s \pmod{t}$.*

Proof. Let a be an element of G, and consider $a, a^2, a^3, \ldots$, which are elements of G by closure. If G has finitely many elements, there must be integers r and s with $0 < s < r$ so that $a^r = a^s$. Then $a^s \cdot a^{r-s} = a^s \cdot 1$ so that $a^{r-s} = 1$ by the cancellation properties of a group (part (1) of Proposition B.1.1). So, $t = r - s$ is a positive integer for which $a^t = 1$.

Now let t be the smallest positive integer for which $a^t = 1$. If n is an integer, we can write $n = tq + r$, for some integers q and r with $0 \leq r < t$, by the division algorithm. By the exponent rules that hold in every group,

$$a^n = a^{tq+r} = a^{tq} \cdot a^r = (a^t)^q \cdot a^r = 1^q \cdot a^r = a^r.$$

But then $1 = a^n = a^r$ if and only if $r = 0$, since otherwise the definition of t as the *smallest* positive integer for which $a^t = 1$ is violated.

Finally, if $a^r = a^s$, we conclude as above that $a^{r-s} = 1$. By the previous statement, this is true if and only if t divides $r - s$, that is, $r \equiv s \pmod{t}$. □

Definition. If a is an element of a group G, and $a^t = 1$ for some positive integer t, we refer to the smallest positive integer with that property as the *order* of a in G, and write $t = \mathrm{ord}(a)$. If no such t exists (which by Proposition B.1.4 can occur only in an infinite group), we say that a has *infinite order* in G. We also refer to the number of elements in G as the *order* of G, written as $|G|$.

Corollary B.1.5. *Let a be an element of a group G.*

1. *If a has infinite order in G, then*

$$\langle a \rangle = \{ \ldots, a^{-3}, a^{-2}, a^{-1}, 1, a, a^2, a^3, \ldots \},$$

with no two of these elements equal.

2. *If a has order t in G, then*

$$\langle a \rangle = \{ 1, a, a^2, \ldots, a^{t-1} \},$$

with t distinct elements.

Proof. As we saw in the proof of Proposition B.1.4, if $a^r = a^s$ with $r \neq s$, then there is a positive integer t so that $a^t = 1$. If a has infinite order in G, this cannot occur, and so each integer power of a is distinct in G. Suppose on the other hand that a has order t in G. Then, as noted in Proposition B.1.4, $a^r = a^s$ if and only if $r \equiv s \pmod{t}$. Hence, only the powers $a^0, a^1, a^2, \ldots, a^{t-1}$ are distinct in G. □

Proposition B.1.6. *Let a be an element of finite order t in a group G. For every integer k, if $\gcd(k, t) = d$, then the order of a^k in G is t/d.*

Proof. We first show that if $\gcd(k, t) = d$, then a^k and a^d have the same order in G, that is, $(a^k)^n = 1$ if and only if $(a^d)^n = 1$. If $\gcd(k, t) = d$, then d divides k, say with $k = ds$ for some integer s. So, if $(a^d)^n = 1$, then

$$(a^k)^n = (a^{ds})^n = ((a^d)^n)^s = 1^s = 1.$$

Conversely, if $\gcd(k, t) = d$, then we can write $d = kq + tr$ for some integers q and r. Now if $(a^k)^n = 1$, then

$$(a^d)^n = (a^{kq+tr})^n = (a^{kq} \cdot (a^t)^r)^n = ((a^k)^n)^q = 1^q = 1.$$

(Here, we use the fact that $a^t = 1$ by definition of the order of a in G.)

Now suppose that d is a positive divisor of t, say with $t = dq$. Then $(a^d)^q = a^{dq} = 1$, so that $\text{ord}(a^d) \leq q$. But if $0 < r < q$, then $0 < dr < dq = t$. Hence, $(a^d)^r \neq 1$, since otherwise the definition of $t = \text{ord}(a)$ is violated. Thus, $\text{ord}(a^d) = q = t/d$, completing the proof. $\qquad\square$

B.2 Cosets and quotient groups

If H is a subgroup of a group G, we can define an equivalence relation on G in terms of H. In some cases, we can make the resulting collection of equivalence classes into a group, which helps us describe properties of the group G.

Definition. Let H be a subgroup of a group G. If a and b are elements of G, we say that a is *congruent* to b modulo H, and write $a \equiv b \pmod{H}$, if $a^{-1}b$ is an element of H.

Proposition B.2.1. *Let H be a subgroup of a group G. Then congruence modulo H, as defined above, is an equivalence relation on G.*

Proof. Let a, b, and c be elements of G.
1. $a \equiv a \pmod{H}$ is true since $a^{-1}a = 1$ is an element of any subgroup H.
2. Suppose that $a \equiv b \pmod{H}$, so that $a^{-1}b$ is an element of H. Then $(a^{-1}b)^{-1}$ is also an element of H. But $(a^{-1}b)^{-1} = b^{-1}(a^{-1})^{-1} = b^{-1}a$ by Proposition B.1.2. Thus, $b \equiv a \pmod{H}$ by definition.
3. Suppose that $a \equiv b \pmod{H}$ and $b \equiv c \pmod{H}$, so that $a^{-1}b$ and $b^{-1}c$ are elements of H. Then $a^{-1}b \cdot b^{-1}c = a^{-1}c$ is in H, since a subgroup is closed under the operation of the group, so that $a \equiv c \pmod{H}$.

Thus congruence has the reflexive, symmetric, and transitive properties. $\qquad\square$

Example. The set $\mathbb{Z}$ of integers is a group under addition. (Note that the identity element is 0 and that the inverse of a typical integer a is its usual negative, $-a$.) The cyclic subgroup $H = \langle 4 \rangle$ consists of all integer multiples of 4. (Here, "exponents" of 4 mean the integers obtained by repeated addition of 4 to itself, and the inverses, that is, negatives, of those integers.) In this context, a "product" $a^{-1}b$ is interpreted as $-a + b$. Thus, $a \equiv b \pmod{H}$ if and only if $b - a$ is a multiple of 4. That is, congruence modulo H is the same as the usual relation of congruence modulo 4 on the set of integers. Any other integer m can be used in place of 4 here.

Proposition B.2.2. *Let H be a subgroup of a group G and let a be an element of G. If b is an element of G, then $a \equiv b \pmod{H}$ if and only if $b = ah$ for some h in H.*

Proof. If $b = ah$, then $a^{-1}b = a^{-1}(ah) = h$ is an element of H, so that $a \equiv b \pmod{H}$. Conversely, if $a \equiv b \pmod{H}$, so that $a^{-1}b = h$ is in H, then $b = ah$. $\qquad\square$

Definition. If H is a subgroup of a group G and a is an element of G, then the set $aH = \{ah \mid h \in H\}$ is called aH the *left coset* of H in G determined by a.

Proposition B.2.2 shows that aH is the equivalence class of a under the relation of congruence modulo m. Left cosets have the following properties.

Proposition B.2.3. *Let H be a subgroup of a group G and let a and b be elements of G. Then the following statements are true:*
1. *a is an element of aH.*
2. *If a is congruent to b modulo H, then $aH = bH$.*
3. *If a is not congruent to b modulo H, then $aH \cap bH = \emptyset$.*
4. *If H is finite, then each left coset aH has the same number of elements as does H.*

Proof. Statements (1)–(3), which imply that the collection of left cosets of H in G form a *partition* of the group G, are standard properties of equivalence classes. We omit further proof. For statement (4), if $H = \{h_1, h_2, \ldots, h_n\}$ is finite, with n elements, then $aH = \{ah_1, ah_2, \ldots, ah_n\}$. These products are distinct, since if $ah_i = ah_j$, then $h_i = h_j$ by cancellation. Thus, aH also contains n elements.　□

The following key theorem of finite groups is an immediate consequence of these properties of left cosets.

Theorem B.2.4 (Lagrange's theorem). *If H is a subgroup of a finite group G, then the order of H divides the order of G.*

Proof. Proposition B.2.3 shows that the set of left cosets forms a partition of G in which each subset has the same number of elements, namely $|H|$, the order of H. If there are k distinct left cosets, it follows that $|G| = k \cdot |H|$, and so $|H|$ divides $|G|$.　□

Corollary B.2.5. *If a is an element of a finite group G, then $\operatorname{ord}(a)$ divides $|G|$.*

Proof. If G is a finite group, then each of its elements has finite order. Corollary B.1.5 shows that if $\operatorname{ord}(a) = t$, then $\langle a \rangle$ contains t elements. But Lagrange's theorem implies that $|\langle a \rangle|$ divides $|G|$, that is, t divides the order of G.　□

Corollary B.2.6. *If G is a group of order n, then $a^n = 1$ for every element a in G.*

Proof. If t is the order of an element a of G, then Corollary B.2.5 shows that t divides n. But then Proposition B.1.4 implies that $a^n = 1$.　□

If G is a group (finite or infinite) and H is a subgroup of G, write the set of all distinct left cosets of H in G as G/H (read as G modulo H, or G mod H). In some cases, G/H is a group under an operation derived from that in G. The following definition is the property required to make this work.

Definition. If N is a subgroup of G, we say that N is *normal* in G if for all n in N and a in G, the element $a^{-1}na$ is in N.

Theorem B.2.7. *Let N be a normal subgroup of a group G and let G/N be the set of all distinct left cosets of N in G. Then the operation $aN \cdot bN = (ab)N$ is well-defined and makes G/N into a group.*

Definition. When N is a normal subgroup of G, then we refer to G/N as the *quotient group* of G modulo N. (The term *factor group* is sometimes used as well.)

Proof. To establish that the operation above is well-defined, we need to show that if $aN = cN$ and $bN = dN$, then $aN \cdot bN = cN \cdot dN$, that is, $(ab)N = (cd)N$. By Proposition B.2.2, $aN = cN$ if and only if $a \equiv c \pmod{N}$, that is, $a^{-1}c$ is an element of N, and we have similar statements for the other coset equations. Thus, we can rephrase our requirement as follows.

If $a^{-1}c$ and $b^{-1}d$ are elements of N, then $(ab)^{-1}(cd)$ is also an element of N.

By Proposition B.1.2 and other properties that hold in every group, we can write

$$(ab)^{-1}(cd) = b^{-1}a^{-1}cd = b^{-1}a^{-1}c(bb^{-1})d = b^{-1}(a^{-1}c)b \cdot b^{-1}d.$$

But if N is normal in G, then $b^{-1}(a^{-1}c)b$ is an element of N, and so $b^{-1}(a^{-1}c)b \cdot b^{-1}d)$ is in N by closure. This establishes that the multiplication operation on G/N is well-defined. Now it is straightforward to show that G/N has all the properties of a group under this operation. For instance, if 1 is the identity element of G, then $1N$ is an identity element for G/N since

$$aN \cdot 1N = (a \cdot 1)N = aN = (1 \cdot a)N = 1N \cdot aN$$

for all aN in G/N. We leave the other verifications as exercises. □

If the operation of G is commutative, then $a^{-1}na = a^{-1}an = n$ for all a in G and n in N. Thus, every subgroup N of an Abelian group G is normal in G. We will restrict our attention to this case in the remainder of Appendix B.

B.3 Finite Abelian groups

In this section, we will use quotient groups to help prove an important classification of Abelian groups of finite order. We begin with a definition and a sequence of lemmas before stating the main result.

Definition. If G is a finite group, then there is a smallest positive integer t so that $a^t = 1$ for all a in G. In this case, we say that t is the *exponent* of G.

If $|G| = n$, then Corollary B.2.6 states that $a^n = 1$ for all a in G. Thus, the exponent of G exists and is no larger than the order of G. Proposition B.1.4 implies that $\mathrm{ord}(a)$ divides the exponent t of G for every a in G.

Lemma B.3.1. *Let a and b be elements of a finite Abelian group G, and suppose that* $\mathrm{ord}(a) = q$ *and* $\mathrm{ord}(b) = r$ *are relatively prime. Then* $\mathrm{ord}(ab) = qr$.

Proof. Let $\mathrm{ord}(ab) = s$. Since G is Abelian, then

$$(ab)^{qr} = (a^q)^r \cdot (b^r)^q = 1^r \cdot 1^q = 1,$$

so that s divides qr by Proposition B.1.4. On the other hand,

$$1 = ((ab)^s)^r = a^{rs} \cdot (b^r)^s = a^{rs},$$

implying that q divides rs. Since $\gcd(q, r) = 1$, it follows that q divides s. Similarly,

$$1 = ((ab)^s)^q = (a^q)^s \cdot b^{qs} = b^{qs},$$

implying that r divides qs, and then r divides s. Now s is a common multiple of q and r, and since $\gcd(q, r) = 1$, it follows that qr divides s. Therefore, $s = qr$. ☐

Lemma B.3.2. *Let a and b be elements of a finite Abelian group G with* $\mathrm{ord}(a) = s$ *and* $\mathrm{ord}(b) = t$. *Then there is an element of G whose order is* $\mathrm{lcm}(s, t)$, *the least common multiple of s and t.*

Proof. For a prime number p, recall that $e_p(n)$ is the largest nonnegative integer e so that p^e divides n. We can define integers q and r by saying that

$$e_p(q) = \begin{cases} e_p(s), & \text{if } e_p(s) \geq e_p(t) \\ 0, & \text{if } e_p(s) < e_p(t) \end{cases} \quad \text{and} \quad e_p(r) = \begin{cases} 0, & \text{if } e_p(s) \geq e_p(t) \\ e_p(t), & \text{if } e_p(s) < e_p(t) \end{cases}$$

for every prime p. Then the following statements are true, using properties of these exponent functions established in Corollary A.3.4.

1. q divides s and r divides t, since $e_p(q) \leq e_p(s)$ and $e_p(r) \leq e_p(t)$ for every prime p.
2. q and r are relatively prime, since either $e_p(q) = 0$ or $e_p(r) = 0$ for every prime p.
3. $qr = \mathrm{lcm}(s, t)$, since $e_p(q) + e_p(r)$ is the larger of $e_p(s)$ and $e_p(t)$ for every prime p.

Now $\mathrm{ord}(a^{s/q}) = q$ and $\mathrm{ord}(b^{t/r}) = r$, as in Proposition B.1.6. Since $\gcd(q, r) = 1$, Lemma B.3.1 implies that $a^{s/q} \cdot b^{t/r}$ is an element of G of order $qr = \mathrm{lcm}(s, t)$. ☐

Lemma B.3.3. *Let G be a finite Abelian group. If t is the maximum order of all elements of G, then t is the exponent of G. Conversely, if t is the exponent of G, then there is an element a in G whose order equals t.*

Proof. Let t be the maximum order of all elements of G, say with a an element of G having order t. If b is in G and $\mathrm{ord}(b) = s$, then Lemma B.3.2 implies that G contains an element of order $\mathrm{lcm}(s, t)$. Since $t \leq \mathrm{lcm}(s, t)$, we must conclude that $\mathrm{lcm}(s, t) = t$ to avoid contradicting the definition of t. But then s divides t by definition, which implies that $b^t = 1$ by Proposition B.1.4. Thus, t is the exponent of G, that is, the smallest positive integer so that $b^t = 1$ for all b in G. (The existence of a, whose order is t, ensures that no smaller value can work.)

Conversely, let t be the exponent of G and let s be the maximum order of all elements of G, say with a an element of G having order s. As above, it follows that the order of each element of G divides s, so that $b^s = 1$ for all b in G. But then, by definition, s is the exponent of G, and so $s = t$. That is, a is an element of G whose order equals the exponent of G. $\qquad\square$

Lemma B.3.4. *Let G be a finite Abelian group. Suppose that t is the exponent of G and that a is an element of G with $\mathrm{ord}(a) = t$ (which must exist by Lemma B.3.3). Let $N = \langle a \rangle$, the cyclic subgroup of G generated by a, and let G/N be the quotient group G mod N. If b is an element of G and bN has order s as an element of G/N, then s divides t, and bN contains an element whose order in G is s.*

By definition, the order of bN in G/N is the smallest positive integer s such that

$$(bN)^s = b^s N = 1N = N.$$

We can also say that the order of bN in G/N is the smallest positive integer s such that b^s is in N.

Proof. Let n be the order of b in G. Note that $(bN)^n = b^n N = 1N$ in G/N, so it follows that s divides n. Likewise, s divides $\mathrm{ord}(c)$ for any element c in bN. By Lemma B.3.3, we know that n divides t, so then s divides t, say that $t = sq$ for some integer q. Now $1N = (bN)^s = b^s N$ implies that b^s is an element of $N = \langle a \rangle$, that is, $b^s = a^m$ for some integer m. Notice that $1 = b^t = (b^s)^q = (a^m)^q = a^{mq}$, implying that $\mathrm{ord}(a) = t$ divides mq. So, now $mq = tr = sqr$, or $m = sr$, for some integer r. But then $b^s = a^m = (a^r)^s$, so that $(ba^{-r})^s = 1$ in G. Since a^{-r} is an element of N, then $c = ba^{-r}$ is an element of bN whose order divides s. But s divides $\mathrm{ord}(c)$ as noted above, and so $\mathrm{ord}(c) = s$. $\qquad\square$

Lemma B.3.4 implies that if $N = \langle a \rangle$ for some element a of maximal order in a finite Abelian group G, then we can select a representative b from each coset in G/N so that b has the same order in G as bN has in G/N. This is not generally true for all subgroups of G. For example, let $G = \mathbb{Z}_4 = \{0, 1, 2, 3\}$, a group under addition modulo 4, and let $N = \langle 2 \rangle = \{0, 2\}$, a normal subgroup of G. We can write $G/N = \{0 + N, 1 + N\}$. (Since the operation in G is addition, it is best to write the typical element of G/N as $a + N$ rather than aN.) Now we find that $1 + N$ has order two in G/N, since

$$(1 + N) + (1 + N) = (1 + 1) + N = 2 + N = 0 + N,$$

the identity element of G/N. But each of the potential representatives for $1 + N$, that is, 1 and 3, have order four in $G = \mathbb{Z}_4$.

Now we can move toward the statement of our main result, using the following terminology.

Definition. Let G be a finite Abelian group, with its operation written as multiplication. We say that a subset $\{a_1, a_2, \ldots, a_k\}$ of G is a *basis* for G of *type* $(n_1, n_2, \ldots, n_k)$ if each a_i has order n_i in G, and if every element of G can be written uniquely as $a_1^{r_1} a_2^{r_2} \cdots a_k^{r_k}$ with $0 \le r_i < n_i$ for $i = 1, 2, \ldots, k$.

Note that if G has a basis of type $(n_1, n_2, \ldots, n_k)$, then $|G| = n_1 n_2 \cdots n_k$. If G is trivial, that is, $G = \{1\}$, then $\{1\}$ is the only basis for G, of type (1). Typically, a finite Abelian group can have more than one basis, and might have bases of different types.

Example. Let $G = \mathbb{Z}_{13}^* = \{1, 2, 3, 4, 5, 6, 7, 8, 9, 10, 11, 12\}$, a group under an operation of multiplication modulo 13. (That is, $a \cdot b = c$ in G if $(ab) \bmod 13 = c$ in the integers.) Then $\{2\}$ is a basis for G of type (12), since the powers of 2,

$$2, 4, 8, 3, 6, 12, 11, 9, 5, 10, 7, 1,$$

include all elements of G exactly once. But the set $\{3, 5\}$ is also a basis for G, of type $(3, 4)$. In the following array, if we number the rows and columns starting with 0, then the entry in row i and column j is $3^i \cdot 5^j$, as calculated in G:

$$\begin{array}{cccc} 1 & 5 & 12 & 8 \\ 3 & 2 & 10 & 11 \\ 9 & 6 & 4 & 7 \end{array}$$

For instance, $3^1 \cdot 5^2 = 75$ has remainder 10 when divided by 13. We see that each element of G appears exactly once in this array.

If a finite Abelian group G has a basis, we can use its type to determine the number of elements of G whose order divides a particular number.

Lemma B.3.5. *Let G be a finite Abelian group. Then for every positive integer t, the set*

$$H_t = \{x \in G \mid x^t = 1\}$$

is a subgroup of G. If G has a basis $\{a_1, a_2, \ldots, a_k\}$ of type $(n_1, n_2, \ldots, n_k)$, then the order of H_t is

$$|H_t| = \gcd(n_1, t) \cdot \gcd(n_2, t) \cdots \gcd(n_k, t).$$

Proof. The identity element 1 has the property that $1^t = 1$, so that 1 is in H_t. If $x^t = 1$, then $(x^{-1})^t = x^{-t} = (x^t)^{-1} = 1^{-1} = 1$. Thus, if x is in H_t, then x^{-1} is in H_t. Finally, if $x^t = 1$ and $y^t = 1$ for elements x and y in an *Abelian* group G, then $(xy)^t = x^t y^t = 1 \cdot 1 = 1$. Thus, H_t is closed, completing the proof that H_t is a subgroup of G.

Note that $(a_1^{r_1} a_2^{r_2} \cdots a_k^{r_k})^t = a_1^{tr_1} a_2^{tr_2} \cdots a_k^{tr_k}$ in an Abelian group. This product equals the identity element 1 if and only if $tr_i \equiv 0 \pmod{n_i}$ for $i = 1, \ldots, k$. Corollary A.2.4 shows that if n_i divides tr_i, then $\frac{n_i}{\gcd(n_i, t)}$ divides r_i. There are precisely $\gcd(n_i, t)$ values of r_i with $0 \le r_i < n_i$ for which this is true. The formula for $|H_t|$ follows by the multiplicative counting principle. $\qquad\square$

Example. Let $G = \mathbb{Z}_{13}^*$, as in the preceding example, and let $t = 6$. Since G has a basis of type (12), then $|H_6| = \gcd(12, 6) = 6$. We saw that G also has a basis of type $(3, 4)$. Using this basis, we have $|H_6| = \gcd(3, 6) \cdot \gcd(4, 6) = 3 \cdot 2 = 6$ as well. One can confirm, by direct calculation or by following the approach in the proof of Lemma B.3.5, that $H_6 = \{1, 3, 4, 9, 10, 12\}$ in this case.

Our main result states that a finite Abelian group G always has a basis, and that the *type* of the basis is unique under the following restriction.

Definition. Suppose that $\{a_1, a_2, \ldots, a_k\}$ is a basis of type $(n_1, n_2, \ldots, n_k)$ for some Abelian group G. We say that this basis has *invariant factor* type if n_{i+1} divides n_i for $i = 1, \ldots, k-1$, with n_i allowed to be 1 only if $i = 1$ (in which case G is trivial).

Example. The group $G = \mathbb{Z}_{13}^*$ has a basis $\{2\}$ of invariant factor type (12). (There is no divisibility condition to check when $k = 1$.) The set $\{3, 5\}$ is a basis, as noted in a previous example, but its type, $(3, 4)$, is not an invariant factor type (4 does not divide 3). Note that the elements of a basis of invariant factor type are not necessarily unique. For instance, one can check that $\{6\}$ is also a basis for $G = \mathbb{Z}_{13}^*$ of type (12).

If G has a basis of invariant factor type $(n_1, n_2, \ldots, n_k)$, we also say that $(n_1, n_2, \ldots, n_k)$ are the *invariant factors* of G. We can now state and prove our main result on the structure of finite Abelian groups.

Theorem B.3.6 (Fundamental theorem of finite Abelian groups). *If G is a finite Abelian group, then G has a basis of invariant factor type, and the invariant factors of G are unique.*

Proof. To show that G has a basis of invariant factor type, we proceed by induction on $|G|$. If $|G| = 1$, then $\{1\}$ is a basis for G with invariant factor type (1). Assume that G is a nontrivial finite Abelian group, and that every Abelian group with fewer than $|G|$ elements has a basis of invariant factor type.

Let t be the exponent of G, and let a be an element of G having order t. Then $N = \langle a \rangle$ is a nontrivial (normal) subgroup of G, so that G/N is an Abelian group with $|G/N| < |G|$. By the inductive hypothesis, G/N has a basis $\{a_1 N, a_2 N, \ldots, a_k N\}$ of invariant factor type $(n_1, n_2, \ldots, n_k)$. Lemma B.3.4 implies that each n_i divides t. Furthermore, we can assume by the same lemma that each class representative a_i has order n_i in G. We now show that $\{a, a_1, a_2, \ldots, a_k\}$ is a basis for G.

If b is an element of G, so that bN is in G/N, then we can write

$$bN = (a_1N)^{r_1} \cdot (a_2N)^{r_2} \cdots (a_kN)^{r_k} = (a_1^{r_1}a_2^{r_2} \cdots a_k^{r_k})N,$$

using our assumption about a basis for G/N and the definition of the operation in G/N. Since $N = \langle a \rangle$, it follows that $b = a^r \cdot (a_1^{r_1}a_2^{r_2} \cdots a_k^{r_k})$ for some integer r. If $b = a^s a_1^{s_1}a_2^{s_2} \cdots a_k^{s_k}$ also, then we find that

$$bN = (a_1N)^{r_1} \cdot (a_2N)^{r_2} \cdots (a_kN)^{r_k} = (a_1N)^{s_1} \cdot (a_2N)^{s_2} \cdots (a_kN)^{s_k},$$

implying, since $\{a_1N, a_2N, \ldots, a_kN\}$ is a basis for G/N, that $r_i \equiv s_i \pmod{n_i}$ for $i = 1, \ldots, k$. Since a_i has order n_i in G, it follows that $a^r = a^s$ by cancellation in G, and so $r \equiv s \pmod{t}$. So, $\{a, a_1, a_2, \ldots, a_k\}$ is a basis for G of invariant factor type $(t, n_1, n_2, \ldots, n_k)$ (or invariant factor type (t) if $n_1 = 1$). By induction, every finite Abelian group G has a basis of invariant factor type.

To show that the invariant factors of G are unique, suppose that G has two bases, one of invariant factor type $(n_1, n_2, \ldots, n_k)$ and the other of invariant factor type $(m_1, m_2, \ldots, m_\ell)$. If $|G| = n$, we have $n_1 \cdot n_2 \cdots n_k = n = m_1 \cdot m_2 \cdots m_\ell$. Letting $t = n_1$, and applying Lemma B.3.5 to each basis, we have that

$$\gcd(n_1, n_1) \cdots \gcd(n_k, n_1) = \gcd(m_1, n_1) \cdots \gcd(m_\ell, n_1).$$

Since each n_i divides n_1, it follows that

$$n = n_1 \cdots n_k = \gcd(m_1, n_1) \cdots \gcd(m_\ell, n_1).$$

But $\gcd(m_i, n_1) \le m_i$ for $1 \le i \le \ell$, so we must conclude that $\gcd(m_1, n_1) = m_1$, that is, m_1 divides n_1. The same argument, using $t = m_1$, shows that n_1 divides m_1. So, $m_1 = n_1$. Repeating this process with $t = n_2$ and $t = m_2$ then shows that those two values must be equal. Continuing in this way, we conclude that the two basis types for G must be identical. □

Corollary B.3.7. *Let $(n_1, n_2, \ldots, n_k)$ be the invariant factors of a finite Abelian group G. Then n_1 is the exponent of G.*

Proof. Let $\{a_1, a_2, \ldots, a_k\}$ be a basis for G of type $(n_1, n_2, \ldots, n_k)$. Then the order of a_1 is n_1, so the exponent of G can be no smaller than n_1. Consider $H_{n_1} = \{x \in G \mid x^{n_1} = 1\}$, a subgroup of G. By Lemma B.3.5,

$$|H_{n_1}| = \gcd(n_1, n_1) \cdot \gcd(n_2, n_1) \cdots \gcd(n_k, n_1) = n_1 \cdot n_2 \cdots n_k = |G|.$$

(Here, we use the fact that each n_i divides n_1.) But then $a^{n_1} = 1$ for all a in G. Together with the previous observation that the exponent of G is no smaller than n_1, we conclude that n_1 is the exponent of G. □

The following two examples illustrate how we might determine the invariant factors of a finite Abelian group in practice.

Example. Suppose that G is an Abelian group with 32 elements. In this case, Theorem B.3.6 implies that G has a basis of type $(2^{e_1}, 2^{e_2}, \ldots, 2^{e_k})$ where $e_1 \geq e_2 \geq \cdots \geq e_k \geq 1$ and $32 = 2^{e_1} \cdot 2^{e_2} \cdots 2^{e_k}$. The number of distinct possibilities is the same as the number of ways of writing $5 = e_1 + e_2 + \cdots + e_k$, with $e_1 \geq e_2 \geq \cdots \geq e_k \geq 1$. We find that there are seven possibilities,

$$5 = 4 + 1 = 3 + 2 = 3 + 1 + 1 = 2 + 2 + 1 = 2 + 1 + 1 + 1 = 1 + 1 + 1 + 1 + 1,$$

corresponding to the following potential collections of invariant factors of G:

$$(32), \quad (16, 2), \quad (8, 4), \quad (8, 2, 2), \quad (4, 4, 2), \quad (4, 2, 2, 2), \quad (2, 2, 2, 2, 2).$$

We can use Lemma B.3.5 to determine which of these seven possibilities is true for a given group G. If G has invariant factors $(2^{e_1}, 2^{e_2}, \ldots, 2^{e_k})$, then $H_2 = \{x \in G \mid x^2 = 1\}$ has order

$$\gcd(2^{e_1}, 2) \cdot \gcd(2^{e_2}, 2) \cdots \gcd(2^{e_k}, 2) = 2^k.$$

For instance, if there are 16 elements x in G for which $x^2 = 1$, then $k = 4$, and the only possible invariant factor collection for G is $(4, 2, 2, 2)$. If $x^2 = 1$ has $2^3 = 8$ solutions, then G has invariant factor type either $(8, 2, 2)$ or $(4, 4, 2)$. Since the first calculation is inconclusive in this case, we may apply Lemma B.3.5 again with $t = 4$. If G has invariant factor type $(8, 2, 2)$, then $x^4 = 1$ has $\gcd(8, 4) \cdot \gcd(2, 4) \cdot \gcd(2, 4) = 4 \cdot 2 \cdot 2 = 16$ solutions, while if G has invariant factor type $(4, 4, 2)$, we find that all 32 elements of G satisfy $x^4 = 1$. The following table summarizes the order of H_t for $t = 2$ and $t = 4$, and shows that each invariant factor type is determined by looking at these two cases:

t	(32)	(16, 2)	(8, 4)	(8, 2, 2)	(4, 4, 2)	(4, 2, 2, 2)	(2, 2, 2, 2, 2)
2	2	4	4	8	8	16	32
4	4	8	16	16	32	32	32

Example. If G is an Abelian group with $200 = 2^3 \cdot 5^2$ elements, possible invariant factor forms are obtained by combining all such forms for $2^3 = 8$ (namely (8), $(4, 2)$, and $(2, 2, 2)$) and for $5^2 = 25$ (which are (25) and $(5, 5)$). If in each case, we match up the highest powers of 2 and 5, and then the next highest powers, and so forth, we obtain six possibilities, each in invariant factor form. These possibilities are listed in the following table, along with the order of $H_t = \{x \in G \mid x^t = 1\}$ for $t = 2$ and $t = 5$. Those two calculations are sufficient to determine the invariant factors of an Abelian group with 200 elements.

t	(200)	(100, 2)	(50, 2, 2)	(40, 5)	(20, 10)	(10, 10, 2)
2	2	4	8	2	4	8
5	5	5	5	25	25	25

C Appendix: Ideals of quadratic domains

In Section 1.6, we asserted that a congruence class of ideal numbers of discriminant Δ can be interpreted as an ideal of a quadratic domain. In this appendix, we prove this claim and compile properties of these ideals that we use throughout the text. In particular, we define an operation of multiplication of ideals and a relation of equivalence of ideals, which we use in the text to establish properties of ideal number composition.

C.1 Ideals of integral domains

An ideal of an integral domain D is a subgroup of D under addition with a "strong closure" property for multiplication. The following definition spells this out more precisely.

Definition. Let D be an integral domain. A subset A of D is called an *ideal* of D if the following statements are true:
1. A contains 0.
2. If v is an element of A, then $-v$ is in A.
3. If v and w are elements of A, then $v + w$ is in A.
4. If v is in A and x is in D, then vx is in A.

Example. The set $\{0\}$ is an ideal of every integral domain D, called the *trivial* ideal of D. If A is an ideal of D, then we say that A is *nontrivial* if $A \neq \{0\}$.

Example. Every integral domain D is an ideal of itself, called the *improper* ideal of D. If A is an ideal of D, then we say that A is a *proper* ideal of D if $A \neq D$.

The following propositions provide some general examples of ideals of an integral domain.

Proposition C.1.1. *If x is an element of an integral domain D and A is an ideal of D, then the set $xA = \{xa \mid a \in A\}$ is an ideal of D.*

Proof. Let a and b be elements of A, so that xa and xb are in xA, and let y be an element of D.
1. $0 = x0$ is in xA, since 0 is in A.
2. $-(xa) = x(-a)$ is in xA, since $-a$ is in A.
3. $xa + xb = x(a + b)$ is in xA, since $a + b$ is in A.
4. $(xa)y = x(ay)$ is in xA, since ay is in A.

Thus, xA has all the properties of an ideal of D. □

As a special case of Proposition C.1.1, if x is an element of D, then xD is an ideal of D. We call xD the *principal ideal* of D generated by x.

https://doi.org/10.1515/9783111319360-010

Proposition C.1.2. *If A and B are ideals of an integral domain D, then the following are true:*
1. *$A \cap B$ is the largest ideal of D contained in both A and B.*
2. *$A + B = \{a + b \mid a \in A \text{ and } b \in B\}$ is the smallest ideal of D containing both A and B.*

Proof. The intersection of A and B is the largest *set* contained in both A and B. Thus, for statement (1), it suffices to show that $A \cap B$ is an ideal of D. We leave this verification as an exercise. For statement (2), we assume that a, a_1, and a_2 are elements of A and that b, b_1, and b_2 are elements of B, with x an element of D.
1. Since 0 is an element of both A and B, then $0 + 0 = 0$ is in $A + B$.
2. $-(a + b) = (-a) + (-b)$ is in $A + B$, since $-a$ is in A and $-b$ is in B.
3. $(a_1 + b_1) + (a_2 + b_2) = (a_1 + a_2) + (b_1 + b_2)$ is in $A + B$, since A and B are closed under addition.
4. $(a + b)x = ax + bx$ is an element of $A + B$, since ax is in A and bx is in B.

For the final claim, note that if a is an element of A, then $a = a + 0$ is in $A + B$. Hence, A is a subset of $A + B$. Similarly, if b is in B, then $b = 0 + b$ is in $A + B$, and so B is a subset of $A + B$. Suppose that C is an ideal of D containing A and B. If a is in A and b is in B, then a and b are both in C, and thus $a + b$ is in C, since an ideal is closed under addition. Thus, the typical element of $A + B$ is an element of C, and so $A + B$ is a subset of C. Therefore, $A + B$ is the smallest ideal of D containing A and B. $\square$

Definition. If A and B are ideals of an integral domain, then the *product* of A and B consists of all finite sums of products of elements of A with elements of B. That is,

$$AB = \{a_1 b_1 + a_2 b_2 + \cdots + a_n b_n \mid a_i \in A \text{ and } b_i \in B\}.$$

Proposition C.1.3. *If A and B are ideals of an integral domain D, then AB is an ideal of D, contained in both A and B. Ideal multiplication is commutative and associative, and $AD = A$ for every ideal A of D.*

Proof. The set AB has the following properties:
1. $0 = 0 \cdot 0$ is in AB since 0 is in A and 0 is in B.
2. The negative of a typical element of AB is

$$-(a_1 b_1 + a_2 b_2 + \cdots + a_n b_n) = (-a_1)b_1 + (-a_2)b_2 + \cdots + (-a_n)b_n,$$

   which is also in AB as A contains the negative of each of its elements.
3. The sum of two finite sums of products is itself a finite sum of products, and so AB is closed under addition.
4. If x is an element of D, then

$$(a_1 b_1 + a_2 b_2 + \cdots + a_n b_n)x = a_1(b_1 x) + a_2(b_2 x) + \cdots a_n(b_n x)$$

   is an element of AB, since B contains each $b_i x$.

Hence, AB is an ideal of D. Each $a_i b_i$ is an element of A by the strong closure property of multiplication in A, and likewise an element of the ideal B. Since ideals are closed under addition, a sum of all such products is also in A and in B. Thus, AB is contained in both A and B.

Since $a_1 b_1 + a_2 b_2 + \cdots + a_n b_n = b_1 a_1 + b_2 a_2 + \cdots + b_n a_n$, then AB is a subset of BA. Likewise, each element of BA is also in AB, and so $AB = BA$. If A, B, and C are ideals of D, then the typical element of $(AB)C$ has the form $v_1 c_1 + v_2 c_2 + \cdots + v_n c_n$, where each v_i is in AB and each c_i is in C. When v_i is written as a sum of products, and the distributive property is applied to each $v_i c_i$, then the typical element of $(AB)C$ is a finite sum of products $(a_i b_i)c_i$. But now, $(a_i b_i)c_i = a_i(b_i c_i)$, with each $b_i c_i$ in BC, and so $(AB)C$ is contained in $A(BC)$. The reverse inclusion is similar. Therefore, $(AB)C = A(BC)$.

Finally, let A be an ideal of D. As noted above, the product AD is contained in A. For the reverse inclusion, if a is in an element of A, then $a \cdot 1 = a$ is in AD as well, since D contains 1. Thus, $AD = A$. $\qquad\square$

We conclude Section C.1 with the following observation.

Proposition C.1.4. *Let w be an element and A an ideal of an integral domain D. Then $wA = wD \cdot A$.*

Proof. The typical element of wA has the form wa where a is in A. But $wa = (w \cdot 1)a$ and, since 1 is in D, then wa is an element of $wD \cdot A$. For the reverse inclusion, the typical element of $wD \cdot A$ can be written as $(wx_1)a_1 + (wx_2)a_2 + \cdots + (wx_n)a_n$. But each $(wx_i)a_i = w(x_i a_i)$ is an element of wA, since A is an ideal of D. A sum of a finite number of terms in the ideal wA is also in wA, so we conclude that $wD \cdot A$ is a subset of wA. $\qquad\square$

C.2 Quadratic domains

We now introduce a type of integral domain in which we will be able to classify all ideals. A complex number v is called a *quadratic integer* if v is a root of a monic quadratic polynomial with integer coefficients, $f(x) = x^2 + bx + c$. If $f(x)$ is irreducible over the rational numbers, then the discriminant of $f(x)$ is also called the *discriminant of* v, written as $\Delta(v) = b^2 - 4c$. A rational number v is a quadratic integer if and only if v is an integer in the usual sense. In this context, we often refer to an element v of $\mathbb{Z}$ as a *rational integer*, and it is convenient to let $\Delta(v) = 0$ in that case. With this terminology, we define the following sets of quadratic integers.

Definition. Let $\Delta = \Delta(d, g)$ be a discriminant. Then the *quadratic domain* of discriminant Δ, denoted as $D(\Delta)$, is the set of all quadratic integers v for which $\Delta(v) = r^2 \Delta$ for some integer r.

We can classify quadratic domains as follows.

Proposition C.2.1. *Let $P(x)$ be the principal polynomial of discriminant Δ and let*

$$z = z_\Delta = \frac{P'(0) + \sqrt{\Delta}}{2}. \tag{C.1}$$

Then the quadratic domain of discriminant Δ is $D(\Delta) = \{q + rz \mid q, r \in \mathbb{Z}\}$.

Before proving Proposition C.2.1, we introduce some useful terminology. An expression of the form $q + rz$ is called a $\mathbb{Z}$-*combination* of the set $\{1, z\}$. Since Δ is not a square, then z is not a rational number, and so $q + rz = s + tz$ if and only if $q = s$ and $r = t$. We define the *conjugate* of z in $D = D(\Delta)$ to be $\overline{z} = \frac{P'(0) - \sqrt{\Delta}}{2} = P'(0) - z$, and the *conjugate* of $v = q + rz$ in D to be $\overline{v} = q + r\overline{z}$. These elements are $\mathbb{Z}$-combinations of $\{1, z\}$ and we verify below that they are elements of D. Note that the conjugate of a rational integer k in D equals k. This is typically not the same as the conjugate of k with respect to Δ, which we defined as $\overline{k} = -k - P'(0)$. In Appendix C, context will make clear to which conjugate of an integer we refer. (See, e. g., the statement of Proposition C.4.3.)

Proof. Suppose first that v is an element of $D(\Delta)$. Then v is a root of a polynomial $x^2 + bx + c$, with b and c rational integers and $b^2 - 4c = r^2\Delta$ for some r in $\mathbb{Z}$. By the quadratic formula, then $v = \frac{-b + r\sqrt{\Delta}}{2}$ or $v = \frac{-b - r\sqrt{\Delta}}{2}$. In the first case, we find that $v = q + rz$ with $q = \frac{-b - rP'(0)}{2}$. In the second case, $v = s + tz$ with $s = \frac{-b + rP'(0)}{2}$ and $t = -r$. Note that b has the same parity as $r^2\Delta$, and so the same parity as $rP'(0)$. Thus, v is a $\mathbb{Z}$-combination of $\{1, z\}$ in every case.

Conversely, if $v = q + rz$ for some rational integers q and r, then we find that v and its conjugate $\overline{v}$ are roots of

$$f(x) = (x - v)(x - \overline{v}) = x^2 - (2q + P'(0)r)x + (q^2 + P'(0)qr + P(0)r^2), \tag{C.2}$$

a monic polynomial with rational integer coefficients. (Note that $z \cdot \overline{z} = \frac{P'(0)^2 - \Delta}{4} = P(0)$, since we have seen that $P'(x)^2 - 4P(x) = \Delta$ for all x.) The discriminant of v is thus

$$\Delta(v) = (2q + P'(0)r)^2 - 4(q^2 + P'(0)qr + P(0)r^2) = r^2(P'(0)^2 - 4P(0)) = r^2\Delta. \tag{C.3}$$

Therefore, v is an element of $D(\Delta)$, as we wanted to show. $\qquad\square$

Proposition C.2.2. *The quadratic domain of discriminant Δ is an integral domain.*

Proof. Let $D = D(\Delta)$ be the quadratic domain of discriminant Δ. Since D is a subset of the field of complex numbers, it suffices to show that D is closed under addition, subtraction, and multiplication, and that D contains 0 and 1, the identity elements for addition and multiplication. For multiplicative closure, note that z is a root of $x^2 - P'(0)x + P(0)$ by equation (C.2). Thus, $z^2 = -P(0) + P'(0)z$ and, more generally,

$$(q + rz)(s + tz) = (qs - P(0)rt) + (qt + rs + P'(0)rt)z. \tag{C.4}$$

Since $D = \{q + rz \mid q, r \in \mathbb{Z}\}$, verifications of the remaining claims are straightforward and are left for exercises. □

We conclude Section C.2 with a proposition on properties of conjugates and a related definition. These will be useful for later calculations.

Proposition C.2.3. *If v and w are elements of a quadratic domain D, then*

$$\overline{v} + \overline{w} = \overline{v + w} \quad and \quad \overline{v} \cdot \overline{w} = \overline{vw}.$$

Proof. Let $v = q + rz$ and $w = s + tz$. Then $v + w = (q + s) + (r + t)z$ and we find that

$$\overline{v} + \overline{w} = (q + r\overline{z}) + (s + t\overline{z}) = (q + s) + (r + t)\overline{z} = \overline{v + w}.$$

For multiplication, note that $\overline{z}$ is a root of $x^2 - P'(0)x + P(0)$, so that $\overline{z}^2 = -P(0) + P'(0)\overline{z}$. Then we find that

$$\overline{v} \cdot \overline{w} = (q + r\overline{z})(s + t\overline{z}) = qs + (qt + rs)\overline{z} + rt\overline{z}^2 = (qs - P(0)rt) + (qt + rs + P'(0)rt)\overline{z}$$

which, using equation (C.4), equals $\overline{vw}$. □

Definition. If $v = q + rz$ in a quadratic domain $D = D(\Delta)$, then we define the *norm* of v to be

$$N(v) = v \cdot \overline{v} = q^2 + qr(z + \overline{z}) + r^2 z \cdot \overline{z} = q^2 + P'(0)qr + P(0)r^2. \tag{C.5}$$

Note that $N(v)$ is a rational integer. By Proposition C.2.3, we have

$$N(vw) = vw \cdot \overline{vw} = (vw)(\overline{v} \cdot \overline{w}) = (v \cdot \overline{v})(w \cdot \overline{w}) = N(v) \cdot N(w)$$

for all v and w in D.

C.3 Classification of ideals of a quadratic domain

Let Δ be a fixed discriminant and let $D = D(\Delta) = \{q + rz \mid q, r \in \mathbb{Z}\}$, with $z = z_\Delta$ defined as in equation (C.1). In this section, we classify all nontrivial ideals of the quadratic domain D. We begin with a lemma.

Lemma C.3.1. *Let B be an ideal of the quadratic domain D of discriminant Δ. Suppose that B contains an element of the form $\ell + mz$ with $m > 0$ and that m is the smallest positive rational integer for which this is true. Then $B = mA$ for some ideal A of D.*

Proof. Let $s + tz$ be an element of B and write $t = mq + r$ with $0 \le r < m$. Then

$$(s + tz) - q(\ell + mz) = (s - q\ell) + rz$$

is in B by closure properties of an ideal. We conclude that $r = 0$ to avoid contradicting the definition of m, and so m divides t. Now by the multiplication formula in equation (C.4), we have that

$$(-P'(0) + z)(s + tz) = (-P'(0)s - P(0)t) + (-P'(0)t + s + P'(0)t)z$$
$$= (-P'(0)s - P(0)t) + sz$$

is in B by the strong closure property for multiplication. By the same reasoning, we conclude that m divides s. Thus, every element x of B can be written as $x = my$ for some y in D. If we let $A = \{\frac{1}{m} \cdot x \mid x \in B\}$, then it is straightforward to verify that A is an ideal of D and that $B = mA$. We leave the details as an exercise. □

In the situation of Lemma C.3.1, we say that m is the *divisor* of B as an ideal of D. If the quadratic domain D is clear from context, we write $m = \mathrm{div}(B)$.

Proposition C.3.2. *Let B be a nontrivial ideal of the quadratic domain $D = D_\Delta$ and let z be defined as in (C.1). Then there is a smallest positive rational integer m so that B contains an element of the form $m(k + z)$, and there is a smallest positive rational integer in B, which equals ma for some rational integer a. In this case, $B = mA$ where*

$$A = \{q(a) + r(k + z) \mid q, r \in \mathbb{Z}\}. \tag{C.6}$$

Proof. If B is nontrivial, then B contains a nonzero element v. Thus, $\bar{v}$ is a nonzero element of D, and then $N(v) = v \cdot \bar{v}$ is a nonzero rational integer in B, as is $-N(v)$, by closure properties of an ideal. Thus, B contains a positive rational integer, and we can take b to be the smallest positive rational integer in B. Now with z in D, it follows that bz is in B. Therefore, B contains an element $\ell + mz$ with m a positive rational integer, and we can assume that m is as small as possible with this property. By Lemma C.3.1, then m divides the coefficients of 1 and of z in every $\mathbb{Z}$-combination of $\{1, z\}$ included in B. In particular, $\ell = mk$ and $b = ma$ for some integers k and a, and we can write $B = mA$ as in Lemma C.3.1.

Notice that A contains $k + z$ and that a is the smallest positive rational integer in A. If n is a rational integer in A, we can write $n = aq + r$ with $0 \le r < a$. Then $r = n - aq$ is an element of A by closure properties of an ideal. If r is positive, this contradicts the definition of a, so instead we conclude that a divides n.

Now consider the set $\{qa + r(k + z) \mid m, n \in \mathbb{Z}\}$, a subset of A. If $s + tz$ is an element of A, then $(s + tz) - t(k + z) = s - tk$ is a rational integer in A. Hence, there is a rational integer q so that $s - tk = aq$. But now

$$s + tz = (s - tk) + t(k + z) = q(a) + t(k + z).$$

We conclude that every element of A is in the set $\{q(a) + r(k + z) \mid q, r \in \mathbb{Z}\}$, as we wanted to show. □

We call an element of the form in equation (C.6) a $\mathbb{Z}$-combination of $\{a, k + z\}$. More generally, an element of $B = mA$ is a $\mathbb{Z}$-combination of $\{ma, m(k + z)\}$. It is straightforward to show that every set of the form in equation (C.6) has the closure properties for addition required of an ideal. However, not every such set is an ideal of a quadratic domain, as the following example illustrates.

Example. Let D be the quadratic domain of discriminant $\Delta = -35$, with $z = \frac{1+\sqrt{-35}}{2}$. Consider the set $A = \{q(2) + r(1 + z) \mid q, r \in \mathbb{Z}\}$ of $\mathbb{Z}$-combinations of $\{2, 1 + z\}$. An element $m + nz$ is in A if and only if $m \equiv n \pmod 2$. Then $1 + z$ is an element of A and z is an element of D, but $(1+z)z = -9 + 2z$ is not in A. (We use equation (C.4), with $P(x) = x^2 + x + 9$, in calculating this product.) Thus, A is not an ideal of D.

The following theorem, our main result for this section, provides a criterion for the set in equation (C.6) to be an ideal of a quadratic domain and connects ideals to ideal numbers.

Theorem C.3.3. *Let $P(x)$ and $D = \{q + rz \mid q, r \in \mathbb{Z}\}$ be the principal polynomial and the quadratic domain respectively of some discriminant Δ. Then a set of the form $A = \{q(a) + r(k + z) \mid q, r \in \mathbb{Z}\}$ is an ideal of D if and only if a divides $P(k)$.*

Proof. Using equation (C.5), we find that

$$N(k + z) = (k + z)(k + \bar{z}) = k^2 + P'(0)k + P(0) = P(k). \tag{C.7}$$

If $A = \{q(a) + r(k + z) \mid q, r \in \mathbb{Z}\}$ is an ideal of D, then $P(k)$ is a rational integer in A, since $k + z$ is in A and $k + \bar{z}$ is in D. But then $P(k) + 0z = (aq + kr) + rz$ implies that $r = 0$ and $P(k) = aq$. Thus, a divides $P(k)$.

Conversely, suppose that $P(k) = ac$ for some integer c. We noted above that A as in equation (C.6) has the properties of addition required for an ideal. Thus, to prove that A is an ideal of D, it will suffice to show that A contains ax and $(k+z)x$ when x is an element of D_ℓ. For the first claim, if $x = s + tz$, then $ax = (s - kt)a + at(k + z)$, an element of A. To help establish the second claim, note that $x = s + tz$ can be rewritten as $u - t(k + \bar{z})$, where $u = s + kt + P'(0)t$ is a rational integer. Then using equation (C.7) and the assumption that $P(k) = ac$, we have

$$(k + z)x = u(k + z) - t(k + z)(k + \bar{z}) = u(k + z) - tP(k) = -tc(a) + u(k + z),$$

likewise an element of A. □

Theorem C.3.3 implies that $A = \{q(a) + r(k + z) \mid q, r \in \mathbb{Z}\}$ is an ideal of $D = D(\Delta)$ if and only if $(a : k)$ is an ideal number of discriminant Δ. Furthermore, if a_1 and k_1 are rational integers, then we find that

$$\{q(a) + r(k + z) \mid q, r \in \mathbb{Z}\} = \{q(a_1) + r(k_1 + z) \mid q, r \in \mathbb{Z}\}$$

if and only if $a_1 = \pm a$ and $k_1 \equiv k \pmod{a}$. Thus, A corresponds to the entire congruence class of $(a : k)$. To emphasize this connection, we may write $A = \langle a : k \rangle$ if the discriminant Δ is apparent from context. If $B = mA$, we similarly write $B = m\langle a : k \rangle$ We call these *ideal number* expressions for the ideals A and B of D.

Example. Let $\Delta = \Delta(-35, 1) = -35$, with principal polynomial $P(x) = x^2 + x + 9$. Since $P(1) = 11$ is not divisible by 2, the set $A = \{q(2) + r(1 + z) \mid q, r \in \mathbb{Z}\}$ is not an ideal of $D = D(-35)$. We saw this more directly in a previous example.

Example. Let $\Delta = \Delta(-15, 1) = -15$, with principal polynomial $P(x) = x^2 + x + 4$. Since $P(16) = 276 = 46 \cdot 6$, then $A = \langle 46 : 16 \rangle$ is an ideal number expression for an ideal A of $D = D(-15)$.

The following term is one that we will need in later sections.

Definition. Let A be a nontrivial ideal of a quadratic domain D. If $m\langle a : k \rangle$ is an ideal number expression for A, define the *norm* of A to be $N(A) = m^2 a$.

Note that $N(A)$ is the product of ma, the smallest positive rational integer in A, and m, the smallest positive coefficient of z in an element of A. The norm of an ideal A of D can also be defined as the number of distinct cosets of A in D.

We conclude Section C.3 with an observation connecting the partial order relation that we defined on $\mathcal{Q}_\Delta$ in Section 2.1 to ideals with divisor $m = 1$ in a quadratic domain. The following proposition is sometimes summarized as: "To contain is to divide."

Proposition C.3.4. *Let A and B be ideals of a quadratic domain D having ideal number expressions $A = \langle a : k \rangle$ and $B = \langle b : \ell \rangle$. Then $\langle a : k \rangle \preceq \langle b : \ell \rangle$ if and only if $B \subseteq A$.*

Proof. We are given that

$$A = \{qa + r(k + z) \mid q, r \in \mathbb{Z}\} \quad \text{and} \quad B = \{qb + r(\ell + z) \mid q, r \in \mathbb{Z}\},$$

and we may assume that a and b are positive. Suppose first that $\langle a : k \rangle \preceq \langle b : \ell \rangle$, that is, a divides b and $\ell \equiv k \pmod{a}$. Then $b = as$ and $\ell = k + at$ for some integers s and t. The typical element of B can then be expressed as

$$qb + r(\ell + z) = q(as) + r(k + at + z) = (qs + rt)a + r(k + z),$$

for some integers q and r. Since $qs + rt$ is an integer, an element of this form is in A. Therefore, B is a subset of A.

For the converse, suppose that B is a subset of A. Since b is in B, then $b + 0z = qa + r(k + z) = (qa + rk) + rz$ for some integers q and r. Comparing coefficients shows that $r = 0$ and so $b = qa$. That is, a divides b. Likewise, since $\ell + z$ is in B, then $\ell + z = qa + r(k + z) = (qa + rk) + rz$ for some integers q and r. Then $r = 1$ and $\ell = qa + k$, which implies that $\ell \equiv k \pmod{a}$. Therefore, $\langle a : k \rangle \preceq \langle b : \ell \rangle$. $\square$

C.4 Ideal number expressions for ideals

Proposition C.3.2 implies that we can explicitly describe the elements of an ideal A of D by finding the smallest positive rational integer in A and an element of A with minimal positive coefficient of z. In this section, we note some methods of finding ideal number expressions for ideals in practice. We begin with an example of a principal ideal, which we generalize in Proposition C.4.1.

Example. Let D be the quadratic domain of discriminant $\Delta = -15$. The principal polynomial of discriminant -15 is $P(x) = x^2 + x + 4$, and so

$$D = \{q + rz \mid q, r \in \mathbb{Z}\}$$

where $z = \frac{P'(0)+\sqrt{\Delta}}{2} = \frac{1+\sqrt{-15}}{2}$. Consider the principal ideal $A = (2 + 3z)D$ of D. Using the multiplication formula of equation (C.4), the typical element of A is

$$(2 + 3z)(s + tz) = (2s - 12t) + (3s + 5t)z,$$

where s and t are integers. Since $\gcd(3, 5) = 1$, then A contains an element of the form $k + z$, such as $16 + z$ when $s = 2$ and $t = -1$. On the other hand, $(2s - 12t) + (3s + 5t)z$ is a rational integer if and only if $3s + 5t = 0$, which occurs only when $s = 5q$ and $t = -3q$ for some integer q. In that case, $2s - 12t = 2(5q) - 12(-3q) = 46q$, and the smallest positive integer of this form is $a = 46$. Therefore,

$$A = \{q(46) + r(16 + z) \mid q, r \in \mathbb{Z}\} = \{(46q + 16r) + rz \mid q, r \in \mathbb{Z}\},$$

the set of $\mathbb{Z}$-combinations of $\{46, 16 + z\}$, for which A is $\langle 46 : 16 \rangle$ is an ideal number expression. (In an example in Section C.3, we saw that $\langle 46 : 16 \rangle$ is in fact an ideal number expression for some ideal of $D = D(-15)$.)

Proposition C.4.1. *Let $D = \{q + rz \mid q, r \in \mathbb{Z}\}$ be the quadratic domain of discriminant Δ. Let $w = q + rz$ be an element of D with $\gcd(q, r) = 1$. Then $A = wD$, the principal ideal of D generated by w, can be written as $A = \langle a : k \rangle$ where $a = |N(w)|$ and $rk \equiv q \pmod{a}$.*

Proof. Let $P(x)$ be the principal polynomial of discriminant Δ. By equation (C.4), the typical element of $A = wD$ has the form

$$(q + rz)(s + tz) = (qs - P(0)rt) + (qt + rs + P'(0)rt)z$$

where s and t are integers. This product is a rational integer if and only if

$$qt + r(s + P'(0)t) = 0.$$

In this case, with $\gcd(q, r) = 1$, it follows that $s = (-q - P'(0)r)\ell$ and $t = r\ell$ for some rational integer ℓ. Then

$$qs - P(0)rt = (-q^2 - P'(0)qr - P(0)r^2)\ell = -N(w)\ell,$$

using the formula for $N(q + rz)$ in equation (C.5). The smallest positive rational integer of this form is $a = |N(w)|$.

If $\gcd(q, r) = 1$, we can write $1 = qm + rn$ for some integers m and n. Then when $s = n - P'(0)m$ and $t = m$, we find that $(q+rz)(s+tz) = k+z$ where $k = qn - P'(0)qm - P(0)rm$. Note that

$$rk = qrn - P'(0)qrm - P(0)r^2m = qrn + q^2m - q^2m - P'(0)qrm - P(0)r^2m$$
$$= q(rn + qm) - (q^2 - P'(0)qr - P(0)r^2) = q - N(w)m,$$

again using the formula for $N(q + rz)$. If $a = |N(w)|$, then $rk \equiv q \pmod{a}$. □

More generally, we can write each nonzero element w of $D = D(\Delta)$ as $w = m(q + rz)$ with m positive and $\gcd(q, r) = 1$. In that case, $A = wD = m\langle a : k\rangle$ where a and k are as given in Proposition C.4.1.

Corollary C.4.2. *Let $A = wD$ for some element w of a quadratic domain D. Then $N(A) = |N(w)|$. That is, the norm of a principal ideal of D is the absolute value of the norm of a generator of that ideal.*

Proof. Let $w = m(q + rz)$ with $\gcd(q, r) = 1$. By Proposition C.4.1 and the preceding note, $A = wD = m\langle a : k\rangle$ where $a = |N(q + rz)|$. The norm of A is then $N(A) = m^2a$, which equals $|N(m(q + rz))|$. □

Example. Returning to the preceding example, let $w = 2 + 3z$ in $D = D(-15)$. With $P(x) = x^2 + x + 4$, we find that $N(2 + 3z) = 2^2 + 2 \cdot 3 + 4 \cdot 3^2 = 46$. We can write $1 = 2m + 3n$ with $m = -1$ and $n = 1$, among other possibilities. Then $A = wD$ contains $k + z$ where

$$k = qn - P'(0)qm - P(0)rm = 2(1) - 2(-1) - 4(3)(-1) = 16.$$

(We could also find $k = 16$ by solving $3k \equiv 2 \pmod{46}$.)

The following definition and propositions will be needed in later sections.

Definition. Let A be an ideal of a quadratic domain D. Then the *conjugate* of A is the set consisting of the conjugates, as defined in D, of the elements of A:

$$\overline{A} = \{\overline{x} \mid x \in A\}. \tag{C.8}$$

Proposition C.4.3. *Let A be an ideal of a quadratic domain $D = D(\Delta)$, and suppose that $A = m\langle a : k\rangle$ is an ideal number expression for A. Then $\overline{A}$ is also an ideal of D, and we can write $\overline{A} = m\langle a : \overline{k}\rangle$ where $\overline{k}$ is the conjugate of k with respect to Δ.*

Proof. The verification that $\overline{A}$ is an ideal of D uses the properties of conjugates in Proposition C.2.3. We leave the details as an exercise.

For the claim about the ideal number expression for $\overline{A}$, let $P(x)$ be the principal polynomial of discriminant Δ. The conjugate of k with respect to Δ is defined as $\overline{k} = -k - P'(0)$. We have also seen that if z is defined as in equation (C.1), then the conjugate of z in D is $\overline{z} = P'(0) - z$. If $A = m\langle a : k \rangle$, then the typical element of A has the form $m(qa + rk) + mrz$ and so the typical element of $\overline{A}$ is

$$m(qa + rk) + mr\overline{z} = m(qa + rk + rP'(0)) - mrz.$$

The smallest positive rational integer of this form is ma when $q = 1$ and $r = 0$. We obtain an element with minimal positive coefficient of z as

$$m(-k - P'(0)) + mz = m(\overline{k} + z)$$

when $q = 0$ and $r = -1$. Thus, $\overline{A}$ has ideal number expression $m\langle a : \overline{k} \rangle$. $\qquad\square$

Proposition C.4.4. *If A and B are ideals of a quadratic domain $D = D(\Delta)$, then $\overline{AB} = \overline{A} \cdot \overline{B}$.*

Proof. The typical element of $\overline{AB}$ is the conjugate in D of a finite sum

$$v_1 w_1 + v_2 w_2 + \cdots v_n w_n$$

in which each v_i is in A and each w_i is in B. By properties of conjugates in Proposition C.2.3, this conjugate can be written as

$$\overline{v_1} \cdot \overline{w_1} + \overline{v_2} \cdot \overline{w_2} + \cdots + \overline{v_n} \cdot \overline{w_n}.$$

Since each $\overline{v_i}$ is in $\overline{A}$ and each $\overline{w_i}$ is in $\overline{B}$, this sum of products is in $\overline{A} \cdot \overline{B}$, and so $\overline{AB}$ is a subset of $\overline{A} \cdot \overline{B}$. The reverse inclusion is similar. $\qquad\square$

C.5 Complete quadratic domains and their subdomains

Throughout this section, let $d \neq 1$ be a fixed square-free integer. For each positive rational integer g,

1. let $\Delta_g = \Delta(d, g)$,
2. let $P_g(x)$ be the principal polynomial of discriminant Δ_g,
3. let $z_g = \dfrac{P'_g(0) + \sqrt{\Delta_g}}{2}$, and
4. let $D_g = \{q + rz_g \mid q, r \in \mathbb{Z}\}$, the quadratic domain of discriminant Δ_g.

(When $g = 1$, we may omit the subscript g on these items.) Recall that D is the set of all quadratic integers v whose discriminant is a square multiple of Δ. But $\Delta_g = g^2\Delta$, so that every square multiple of Δ_g is also a square multiple of Δ. It follows that D_g is a subdomain of D. In fact, recalling from Lemma 1.1.3 that

$$P_g(gx) = g^2 P(x) \quad \text{and} \quad P'_g(gx) = gP'(x)$$

for all x, then $z_g = \frac{P'_g(0) + \sqrt{\Delta_g}}{2} = \frac{gP'(0) + g\sqrt{\Delta}}{2} = gz$ and

$$D_g = \{q + rz \in D \mid g \text{ divides } r\}.$$

We refer to $D = D_1$ as a *complete* quadratic domain and to each D_g with $g > 1$ as a *subdomain* of D. More generally, D_g is a subdomain of D_h if and only if h divides g.

An ideal of one quadratic domain might be a subset of some other quadratic domain and, in that case, may or may not be an ideal of that domain. In this situation, there can be some ambiguity in terminology that we associate to an ideal, such as the divisor or norm of an ideal. (More precisely, it emphasizes that these terms depend not just on an ideal as a set but on the quadratic domain of which it is an ideal.) Our next proposition illustrates this.

Proposition C.5.1. *Let $d \neq 1$ be a square-free integer and, for each positive integer g, let D_g be the quadratic domain of discriminant $\Delta_g = \Delta(d, g)$. Suppose that A is a nontrivial ideal of D_g for some g and that the divisor of A in D_g is m. Then A is also an ideal of D_{gm}, and the divisor of A in D_{gm} is 1. If $A = m\langle a : k\rangle$ is an ideal number expression for A as an ideal of D_g, then $A = \langle ma : mk\rangle$ as an ideal of D_{gm}.*

Proof. Let A be an ideal of D_g, with $A = m\langle a : k\rangle$ as an ideal number expression. Since $mz_g = m(gz) = (gm)z = z_{gm}$, then A consists of all numbers of the form

$$m(q(a) + r(k + z_g)) = q(ma) + r(mk + mz_g) = q(ma) + r(mk + z_{gm}),$$

where q and r are rational integers. Thus, A is a subset of D_{gm} and is also an ideal of D_{gm}, as we now show. The set A has the closure properties for addition required of an ideal. (These do not depend on the domain of which A is a subset.) Given any v in A and x in D_g, we know that vx is in A. Since every x in D_{gm} is also in D_g, then A continues to have the strong closure property for multiplication when viewed as a subset of D_{gm}. With $mk + z_{gm}$ in A, and with ma the smallest positive rational integer in A, we see that the divisor of A as an ideal of D_{gm} is 1 and that $A = \langle ma : mnk\rangle$ is an ideal number expression of A in D_{gm}. □

Example. Let $d = -35$. If $g = 3$, then the principal polynomial of discriminant $\Delta_3 = \Delta(-35, 3) = -315$ is $P_3(x) = x^2 + 3x + 81$. Here, $a = 9$ divides $P_3(-3) = 81$, so

$$A = 5\langle 9 : -3\rangle = \{5(q(9) + r(-3 + z_3)) \mid q, r \in \mathbb{Z}\}$$

is an ideal of $D_3 = \{q + rz_3 \mid q, r \in \mathbb{Z} \text{ and } z_3 = \frac{3 + \sqrt{-315}}{2}\}$, the quadratic domain of discriminant -315. We can also write

$$A = \{q(45) + r(-15 + 5z_3) \mid q, r \in \mathbb{Z}\} = \{q(45) + r(-15 + z_{15}) \mid q, r \in \mathbb{Z}\},$$

where $z_{15} = \frac{15+\sqrt{-7875}}{2} = 5z_3$. Proposition C.5.1 shows that $A = \langle 45 : -15 \rangle$ as an ideal of D_{15}, the quadratic domain of discriminant $\Delta_{15} = \Delta(-35, 15) = -7875$. To verify more directly that A is an ideal of this domain, note that $P_{15}(x) = x^2 + 15x + 2025$ is the principal polynomial of discriminant Δ_{15} and that 45 divides $P_{15}(-15) = 2025$.

Proposition C.5.1 implies that if A is an ideal of some quadratic domain D_g, then there is a largest positive integer ℓ so that A is an ideal of D_ℓ. In that case, the divisor of A in D_ℓ is 1. If h divides ℓ, then it is always true that A is a subset of D_h. But A may or may not be an ideal of D_h, depending on whether A retains the strong closure property for multiplication. (We know that A contains vx if v is in A and x is D_ℓ, but since D_ℓ is a subset of D_h, this is not necessarily still true if x is an arbitrary element of D_h.) The following general definition helps us determine which is the case.

Definition. Let A be an ideal of D_g, the quadratic domain of discriminant $\Delta_g = \Delta(d, g)$, and suppose that A has an ideal number expression $m\langle a : k \rangle$ in D_g. Let $P_g(k) = ac$ and $P'_g(k) = b$, where $P_g(x)$ is the principal polynomial of discriminant Δ_g. Then the *content* of A in D_g is defined as $\gcd(a, b, c)$.

The content of an ideal $A = m\langle a : k \rangle$ in D_g is the same as the content of the ideal number $(a : k)$ in $\mathcal{I}_{\Delta_g}$. By Proposition 2.1.3, this content is the same for all members of the congruence class of $(a : k)$, so the content of an ideal A of D is well-defined. Furthermore, as shown in Lemma 1.5.2, the content of an ideal of D_g divides g, the index of $\Delta_g = \Delta(d, g)$. However, as with the divisor of an ideal, the content of A can vary if A is an ideal of more than one quadratic domain.

Example. In the preceding example, we saw that $5\langle 9 : -3 \rangle$ and $\langle 45 : -15 \rangle$ are ideal number expressions for the same set A as an ideal of the quadratic domains (D_3 and D_{15}) of discriminant $\Delta_3 = -315$ and $\Delta_{15} = -7875$, respectively. The principal polynomials of these discriminants are

$$P_3(x) = x^2 + 3x + 81 \quad \text{and} \quad P_{15}(x) = x^2 + 15x + 2025.$$

With $P_3(-3) = 81 = 9 \cdot 9$ and $P'_3(-3) = -3$, we find that the content of A as an ideal of D_3 is $\gcd(9, -3, 9) = 3$. On the other hand, since $P_{15}(-15) = 2025 = 45 \cdot 45$ and $P'_{15}(-15) = -15$, then $\gcd(45, -15, 45) = 15$ is the content of A as an ideal of D_{15}.

Proposition C.5.2. *Let $d \neq 1$ be a square-free integer and, for each positive integer g, let D_g be the quadratic domain of discriminant $\Delta_g = \Delta(d, g)$. Suppose that A is a nontrivial ideal of D_ℓ for some positive integer ℓ, and that the divisor and content of A in D_ℓ are 1 and m, respectively. Then $\ell = mh$ for some integer h, and A is un ideal of D_h having divisor m and content 1.*

Proof. Let A be an ideal of D_ℓ with divisor 1, so that A has an ideal number expression $\langle a_1 : k_1 \rangle$ in D_ℓ. Let $P_\ell(x)$ be the principal polynomial of discriminant Δ_ℓ, and let $P_\ell(k_1) = a_1 c_1$ and $P'_\ell(k_1) = b_1$. If, as given, A has content m in D_ℓ, then $a_1 = ma$, $b_1 = mb$, and

$c_1 = mc$ for some integers a, b, and c. As noted above, m divides ℓ, so we can let $\ell = mh$ for some integer h. Since $b^2 - 4ac = \Delta_h$, we find that $k = \frac{b - P'_h(0)}{2}$ is an integer and $k_1 = mk$. Now we find that

$$ma \cdot mc = a_1 c_1 = P_\ell(k_1) = P_\ell(mk) = m^2 P_h(k)$$

and

$$mb = b_1 = P'_\ell(k_1) = P'_\ell(mk) = m P'_h(k).$$

Since a divides $P_h(k)$, then $\langle a : k \rangle$ is an ideal of D_h, as is $m\langle a : k \rangle$. In fact,

$$m\langle a : k \rangle = \{m(q(a) + r(k + z_h)) \mid q, r \in \mathbb{Z}\} = \{q(a_1) + r(k_1 + z_\ell) \mid q, r \in \mathbb{Z}\} = A.$$

That is, A is an ideal of D_h with divisor m and content $\gcd(a, b, c) = 1$. ☐

Example. In previous examples, we saw that $\langle 45 : -15 \rangle$ is an ideal number expression for an ideal A in the quadratic domain D_{15} of discriminant $-7875 = \Delta(-35, 15)$ and the content of A in D_{15} is $m = 15$. Proposition C.5.2 shows that A can also be written as $15\langle 3 : -1 \rangle$ in the quadratic domain D of discriminant $\Delta = \Delta(-35, 1) = -35$. Using Proposition C.5.1, we also have that $A = 3(5\langle 3 : -1 \rangle) = 3\langle 15 : -5 \rangle$ in D_5, the quadratic domain of discriminant $\Delta_5 = \Delta(-35, 5) = -875$, as well as the other expressions for A in previous examples.

We summarize the implications of Propositions C.5.1 and C.5.2 in the following corollary, without additional proof.

Corollary C.5.3. *Let $d \neq 1$ be a square-free integer and, for each positive integer g, let D_g be the quadratic domain of discriminant $\Delta_g = \Delta(d, g)$. Suppose that A is a nontrivial ideal of D_g for some positive integer g. Then there is a largest positive integer ℓ and a smallest positive integer h so that A is an ideal of D_ℓ and an ideal of D_h. In this case, $\ell = mh$ for some integer m, and A is an ideal of D_g if and only if h divides g and g divides ℓ. If n is a positive divisor of m, then the content and divisor of A as an ideal of D_{nh} are n and $\frac{m}{n}$, respectively.*

C.6 Multiplication of ideals of a quadratic domain

In Section C.1, we introduced an operation of multiplication on the set of ideals of an arbitrary integral domain. Our goal in Section C.6 is to develop a formula for ideal multiplication in a quadratic domain using ideal number expressions. We begin by recalling terminology introduced in previous sections.

Throughout this section, let $D = \{q + rz \mid q, r \in \mathbb{Z}\}$ be the quadratic domain and let $P(x)$ be the principal polynomial of some fixed discriminant $\Delta = \Delta(d, g)$. Every nontrivial ideal A of D can be expressed as

$$A = m\langle a : k \rangle = \{m(qa + rk) + mrz \mid q, r \in \mathbb{Z}\} \tag{C.9}$$

for some positive rational integers m and a and some rational integer k. We use the following terms in reference to A as an ideal of D:

1. The integer m is called the *divisor* of A in D, written as $m = \text{div}(A)$ when D is clear from context. This divisor can be described as the smallest positive coefficient m of z in an element of A.

2. The *norm* of A is $N(A) = m^2 a$, the product of the smallest positive rational integer in A and the smallest positive coefficient of z in an element of A.

3. A set of the form in equation (C.9) is an ideal of D if and only if a divides $P(k)$. If $P(k) = ac$ and $P'(k) = b$, then $\gcd(a, b, c)$ is called the *content* of A in D. If D is clear from context, we write the content of A in D as $\text{cont}(A)$.

As we saw with examples in Section C.4, we can determine an ideal number expression for an ideal A of a quadratic domain D by finding the smallest positive rational integer in A and an element of minimal positive coefficient of z in A. Our formula for a product of ideals arises in this way, as the following example illustrates.

Example. Let D be the quadratic domain of discriminant $\Delta = -35$, with $P(x) = x^2 + x + 9$ as its principal polynomial. Since $P(-1) = 9$ and $P(-3) = 15$, then $A = \langle 9 : -1 \rangle$ and $B = \langle 15 : -3 \rangle$ are ideals of D. The typical product of a $\mathbb{Z}$-combination of $\{9, -1 + z\}$ with a $\mathbb{Z}$-combination of $\{15, -3 + z\}$ is a $\mathbb{Z}$-combination of the set of products

$$\{9(15), 9(-3 + z), 15(-1 + z), (-1 + z)(-3 + z)\} = \{135, -27 + 9z, -15 + 15z, -6 - 3z\}.$$

(We use the product formula of equation (C.4) in the final calculation.) An element of AB is a finite sum of products of elements of A with elements of B, so is likewise a $\mathbb{Z}$-combination of this set. Hence, we can describe the product AB as the set of all

$$(135q - 27r - 15s - 6t) + (9r + 15s - 3t)z, \tag{C.10}$$

where $q, r, s,$ and t are integers.

Here, $n = \gcd(9, 15, -3) = 3$ is the smallest positive coefficient of z in an element of AB, for instance, with $6 + 3z = 3(2 + z)$ in AB when $t = -1$ and $q = r = s = 0$. An element of the form in (C.10) is a rational integer only when $9r + 15s - 3t = 0$, that is, when $t = 3r + 5s$. In that case,

$$135q - 27r - 15s - 6t = 135q - 27r - 15s - 6(3r + 5s) = 135q - 45r - 45s = 45(3q - r - s),$$

and the smallest positive integer of that form is $45 = 3(15)$, say when $r = -1, q = s = 0$, and $t = -3$. We conclude that $AB = 3\langle 15 : 2 \rangle$.

If $C = \langle 15 : 2 \rangle$, then similar calculations show that AC is the set of all

$$(135q + 18r - 15s - 11t) + (9r + 15s + 2t)z. \tag{C.11}$$

Here, $\gcd(9, 15, 2) = 1$, and we find that AC contains $62 + z$, say when $q = 0$, $r = 2$, $s = -1$, and $t = -1$. An element of the form in (C.11) is a rational integer when $9r + 15s + 2t = 0$. If so, then $-15s = 9r + 2t$ and

$$135q + 18r - 15s - 11t = 135q + 18r + (9r + 2t) - 11t = 135q + 27r - 9t = 9(15q + 3r - t).$$

Here, however, some caution is required, as we must select r and t so that $9r + 2t$ is divisible by 15. (That is, s must be an integer, although it does not appear in this expression.) We find that $t \equiv 3r \pmod{15}$, so that $t = 3r + 15x$ for some integer x. Therefore, $9(15q + 3r - t) = 9(15q - 15x) = 135(q - x)$. The smallest positive rational integer of this form is 135, and we conclude that $AC = \langle 135 : 62 \rangle$.

Following the method of the preceding example, we can establish a general formula for multiplication of ideals of $D(\Delta)$ written as ideal number expressions. If $A_1 = mA$ and $B_1 = nB$, then $A_1 B_1 = (mn)AB$. Thus, we can concentrate on multiplying ideals of the form $A = \langle a : k \rangle$ and $B = \langle b : \ell \rangle$, that is, with divisor 1. If $P(x)$ is the principal polynomial of discriminant Δ, then finite sums of products of elements of A with elements of B produce all $\mathbb{Z}$-combinations of the set of products

$$\{ab, a(\ell + z), b(k + z), (k + z)(\ell + z)\} = \{ab, a\ell + az, bk + bz, v + uz\},$$

where $u = k + \ell + P'(0)$ and $v = k\ell - P(0)$. The typical element of AB thus has the form

$$(abq + a\ell r + bks + vt) + (ar + bs + ut)z, \tag{C.12}$$

for some integers q, r, s, and t. As the example illustrates, we can determine an ideal number expression for AB by finding the minimal positive value of $ar + bs + ut$ and describing all rational integers given by the formula in (C.12). We begin with some special cases in the following propositions.

Proposition C.6.1. *Let D be the quadratic domain of discriminant $\Delta = \Delta(d, g)$. Then for every positive divisor a of g, there is a unique ideal of D whose norm and content both equal a, namely $\langle a : 0 \rangle$. If a and b are positive divisors of g, with greatest common divisor m and least common multiple n, then*

$$\langle a : 0 \rangle \cdot \langle b : 0 \rangle = \gcd(a, b)\langle \mathrm{lcm}(a, b) : 0 \rangle = m\langle n : 0 \rangle. \tag{C.13}$$

Proof. Let $P(x)$ be the principal polynomial of discriminant $\Delta = \Delta(d, g)$. Then g divides $P'(0)$ and g^2 divides $P(0)$, the linear and constant coefficients of $P(x)$ respectively. Thus, if a divides g, then likewise a^2 divides $P(0)$ and a divides $P'(0)$. It follows that $\langle a : 0 \rangle$ is an ideal of D, and that if $P(0) = ac$ and $P'(0) = b$, then a divides both b and c. Hence, the content of $\langle a : 0 \rangle$ is $\gcd(a, b, c) = a$. If $\langle a : k \rangle$ is an ideal number expression for an ideal with content m in some quadratic domain, then the proof of Proposition C.5.2 show that m divides k. Thus, $\langle a : 0 \rangle$ can be the only ideal of D with norm and content a.

Now let a and b be positive divisors of g, so that $\langle a : 0 \rangle$ and $\langle b : 0 \rangle$ are ideals of D. By equation (C.12), the typical element of $\langle a : 0 \rangle \cdot \langle b : 0 \rangle$ is

$$(abq - P(0)t) + (ar + bs + P'(0)t)z$$

for some integers q, r, s, and t. Since a and b both divide g, and thus divide $P'(0)$, the smallest positive coefficient of z in such an element is $\gcd(a, b) = m$. Furthermore, we can select r and s so that $ar + bs = m$, and then we find that $0 + mz$ is an element of this product, when $q = t = 0$. With a and b both dividing g, then ab divides $P(0)$, and ab divides every rational integer in the product. We obtain ab itself when $q = 1$ and $r = s = t = 0$. Since $n = \text{lcm}(a, b) = \frac{ab}{\gcd(a,b)}$, we conclude that $\langle a : 0 \rangle \cdot \langle b : 0 \rangle = m\langle n : 0 \rangle$, as we wanted to show. $\qquad\square$

Proposition C.6.2. *Let $A = m\langle a : k \rangle$ be an ideal of a quadratic domain $D = D_\Delta$ and let $\overline{A}$ be its conjugate. If A has content n as an ideal of D, then*

$$A \cdot \overline{A} = N(A) \cdot \langle n : 0 \rangle.$$

Proof. Let $P(x)$ be the principal polynomial of discriminant Δ, and write $P(k) = ac$ and $P'(k) = b$, so that the content of A is $\gcd(a, b, c) = n$. By Proposition C.4.3, we can write $\overline{A} = m\langle a : \overline{k} \rangle$, where $\overline{k} = -k - P'(0)$ is the conjugate of k with respect to Δ. Let

$$u = k + \overline{k} + P'(0) = 0$$

and

$$v = k \cdot \overline{k} - P(0) = -k^2 - P'(0)k - P(0) = -P(k) = -ac.$$

Then the typical element of $\langle a : k \rangle \cdot \langle a : \overline{k} \rangle$, by equation (C.12), has the form

$$(a^2q + a\overline{k}r + aks - act) + (ar + as)z,$$

with q, r, s, and t integers. The smallest positive coefficient of z in such an expression is a, for instance, with $ak + az$ when $s = 1$ and $q = r = t = 0$. Note that $\overline{k} - k = -2k - P'(0) = -P'(k) = -b$. Hence, when $s = -r$, then

$$(a^2q + a\overline{k}r + aks - act) + (ar + as)z = a(aq - br - ct)$$

is a rational integer in $\langle a : k \rangle \cdot \langle a : \overline{k} \rangle$. Here, $a \cdot \gcd(a, b, c) = an$ is the smallest positive possibility, and thus

$$A \cdot \overline{A} = m\langle a : k \rangle \cdot m\langle a : \overline{k} \rangle = m^2 a \cdot \langle n : k \rangle = N(A) \cdot \langle n : k \rangle.$$

Finally, note that $P(k) = ac = n \cdot \frac{a}{n}c$ and $P'(k) = b$, with $\frac{a}{n}c$ and b both divisible by n. Thus, the ideal $\langle n : k \rangle$ has content n and, as noted in Proposition C.6.1, equals $\langle n : 0 \rangle$. Therefore, $A \cdot \overline{A} = N(A) \cdot \langle n : 0 \rangle$, as we wanted to show. □

Corollary C.6.3. *If $A = wD$ is the principal ideal of the quadratic domain D generated by w, then A has content* 1.

Proof. If A is an ideal of a quadratic domain D having content n, then Proposition C.6.2 implies that the smallest positive rational integer in A is $N(A) \cdot n$. By Corollary C.4.2, if $A = wD$ for some element w, then $N(A) = |N(w)|$. But $N(w) = w \cdot \overline{w}$ is an element of $A = wD$, since $\overline{w}$ is an element of D. Thus, n, the content of $A = wD$, must equal 1. □

Proposition C.6.4. *Let A and B be ideals of a quadratic domain D. Then*

$$N(AB) = N(A) \cdot N(B) \cdot \gcd(\mathrm{cont}(A), \mathrm{cont}(B))$$

and

$$\mathrm{cont}(AB) = \mathrm{lcm}(\mathrm{cont}(A), \mathrm{cont}(B)).$$

Proof. Let $n_1 = \mathrm{cont}(A)$, $n_2 = \mathrm{cont}(B)$, and $n = \mathrm{cont}(AB)$. We have that

$$AB \cdot \overline{AB} = (A \cdot \overline{A})(B \cdot \overline{B})$$

by Proposition C.4.4 and other properties of ideal multiplication. On one hand,

$$AB \cdot \overline{AB} = N(AB) \cdot \langle n : 0 \rangle$$

by Proposition C.6.2. Using Proposition C.6.2 and equation (C.13), we also have

$$(A \cdot \overline{A})(B \cdot \overline{B}) = N(A)\langle n_1 : 0 \rangle \cdot N(B)\langle n_2 : 0 \rangle$$
$$= N(A) \cdot N(B) \cdot \gcd(n_1, n_2) \cdot \langle \mathrm{lcm}(n_1, n_2) : 0 \rangle.$$

We conclude that $N(AB) = N(A) \cdot N(B) \cdot \gcd(n_1, n_2)$ by comparing the minimal positive co-efficient of z in these ideals. Likewise, $n = \mathrm{lcm}(n_1, n_2)$ by comparing the minimal positive rational integer in both ideals. □

Finally, we obtain the following formula for multiplication of ideals of quadratic domains, written as ideal number expressions. As before, we can restrict our attention to ideals of the form $\langle a : k \rangle$ as part of the product.

Theorem C.6.5. *Let $A = \langle a : k \rangle$ and $B = \langle b : \ell \rangle$ be ideals of the quadratic domain D_Δ. Let $P(x)$ be the principal polynomial of discriminant Δ, and let*

$$u = k + \ell + P'(0) \quad and \quad v = k\ell - P(0).$$

Then $AB = n\langle c : m\rangle$ where n, c, and m are defined as follows:
1. $n = \gcd(a, b, u)$.
2. $c = \frac{a}{n} \cdot \frac{b}{n} \cdot \gcd(\text{cont}(A), \text{cont}(B))$.
3. *If $n = ar + bs + ut$ for some integers r, s, and t, then*

$$m = \frac{1}{n}(a\ell r + bks + vt).$$

Proof. As in equation (C.12), the typical element of AB has the form

$$(abq + a\ell r + bks + vt) + (ar + bs + ut)z,$$

for some integers q, r, s, and t, where $z = \frac{P'(0) + \sqrt{\Delta}}{2}$. The smallest positive coefficient of z is $n = \gcd(a, b, u)$. If we select r, s, and t so that $ar + bs + ut = n$, and let $q = 0$, then $(a\ell r + bks + vt) + nz$ is an element of AB. This has the form $n(m + z)$ with m defined as in (3) above. Thus, AB can be written as $n\langle c : m\rangle$ for some integer c. If so, then $N(AB) = n^2 c$. But by Proposition C.6.4,

$$N(AB) = N(A) \cdot N(B) \cdot \gcd(\text{cont}(A), \text{cont}(B)) = ab \cdot \gcd(\text{cont}(A), \text{cont}(B)).$$

The formula for c in (2) follows by comparing these two expressions. □

Example. Let $\Delta = \Delta(-35, 3) = -315$, with $P(x) = x^2 + 3x + 81$ as its principal polynomial, and let $D = D_{-315}$. Since $P(3) = 99 = 3 \cdot 33$, then $A = \langle 3 : 3\rangle$ is an ideal of D. The content of A is

$$\text{cont}(A) = \gcd(3, 9, 33) = 3.$$

Consider the product $A^2 = A \cdot A$. Here, $u = 3 + 3 + 3 = 9$, so that $n = \gcd(3, 3, 9) = 3 = 3(1) + 3(0) + 9(0)$. We find that $c = \frac{3}{3} \cdot \frac{3}{3} \cdot \gcd(3, 3) = 3$ and $m = \frac{1}{3}(3 \cdot 3 \cdot 1) = 3$. Thus, $\langle 3 : 3\rangle \cdot \langle 3 : 3\rangle = 3\langle 3 : 3\rangle$.

C.7 Composition of primitive ideals

In this section, we introduce a revised multiplication operation on a subset of the set of all ideals of a given quadratic domain. This operation, which we will call *composition*, is precisely the same as the composition operation that we introduced on $\mathcal{G}_\Delta$, the collection of congruence classes of primitive ideal numbers of discriminant Δ, in Chapter 3. In this way, we establish the properties of that operation that are claimed in that chapter.

Definition. Let A be a nontrivial ideal of a quadratic domain D. We say that A is *primitive* in D if the divisor of A in D and the content of A in D both equal 1.

In many texts, an ideal A of D is called primitive simply if its divisor is 1. The definition above is more consistent with our use of the term primitive for ideal numbers and quadratic forms. As we saw in Section C.5, the divisor of an ideal A might be 1 in some quadratic domain but larger than 1 in a different domain. On the other hand, using Corollary C.5.3, an ideal A of D is primitive, according to the preceding definition, if and only if D is the only quadratic domain of which A is an ideal.

Every primitive ideal A of a quadratic domain $D = D(\Delta)$ has an ideal number expression of the form $A = \langle a : k \rangle$. If $P(x)$ is the principal polynomial of discriminant Δ, with $P(k) = ac$ and $P'(k) = b$, then $\gcd(a, b, c) = 1$. Thus, this collection of ideals can be identified with the set of primitive elements in $\mathcal{Q}_\Delta$, and we will denote that collection of ideals as $\mathcal{G}_\Delta$ in this section.

If A and B are elements of $\mathcal{G}_\Delta$, then Proposition C.6.4 shows that

$$\mathrm{cont}(AB) = \mathrm{lcm}(\mathrm{cont}(A), \mathrm{cont}(B)) = \mathrm{lcm}(1, 1) = 1.$$

But the set $\mathcal{G}_\Delta$ is not closed under ideal multiplication, as the product of two ideals with divisor 1 might have a divisor greater than 1. See, for instance, the first example of ideal multiplication in Section C.6, in which we found that

$$\langle 9 : -1 \rangle \cdot \langle 15 : -3 \rangle = 3 \langle 15 : 2 \rangle$$

in the collection of ideals of discriminant $\Delta = -35$. However, the following slight variation on ideal multiplication gives us a well-defined operation on $\mathcal{G}_\Delta$.

Definition. Let A and B be primitive ideals of the quadratic domain $D = D_\Delta$. If n is the divisor, in D, of the product AB, then define the *composite* of A and B to be $A \circ B = \frac{1}{n} AB$. Put another way, if $A = \langle a : k \rangle$ and $B = \langle b : \ell \rangle$ as ideal number expressions, then let

$$\langle a : k \rangle \circ \langle b : \ell \rangle = \langle c : m \rangle \quad \text{if and only if} \quad \langle a : k \rangle \cdot \langle b : \ell \rangle = n \langle c : m \rangle.$$

Theorem C.7.1. *Let Δ be a discriminant and let $\mathcal{G}_\Delta$ be the set of primitive ideals of the quadratic domain $D = D(\Delta)$. Then $\mathcal{G}_\Delta$ is an Abelian group under the operation of composition defined above.*

Proof. In Proposition C.1.3, we saw that ideal multiplication is commutative and associative. It is not difficult to see that these properties carry over to the ideal composition operation. To demonstrate the associative property, for example, let A_1, A_2, and A_3 be primitive ideals of D. Suppose that $A_1 A_2 = n B_3$ and that $A_2 A_3 = m B_1$ for some positive integers n and m and primitive ideal B_3 and B_1. Suppose further that $B_3 A_3 = q C_3$ and that $A_1 B_1 = r C_1$ for some positive integers q and r and primitive ideals C_3 and C_1. Then $(A_1 A_2) A_3 = (nq) C_3$ and $A_1 (A_2 A_3) = (mr) C_1$. Since these products are equal, we find, by comparing the minimal positive coefficient of z in each for instance, that $nq = mr$, and then that $C_3 = C_1$. But now we find that in $\mathcal{G}_\Delta$,

$$(A_1 \circ A_2) \circ A_3 = B_3 \circ A_3 = C_3 = C_1 = A_1 \circ B_1 = A_1 \circ (A_2 \circ A_3).$$

Thus composition is associative.

The improper ideal D of D, which has ideal number expression $\langle 1 : 0 \rangle$, is primitive. We saw in Proposition C.1.3 that $AD = A$ for every ideal A of D. When A is primitive, it follows that $A \circ D = A$ as well. Thus, D serves as an identity element for composition in $\mathcal{G}_\Delta$.

Finally, if $A = \langle a : k \rangle$ is a primitive ideal of D, then $\overline{A} = \langle a : \overline{k} \rangle$ is also primitive in D. (If cont$(A) = \gcd(a, b, c)$, then cont$(\overline{A}) = \gcd(a, -b, c)$.) Proposition C.6.2 shows that $A \cdot \overline{A} = N(A)\langle 1 : 0 \rangle = N(A)D$ when the content of A is 1. Thus, $A \circ \overline{A} = D$ by definition. Therefore, the conjugate of A serves as an inverse for a primitive ideal A under composition. This completes the proof that $\mathcal{G}_\Delta$ has all the properties of an Abelian group under composition. ☐

We conclude Section C.7 with the following corollary ensuring that the composition operation on primitive ideal numbers presented in Chapter 3 is well-defined.

Corollary C.7.2. *Let $\langle a : k \rangle$ and $\langle b : \ell \rangle$ be congruence classes of primitive ideal numbers of discriminant Δ. Let $u = k + \ell + P'(0)$ and $v = k\ell - P(0)$, where $P(x)$ is the principal polynomial of discriminant Δ. Suppose that $n = \gcd(a, b, u) = ar + bs + ut$ for some integers r, s, and t. Then the composition operation on $\mathcal{G}_\Delta$,*

$$\langle a : k \rangle \circ \langle b : \ell \rangle = \langle c : m \rangle$$

where

$$c = \frac{a}{n} \cdot \frac{b}{n} \quad and \quad m = \frac{a\ell r + bks + vt}{n} \tag{C.14}$$

is well-defined. That is, the result is not affected by the choice of the representatives of the congruence classes of $\langle a : k \rangle$ and $\langle b : \ell \rangle$ nor by the choice of the integers r, s, and t.

Proof. Let A and B be the ideals of $D(\Delta)$ having ideal number expressions $\langle a : k \rangle$ and $\langle b : \ell \rangle$, respectively. Any ideal number in the congruence class $\langle a : k \rangle$ represents the same ideal A and similarly for any ideal number in the congruence class $\langle b : \ell \rangle$. The composition formula in equation (C.14) is adapted from the ideal multiplication formula in Theorem C.6.5, and produces the composite ideal $A \circ B$. (Since A and B are primitive, the content of each is 1. The divisor n of AB is not part of the final expression.) The resulting congruence class $\langle c : m \rangle$ must be an ideal number expression for $A \circ B$, which is uniquely determined by A and B. Thus, the operation is not affected by any choices made in its calculation. ☐

C.8 Equivalence of ideals

We now introduce a relation of equivalence on the ideals of a fixed quadratic domain. In Section C.10, we will establish a close connection between this relation and the equivalence of ideal numbers defined in Chapter 4.

Definition. Let A and B be ideals of a quadratic domain D. We say that A is *equivalent* to B, written as $A \sim B$, if there is a nonzero rational integer m and a nonzero element w of D so that $mA = wB$.

The trivial ideal of D is equivalent only to itself. We will restrict our attention to nontrivial ideals in what follows.

Proposition C.8.1. *Ideal equivalence is an equivalence relation on the set of all ideals of a quadratic domain D.*

Proof. Let A, B, and C be ideals of a quadratic domain D.
1. Since $1 \cdot A = 1 \cdot A$, then $A \sim A$. That is, equivalence is reflexive.
2. Suppose that $mA = wB$ with $m \neq 0$ in $\mathbb{Z}$ and $w \neq 0$ in D. If $\overline{w}$ is the conjugate of w in D, then $\overline{w}(mA) = \overline{w}(wB)$, so that $N(w)B = (m\overline{w})A$. Here, $N(w)$ is a nonzero rational integer, and we conclude that $B \sim A$, so that equivalence is symmetric.
3. If $mA = vB$ and $nB = wC$, with m and n nonzero in $\mathbb{Z}$ and v and w nonzero in D, then $(mn)A = n(mA) = n(vB) = v(nB) = v(wC) = (vw)C$. Thus, $A \sim C$ and equivalence is transitive.

Therefore, ideal equivalence has the properties of an equivalence relation. □

We denote the equivalence class of an ideal A under this relation as $[A]$.

Proposition C.8.2. *Let A be an ideal of a quadratic domain D and let w be a nonzero element of D. Then wA is equivalent to A. Furthermore, A is equivalent to the improper ideal D if and only if A is a principal ideal of D.*

Proof. For the first claim, note that $1 \cdot (wA) = wA$, which is sufficient to show that $wA \sim A$. In particular, then $wD \sim D$, so that every principal ideal is equivalent to D as an ideal of itself. Conversely, suppose that $A \sim D$, say with $mA = wD$ for some $m \neq 0$ in $\mathbb{Z}$ and $w \neq 0$ in D. Since 1 is an element of D, then $mx = w$ for some x in A. But then $mA = mxD$, and we conclude that $A = xD$ is a principal ideal of D. □

Proposition C.8.2 shows that each equivalence class of nontrivial ideals of D contains a representative A for which $\mathrm{div}(A) = 1$. We also have the following result leading to a definition of primitive ideal classes.

Proposition C.8.3. *Let A and B be nontrivial ideals of a quadratic domain D. If A is equivalent to B, then the content of A in D equals the content of B in D.*

Proof. Suppose that A is equivalent to B, say with $mA = wB$. By Proposition C.1.4, we can rewrite this equation as $mD \cdot A = wD \cdot B$. The content of these two products of ideals must be equal. Thus, using Proposition C.6.4,

$$\text{cont}(mD \cdot A) = \text{lcm}(\text{cont}(mD), \text{cont}(A))$$
$$= \text{lcm}(\text{cont}(wD), \text{cont}(B)) = \text{cont}(wD \cdot B).$$

But by Corollary C.6.3, the content of a principal ideal is 1. Therefore,

$$\text{cont}(A) = \text{lcm}(\text{cont}(mD), \text{cont}(A)) = \text{cont}(mA)$$
$$= \text{cont}(wB) = \text{lcm}(\text{cont}(wD), \text{cont}(B)) = \text{cont}(B),$$

as we wanted to show. □

Thus, if A is a nontrivial ideal of a quadratic domain D, we can refer to the content of its equivalence class, cont[A], without ambiguity. We will say that [A] is *primitive* if cont[A] = 1. We can assume that A itself is primitive in this case. That is, we can select a representative for [A] having divisor 1.

Theorem C.8.4. *Let D be the quadratic domain of discriminant Δ and let $\mathcal{IC}_\Delta$ be the set of primitive classes of ideals of D. Then $\mathcal{IC}_\Delta$ is an Abelian group under the following composition operation:*

$$[A] \circ [B] = [A \circ B]. \tag{C.15}$$

Proof. Suppose that A, B, A_1, and B_1 are primitive ideals of D, and that $A \sim A_1$ and $B \sim B_1$, say with $mA = vA_1$ and $nB = wB_1$ for nonzero rational integers m and n and nonzero elements v and w of D. Then

$$mn(AB) = (mA)(nB) = (vA_1)(wB_1) = vw(A_1B_1),$$

so that $AB \sim A_1B_1$. Likewise, $A \circ B \sim A_1 \circ B_1$, using Proposition C.8.2, since AB and A_1B_1 are rational integer multiples of $A \circ B$ and $A_1 \circ B_1$, respectively. Thus, the operation in equation (C.15) is well-defined. It is then an immediate consequence of Theorem C.7.1 that $\mathcal{IC}_\Delta$ has all the properties of an Abelian group under this composition operation. □

We refer to $\mathcal{IC}_\Delta$ as the *ideal class group* of discriminant Δ.

C.9 Ordered bases of ideals

In this section, we define ordered bases and their associated quadratic forms, which we can associate to each nontrivial ideal of D. We will see in Section C.10 that this provides a connection between equivalence of quadratic forms (or ideal numbers) and equivalence of ideals of a quadratic domain, as defined in Section C.8. Throughout this section, we let $D = D(\Delta) = \{q + rz \mid q, r \in \mathbb{Z}\}$ be the quadratic domain of discriminant Δ, with $z = z_\Delta$ defined as in equation (C.1). For each element $w = q + rz$ in D, let $\overline{w} = q + r\overline{z}$ be the conjugate of w in D. Note that $w = \overline{w}$ precisely when w is a rational integer. If u and v are elements of D, and m and n are rational integers, then we refer to the element $mu + nv$ in D as a $\mathbb{Z}$-combination of u and v.

Definition. Let A be a nontrivial ideal of the quadratic domain $D = D(\Delta)$. An *ordered basis* for A is a 1×2 matrix $S = [u \quad v]$ whose entries are elements of A satisfying the following properties:

1. Every element of A can be written as a $\mathbb{Z}$-combination of u and v.
2. $mu + nv = 0$ only if $m = 0$ and $n = 0$.
3. $\overline{u} \cdot v - u \cdot \overline{v} = N(A)\sqrt{\Delta}$, where $N(A)$ is the norm of A as an ideal of D.

Proposition C.9.1. *Let $D = D(\Delta) = \{q + rz \mid q, r \in \mathbb{Z}\}$ be a quadratic domain. If an ideal A of D has an ideal number expression $A = \langle a : k \rangle$ with $a > 0$, then $[a \quad k + z]$ is an ordered basis for A.*

Proof. By definition, $A = \langle a : k \rangle = \{m(a) + n(k + z) \mid m, n \in \mathbb{Z}\}$, so that each element of A is a $\mathbb{Z}$-combination of a and $k + z$. If

$$m(a) + n(k + z) = (ma + nk) + nz = 0 = 0 + 0z,$$

then $n = 0$ and $0 = ma + nk = ma$. Since a is positive, then $m = 0$. Now if $u = a = a + 0z$ and $v = k + z$, then

$$\overline{u} \cdot v - u \cdot \overline{v} = a(k + z) - a(k + \overline{z}) = a(z - \overline{z}) = N(A)\sqrt{\Delta},$$

since the norm of $A = \langle a : k \rangle$ is a, and $z - \overline{z} = \frac{P'(0) + \sqrt{\Delta}}{2} - \frac{P'(0) - \sqrt{\Delta}}{2} = \sqrt{\Delta}$. $\square$

Proposition C.9.2. *Let $[u \quad v]$ be an ordered basis for an ideal A of the quadratic domain $D(\Delta)$ and let $u_1 = qu + rv$ and $v_1 = su + tv$ for some rational integers q, r, s, and t. Then $[u_1 \quad v_1]$ is an ordered basis for A if and only if $qt - rs = 1$.*

Proof. Suppose first that $qt - rs = 1$. If $u_1 = qu + rv$ and $v_1 = su + tv$, then

$$tu_1 - rv_1 = qtu + rtv - rsu - rtv = (qt - rs)u = u$$

and

$$-su_1 + qv_1 = -qsu - rsv + qsu + qtv = (qt - rs)v = v.$$

Then $mu + nv = (mt - ns)u_1 + (-mr + nq)v_1$, which shows that every element of A is a $\mathbb{Z}$-combination of u_1 and v_1. If $0 = mu_1 + nv_1 = (mq + ns)u + (mr + nt)v$, with $[u \quad v]$ an ordered basis for A, then $mq + ns = 0$ and $mr + nt = 0$. But now

$$0 = t(mq + ns) - s(mr + nt) = mqt + nst - mrs - nst = m(qt - rs) = m$$

and

$$0 = -r(mq + ns) + q(mr + nt) = -mqr - nrs + mqr + nqt = n(qt - rs) = n.$$

Finally,

$$\overline{u_1} \cdot v_1 - u_1 \cdot \overline{v_1} = (q\overline{u} + r\overline{v})(su + tv) - (qu + rv)(s\overline{u} + t\overline{v})$$
$$= qs(u \cdot \overline{u}) + qt(\overline{u} \cdot v) + rs(u \cdot \overline{v}) + rt(v \cdot \overline{v})$$
$$\quad - qs(u \cdot \overline{u}) - qt(u \cdot \overline{v}) - rs(\overline{u} \cdot v) - rt(v \cdot \overline{v})$$
$$= (qt - rs)(\overline{u} \cdot v - u \cdot \overline{v}) = N(A)\sqrt{\Delta}, \qquad\qquad \text{(C.16)}$$

from the assumption that $[u \quad v]$ is an ordered basis for A.

Conversely, if $[u_1 \quad v_1]$ is an ordered basis for A, where $u_1 = qu + rv$ and $v_1 = su + tv$, then

$$N(A)\sqrt{\Delta} = \overline{u_1} \cdot v_1 - u_1 \cdot \overline{v_1} = (qt - rs)(\overline{u} \cdot v - u \cdot \overline{v}) = (qt - rs)N(A)\sqrt{\Delta},$$

following the same calculations as in equation (C.16). Since $N(A)\sqrt{\Delta} \neq 0$, we conclude that $qt - rs = 1$. □

We can summarize Proposition C.9.2 by saying that if $S = [u \quad v]$ is an ordered basis for an ideal A of a quadratic domain D, and if $U = \left[\begin{smallmatrix} q & s \\ r & t \end{smallmatrix}\right]$ is a matrix with integer entries, then

$$S_1 = S \cdot U = [u \quad v] \cdot \begin{bmatrix} q & s \\ r & t \end{bmatrix} = [qu + rv \quad su + tv]$$

is an ordered basis for A if and only if U is a *unimodular* matrix, that is, U has determinant 1.

Proposition C.9.3. *Let $[u \quad v]$ be an ordered basis for a nontrivial ideal A of the quadratic domain D of discriminant Δ. Let $N(A)$ be the norm of A and, for each w in D, let $\overline{w}$ be the conjugate of w in D. Then*

$$a = \frac{u \cdot \overline{u}}{N(A)}, \quad b = \frac{\overline{u} \cdot v + u \cdot \overline{v}}{N(A)}, \quad c = \frac{v \cdot \overline{v}}{N(A)} \qquad\qquad \text{(C.17)}$$

are rational integers satisfying $b^2 - 4ac = \Delta$.

Proof. Assume first that the divisor of A is 1, so that $N(A)$ is the smallest positive rational integer in A. Recall in this case that a rational integer n is an element of A if and only if $N(A)$ divides n. Now $u \cdot \overline{u}$ is an element of A since u is in A and $\overline{u}$ is in D. Furthermore, since $\overline{u \cdot \overline{u}} = \overline{u} \cdot u = u \cdot \overline{u}$, then $u \cdot \overline{u}$ is a rational integer. Thus, $N(A)$ divides $u \cdot \overline{u}$, and $a = \frac{u \cdot \overline{u}}{N(A)}$ is a rational integer. In a similar way, we find that $\overline{u} \cdot v + u \cdot \overline{v}$ and $v \cdot \overline{v}$ are rational integers in A, and so b and c as defined in equation (C.17) are rational integers. Finally, we find that

$$b^2 - 4ac = \frac{1}{(N(A))^2}\left((u \cdot v + u \cdot \overline{v})^2 - 4(u \cdot \overline{u})(v \cdot \overline{v})\right)$$

$$= \frac{1}{(N(A))^2}(\overline{u} \cdot v - u \cdot \overline{v})^2 = \Delta,$$

since $\overline{u} \cdot v - u \cdot \overline{v} = N(A)\sqrt{\Delta}$.

More generally, if $A = mB$ for some ideal B of D, then an ordered basis for A has the form $[u \quad v] = [mu_1 \quad mv_1]$, where $[u_1 \quad v_1]$ is an ordered basis for B. Since $N(A) = m^2 N(B)$ in this case, we find that a, b, and c as given in equation (C.17) are precisely as those values would be for the ordered basis $[u_1 \quad v_1]$ of B. In particular, they are rational integers and $b^2 - 4ac = \Delta$. □

Proposition C.9.3 implies that each ordered basis S of an ideal A of the quadratic domain $D = D(\Delta)$ produces a quadratic form $f_S(x, y) = ax^2 + bxy + cy^2$ of discriminant Δ, where a, b, and c are defined as in equation (C.17). We refer to f_S as the *quadratic form of the ordered basis S* for A. The following general example is used in Section C.10.

Example. Let $\langle a : k \rangle$, with $a > 0$, be an ideal number expression for an ideal A of the quadratic domain $D = D(\Delta)$. Then $N(A) = a$ and $S = [a \quad k + z]$ is an ordered basis for A, as we saw in Proposition C.9.1. If $u = a$ and $v = k + z$, then we use the calculations that

$$z + \overline{z} = P'(0) \quad \text{and} \quad z \cdot \overline{z} = \frac{P'(0)^2 - \Delta}{4} = P(0)$$

to establish the following three equations:

$$\frac{u \cdot \overline{u}}{N(A)} = \frac{a \cdot a}{a} = a,$$

$$\frac{\overline{u} \cdot v + u \cdot \overline{v}}{N(A)} = \frac{a(k + z) + a(k + \overline{z})}{a} = 2k + (z + \overline{z}) = P'(k),$$

$$\frac{v \cdot \overline{v}}{N(A)} = \frac{(k + z)(k + \overline{z})}{a} = \frac{k^2 + (z + \overline{z})k + (z \cdot \overline{z})}{a} = \frac{P(k)}{a}.$$

If we let $P(k) = ac$ and $P'(k) = b$, then $f_S(x, y) = ax^2 + bxy + cy^2$. Notice that $f_S = (a : k)$ in ideal number notation.

Proposition C.9.4. *Let f_S be the quadratic form of the ordered basis $S = [u \quad v]$ for an ideal A of a quadratic domain D. Let U be a unimodular matrix. Then the quadratic form of the ordered basis $S_1 = S \cdot U$ is the same as $f_S \cdot U$, as that form is defined in Theorem 4.1.2.*

Proof. Let $U = [\begin{smallmatrix} q & s \\ r & t \end{smallmatrix}]$, so that $S_1 = [u_1 \quad v_1]$ with $u_1 = qu + rv$ and $v_1 = su + tv$. If f_S has coefficients a, b, and c as in equation (C.17), then we find that

$$u_1 \cdot \overline{u_1} = (qu + rv)(q\overline{u} + r\overline{v})$$
$$= q^2 u\overline{u} + qr(\overline{u}v + u\overline{v}) + r^2 v\overline{v} = N(A)(aq^2 + bqr + cr^2).$$

Similar calculations show that

$$\overline{u_1} \cdot v_1 + u_1 \cdot \overline{v_1} = N(A)(2aqs + b(qt + rs) + 2crt)$$

and

$$v_1 \cdot \overline{v_1} = N(A)(qs^2 + bst + ct^2).$$

Therefore, the quadratic form of the ordered basis S_1 has the same coefficients as $f_S \cdot U$, as given in Theorem 4.1.2. $\qquad\square$

We conclude Section C.9 with an observation used in Section C.10.

Proposition C.9.5. *Let A be a nontrivial ideal of a quadratic domain D and let w be a nonzero element of D. If $f_S(x,y) = ax^2 + bxy + cy^2$ is the quadratic form of the ordered basis $S = [u \quad v]$ for A, then the following statements are true:*

1. *If $N(w)$ is positive, then $T = [wu \quad wv]$ is an ordered basis for wA, and the quadratic form of T is $f_T(x,y) = ax^2 + bxy + cy^2$.*
2. *If $N(w)$ is negative, then $T = [-wu \quad wv]$ is an ordered basis for wA, and the quadratic form of T is $f_T(x,y) = -ax^2 + bxy - cy^2$.*

Proof. The typical element of wA is wx where x is in A. If $x = mu + nv$, then

$$wx = w(mu + nv) = m(wu) + n(wv) = -m(-wu) + n(wv).$$

Thus, every element of wA is a $\mathbb{Z}$-combination of wu and wv, or of $-wu$ and wv. If $m(wu) + n(wv) = 0$, then $w(mu + nv) = 0$. Since $w \neq 0$, then $mu + nv = 0$, and $m = 0 = n$. We draw the same conclusion if $m(-wu) + n(wv) = 0$.

Since wA equals the product $wD \cdot A$, we have that $N(wA) = |N(w)| \cdot N(A)$, as a consequence of Proposition C.6.4 and Corollaries C.6.3 and C.4.2. Note that

$$\overline{wu} \cdot wv - wu \cdot \overline{wv} = (w \cdot \overline{w})(\overline{u} \cdot v - u \cdot \overline{v}) = N(w) \cdot N(A) \sqrt{\Delta}. \qquad\text{(C.18)}$$

If $N(w) > 0$, then $N(w) \cdot N(A) = N(wA)$, and equation (C.18) implies that $[wu \quad wv]$ is an ordered basis of wA. If $N(w) < 0$, then $N(w) \cdot N(A) = -N(wA)$, and equation (C.18) can be rewritten as

$$\overline{wu} \cdot wv - (-wu) \cdot \overline{wv} = N(wA)\sqrt{\Delta},$$

implying that $[-wu \quad wv]$ is an ordered basis of wA.

Finally, let $f_S(x, y) = ax^2 + bxy + cy^2$ be the quadratic form of the ordered basis $S = [u \quad v]$ for A. Then we find that

$$wu \cdot \overline{wu} = N(w) \cdot N(A) \cdot a = (-wu) \cdot (\overline{-wu})$$

and

$$wv \cdot \overline{wv} = N(w) \cdot N(A) \cdot c,$$

while

$$\overline{wu} \cdot wv + wu \cdot \overline{wv} = N(w) \cdot N(A) \cdot b$$

and

$$\overline{-wu} \cdot wv + (-wu) \cdot \overline{wv} = -N(w) \cdot N(A) \cdot b.$$

If $N(w) > 0$, so that $T = [wu \quad wv]$ is an ordered basis for wA and $N(wA) = N(w) \cdot N(A)$, it follows that the quadratic form of T is $f_T(x, y) = ax^2 + bxy + cy^2$. If $N(w) < 0$, so that $T = [-wu \quad wv]$ is an ordered basis for wA and $N(wA) = -N(w) \cdot N(A)$, we see that the quadratic form of T is $f_T(x, y) = -ax^2 + bxy - cy^2$. □

C.10 Ideal equivalence and ideal number equivalence

In Chapter 4, we defined a relation of equivalence on the set $\mathcal{I}_\Delta$ of ideal numbers of discriminant Δ by saying that $(a : k) \sim (a_1 : k_1)$ if there is a unimodular matrix U so that $(a_1 : k_1) = (a : k) \cdot U$, as defined in Theorem 4.1.2. We now compare this relation to the equivalence of ideals of quadratic domains defined in Section C.8. In this way, we will establish properties of the class group of ideal numbers.

Theorem C.10.1. *If $(a : k)$ is an ideal number of discriminant Δ, then let $A = A(a : k)$ be the ideal of the quadratic domain $D = D(\Delta)$ with the ideal number expression $\langle a : k \rangle$, that is,*

$$A = \{q(a) + r(k + z) \mid q, r \in \mathbb{Z}\}.$$

If $(a : k)$ and $(a_1 : k_1)$ are equivalent ideal numbers in $\mathcal{I}_\Delta$, then $A(a : k) = \langle a : k \rangle$ and $A(a_1 : k_1) = \langle a_1 : k_1 \rangle$ are equivalent ideals of D.

Proof. Let T and V be the translation and involution matrices defined in equation (4.1). It suffices to show that the ideals of D corresponding to $(a : k) \cdot T$ and to $(a : k) \cdot V$ are equivalent to $A = \langle a : k \rangle$. Since $(a : k) \cdot T = (a : k + a)$ and

$$\langle a : k + a \rangle = \{q(a) + r(a + k + z) \mid q, r \in \mathbb{Z}\}$$
$$= \{(q + r)(a) + r(k + z) \mid q, r \in \mathbb{Z}\} = \langle a : k \rangle = A,$$

the first claim is immediately true. For the second, let $P(x)$ be the principal polynomial of discriminant Δ, let $P(k) = ac$, and let $\overline{k} = -k - P'(0)$ be the conjugate of k with respect to Δ. Then $(a : k) \cdot V = (c : \overline{k})$. We will establish that $A = \langle a : k \rangle$ is equivalent to $C = \langle c : \overline{k} \rangle$ by showing that $cA = (k + z)C$. Notice that

$$(k + z)(\overline{k} + z) = k\overline{k} - P(0) + (k + \overline{k} + P'(0))z$$
$$= -k^2 - P'(0)k - P(0) = -P(k) = -ac,$$

using equation (C.4) and the definition of the conjugate of k with respect to Δ. Thus, we find that ca and $c(k+z)$ are in $(k+z)C$. This implies that cA, the set of all $\mathbb{Z}$-combinations of ca and $c(k + z)$, is a subset of $(k + z)C$. Similarly, $(k + z)c$ and $(k + z)(\overline{k} + z) = c(-a)$ are elements of cA, reversing the inclusion. Therefore, $cA = (k + z)C$, as we wanted to show. □

In the reverse direction, more caution is required, in that $\langle a : k \rangle$ equals $\langle -a : k \rangle$, whereas $(a : k)$ and $(-a : k)$ are typically not equivalent as ideal numbers. Here, we will establish the following as a partial converse of Theorem C.10.1.

Theorem C.10.2. *Let A and A_1 be ideals of a quadratic domain $D = D(\Delta)$, written as the ideal number expressions $A = \langle a : k \rangle$ and $A_1 = \langle a_1 : k_1 \rangle$ with a and a_1 positive. Suppose that $mA = wA_1$ where m is a nonzero rational integer and w is a nonzero element of D, so that A is equivalent to A_1. Then the following statements are true:*
1. *If $N(w) > 0$, then $(a : k) \sim (a_1 : k_1)$ in $\mathcal{I}_\Delta$.*
2. *If $N(w) < 0$, then $(a : k) \sim (-a_1 : k_1)$ in $\mathcal{I}_\Delta$.*

The proof of Theorem C.10.2 uses properties of ordered bases of ideals compiled in Section C.9.

Proof. Let $A = \langle a : k \rangle$ and $A_1 = \langle a_1 : k_1 \rangle$ be ideals of D, with a and a_1 positive. Let $mA = wA_1$ with m a nonzero rational integer and w a nonzero element of D. Suppose first that $N(w)$ is positive. By Propositions C.9.1 and C.9.5, then $S = [ma \quad m(k + z)]$ and $S_1 = [wa_1 \quad w(k_1+z)]$ are ordered bases for mA and wA_1, respectively. If these ideals are equal, then there is a unimodular matrix U so that $S_1 = S \cdot U$, as implied by Proposition C.9.2. If we let f and f_1 be the quadratic forms of the ordered bases S and S_1, respectively, then

Proposition C.9.4 shows that $f_1 = f \cdot U$. But by Proposition C.9.5 and the example following Proposition C.9.3, $f = (a : k)$ and $f_1 = (a_1 : k_1)$ in ideal number notation. We conclude that $(a : k) \sim (a_1 : k_1)$ as quadratic forms, and also as ideal numbers in $\mathcal{I}_\Delta$.

Now suppose that $N(w)$ is negative. Then $S = [ma \quad m(k + z)]$ and $S_1 = [-wa_1 \quad w(k_1 + z)]$ are ordered bases for mA and wA_1, respectively, with $S_1 = S \cdot U$ for some unimodular matrix U. If f and f_1 are the quadratic forms of the ordered bases S and S_1, respectively, then $f_1 = f \cdot U$. But now Proposition C.9.5 shows that $f = (a : k)$ while $f_1 = (-a_1 : k_1)$, the negative conjugate of $(a_1 : k_1)$. Thus, $(a : k) \sim (-a_1 : k_1)$ as quadratic forms and as ideal numbers in $\mathcal{I}_\Delta$. $\square$

Proposition C.10.3. *For every discriminant Δ, the operation of composition on the class group $\mathcal{C}_\Delta$ is well-defined.*

Proof. Suppose that

$$(a_1 : k_1) \sim (a_2 : k_2) \quad \text{and} \quad (b_1 : \ell_1) \sim (b_2 : \ell_2)$$

for certain primitive ideal numbers in $\mathcal{I}_\Delta$. Let $A_1, A_2, B_1,$ and B_2 be the ideals of the quadratic domain $D = D(\Delta)$ that correspond to these respective ideal numbers. By Theorem C.10.1, we have that $A_1 \sim A_2$ and $B_1 \sim B_2$, say with $mA_1 = vA_2$ and $nB_1 = wB_2$ for some nonzero rational integers m and n and some nonzero elements v and w of D. Now

$$mA_1 \cdot nB_1 = (mn)(A_1B_1) = (vw)(A_2B_2) = (vA_2)(wB_2),$$

which shows that $A_1B_1 \sim A_2B_2$. Since ideal multiplication (or composition, as in Section C.7) and composition of primitive ideal numbers are given by the same formula, we can then conclude that

$$(a_1 : k_1) \circ (b_1 : \ell_1) \sim (a_2 : k_2) \circ (b_2 : \ell_2)$$

in $\mathcal{I}_\Delta$. (The fact that $N(vw) = N(v) \cdot N(w)$, along with Theorem C.10.2, ensure that there are no unwanted changes of sign in the norms of these ideal numbers.) Thus, composition of classes of primitive ideal numbers in $\mathcal{C}_\Delta$ is well-defined. $\square$

D Appendix: Continued fractions

A continued fraction is an expression that we can associate with an arbitrary real number, with important applications to approximations of real numbers by rational numbers. When v is a real *quadratic number*, that is, a real root of a quadratic polynomial having integer coefficients, the continued fraction of v gives us information about representations of integers by indefinite quadratic forms. In Chapter 7, we introduce some of these concepts in the context of ideal numbers of positive discriminant. In this appendix, we compile some general results about continued fractions that we will need in that process.

D.1 Continued fractions of rational numbers

To motivate the definition of continued fractions, consider the following alternative expressions for the rational number $v = \frac{553}{82}$:

$$\frac{553}{82} = 6 + \frac{61}{82} = 6 + \frac{1}{82/61} = 6 + \frac{1}{1 + \frac{21}{61}} = 6 + \frac{1}{1 + \frac{1}{61/21}} = 6 + \frac{1}{1 + \frac{1}{2 + \frac{19}{21}}}$$

$$= 6 + \frac{1}{1 + \frac{1}{2 + \frac{1}{21/19}}} = 6 + \frac{1}{1 + \frac{1}{2 + \frac{1}{1 + \frac{2}{19}}}} = 6 + \frac{1}{1 + \frac{1}{2 + \frac{1}{1 + \frac{1}{19/2}}}}$$

$$= 6 + \frac{1}{1 + \frac{1}{2 + \frac{1}{1 + \frac{1}{9 + \frac{1}{2}}}}}.$$

Here, we recursively define rational numbers v_n and integers q_n as follows. Let $v_0 = v$ and for each integer $n \geq 0$, let $q_n = \lfloor v_n \rfloor$, the greatest integer less than or equal to v_n. Then, if $q_n \neq v_n$, let $v_{n+1} = \frac{1}{v_n - q_n}$. We find that

$$v = q_0 + \frac{1}{v_1} = q_0 + \frac{1}{q_1 + \frac{1}{v_2}} = q_0 + \frac{1}{q_1 + \frac{1}{q_2 + \frac{1}{v_3}}} = q_0 + \frac{1}{q_1 + \frac{1}{q_2 + \frac{1}{q_3 + \frac{1}{v_4}}}}$$

and so forth, until eventually $v_n = q_n$, where the process terminates. Note that v_0 can be positive or negative, but that $v_1, v_2, \ldots$ are all larger than 1, if they are defined. Likewise, q_0 can be any integer but q_i is positive for all $i \geq 1$. The sequence of integers q_i is sufficient to determine the rational number v. We will write $v = \langle q_0, q_1, \ldots, q_n \rangle$ and refer to this sequence as a *continued fraction expansion* of the rational number v. Thus, for example, we can summarize the equations above by saying that $\frac{553}{82} = \langle 6, 1, 2, 1, 9, 2 \rangle$. We broaden this definition as follows, in a way that is convenient for calculation.

https://doi.org/10.1515/9783111319360-011

Definition. Let $q_0, q_1, q_2, \ldots$ be a sequence of integers with $q_i > 0$ if $i > 0$. If $w \geq 1$ is a real number, we define an expression of the form $\langle q_0, q_1, \ldots, q_n, w \rangle$ recursively by saying that

$$\langle q_0, q_1, \ldots, q_n, w \rangle = \left\langle q_0, q_1, \ldots, q_n + \frac{1}{w} \right\rangle \quad \text{if } n \geq 0.$$

We also let $\langle z \rangle = z$ for every real number z. We refer to $\langle q_0, q_1, \ldots, q_n, w \rangle$ as a *finite continued fraction*. We say that this continued fraction is *simple* if w is a positive integer.

Example. For instance,

$$\langle 6, 1, 2, 1, 9, 2 \rangle = \left\langle 6, 1, 2, 1, 9 + \frac{1}{2} \right\rangle = \left\langle 6, 1, 2, 1, \frac{19}{2} \right\rangle = \left\langle 6, 1, 2, 1 + \frac{2}{19} \right\rangle,$$

and so forth, eventually arriving at $\langle 6, 1, 2, 1, 9, 2 \rangle = \frac{553}{82}$, as above.

The continued fraction expansion of a rational number is not unique, as we see with the following example.

Example. Suppose we want to find a rational number v having the continued fraction expression $\langle 1, 2, 3, 1 \rangle$. Applying the preceding definition, we have

$$\langle 1, 2, 3, 1 \rangle = \left\langle 1, 2, 3 + \frac{1}{1} \right\rangle = \langle 1, 2, 4 \rangle = \left\langle 1, 2 + \frac{1}{4} \right\rangle$$

$$= \left\langle 1, \frac{9}{4} \right\rangle = \left\langle 1 + \frac{4}{9} \right\rangle = \left\langle \frac{13}{9} \right\rangle = \frac{13}{9}.$$

But note that if we apply the same approach as in our first example to $v = \frac{13}{9}$, we have

$$\frac{13}{9} = 1 + \frac{4}{9} = 1 + \frac{1}{9/4} = 1 + \frac{1}{2 + \frac{1}{4}},$$

which we could also write as $\langle 1, 2, 4 \rangle$. (This alternative expression appeared in the calculations for this example.)

It is convenient to allow the ambiguity illustrated by this example for continued fraction expressions of rational numbers. We summarize this observation as follows.

Fact. If v is a rational number, then v has two distinct expressions as a simple continued fraction:

$$v = \langle q_0, q_1, \ldots, q_n, 1 \rangle \quad \text{and} \quad v = \langle q_0, q_1, \ldots, q_n + 1 \rangle,$$

where each q_i is an integer, with $q_i > 0$ if $i > 0$.

Example. We can write $\frac{553}{82}$ as either $\langle 6, 1, 2, 1, 9, 2 \rangle$ or $\langle 6, 1, 2, 1, 9, 1, 1 \rangle$.

D.2 Convergents

The preceding examples show that we can calculate a finite continued fraction recursively, working from right to left. We may also need to work from left to right, particularly when we introduce infinite continued fractions. To do so, we associate a pair of sequences to an arbitrary sequence of positive integer as follows.

Definition. Let $q_0, q_1, q_2, \ldots$ be a sequence of integers with $q_i > 0$ if $i > 0$. Then we define s_i and t_i for $i \geq -2$ by the following recursive definitions:

$$s_{-2} = 0, \quad s_{-1} = 1, \quad s_i = s_{i-2} + s_{i-1} q_i \quad \text{for } i \geq 0 \tag{D.1}$$

and

$$t_{-2} = 1, \quad t_{-1} = 0, \quad t_i = t_{i-2} + t_{i-1} q_i \quad \text{for } i \geq 0. \tag{D.2}$$

We refer to s_i and t_i as the *numerator* and *denominator* sequences, respectively, of the *convergents* of the q_i sequence.

In examples in this section, we apply these definitions only to the terms of a finite simple continued fraction, $\langle q_0, q_1, \ldots, q_n \rangle$, in which case the s_i and t_i sequences terminate as well. Note that $s_0 = q_0$ and $t_0 = 1$. For $i \geq 0$, the denominator sequence contains only positive integers and is strictly increasing, with the possible exception that $t_1 = t_0 = 1$ if $q_1 = 1$. In any case, $t_i \geq i$ for all $i \geq 0$.

Example. We can compile the numerator and denominator sequences most easily using tables, such as the following separate tables for $\langle 6, 1, 2, 1, 9, 2 \rangle$ and $\langle 6, 1, 2, 1, 9, 1, 1 \rangle$, from a previous example:

i	0	1	2	3	4	5	i	0	1	2	3	4	5	6
q	6	1	2	1	9	2	q	6	1	2	1	9	1	1
s	6	7	20	27	263	553	s	6	7	20	27	263	290	553
t	1	1	3	4	39	82	t	1	1	3	4	39	43	82

We obtain the entries in the s row by multiplying the value of q immediately above the entry of interest by the value of s one place to the left and adding the value of s two places to the left. We will typically omit the initial terms, that is, for $i = -2$ and $i = -1$. For the numerator sequence, these are 0 and 1. The t row is calculated similarly, but with initial terms 1 and 0 instead. Notice that the s and t sequences of these two continued fractions are identical, aside from the next-to-last terms in the longer pair of sequences. The final terms in both examples are the numerator and denominator of the rational number for which these are continued fraction expansions.

As this example suggests, the numerator and denominator sequences give us an alternative method of calculating a finite continued fraction. We establish this claim after the following lemma.

Lemma D.2.1. *Let $q_0, q_1, q_2, \ldots$ be a sequence of integers with $q_i > 0$ if $i > 0$, and let s_i and t_i be defined for $i \geq -2$ as in equations (D.1) and (D.2). Then for every nonnegative integer n and real number $w \geq 1$,*

$$\langle q_0, q_1, \ldots, q_n, w \rangle = \frac{s_n w + s_{n-1}}{t_n w + t_{n-1}}.$$

Proof. We proceed by induction on n. If $n = 0$, then

$$\langle q_0, w \rangle = \left\langle q_0 + \frac{1}{w} \right\rangle = \frac{q_0 w + 1}{w} = \frac{s_0 w + s_{-1}}{t_0 w + t_{-1}}$$

since, as noted previously, $s_0 = q_0$ and $t_0 = 1$. Now let n be larger than 0 and suppose that the claim has been established for all nonnegative integers smaller than n. Let $z = q_n + \frac{1}{w}$, a real number larger than 1. Then

$$\langle q_0, q_1, \ldots, q_n, w \rangle = \langle q_0, q_1, \ldots, q_{n-1}, z \rangle = \frac{s_{n-1} z + s_{n-2}}{t_{n-1} z + t_{n-2}}$$

by the inductive hypothesis. But

$$s_{n-1} z + s_{n-2} = s_{n-1} \left(q_n + \frac{1}{w} \right) + s_{n-2}$$

$$= (s_{n-1} q_n + s_{n-2}) + \frac{1}{w} \cdot s_{n-1} = \frac{1}{w}(s_n w + s_{n-1}),$$

using the recursive definition of s_n. Similarly, $t_{n-1} z + t_{n-2} = \frac{1}{w}(t_n w + t_{n-1})$. Thus, we conclude that

$$\frac{s_{n-1} z + s_{n-2}}{t_{n-1} z + t_{n-2}} = \frac{(s_n w + s_{n-1})/w}{(t_n w + t_{n-1})/w} = \frac{s_n w + s_{n-1}}{t_n w + t_{n-1}},$$

which proves the claim by induction. $\qquad\square$

Proposition D.2.2. *Let $q_0, q_1, q_2, \ldots$ be a sequence of integers with $q_i > 0$ if $i > 0$, and let s_i and t_i be defined for $i \geq -2$ as in equations (D.1) and (D.2). Then for all $n \geq 0$,*

$$\langle q_0, q_1, \ldots, q_n \rangle = \frac{s_n}{t_n}.$$

Proof. Since $s_0 = q_0$ and $t_0 = 1$, then $\langle q_0 \rangle = q_0 = \frac{s_0}{t_0}$. For $n > 0$, we can apply Lemma D.2.1 with q_n in place of w, and $n - 1$ in place of n. We find that

$$\langle q_0, q_1, \ldots, q_{n-1}, q_n \rangle = \frac{s_{n-1}q_n + s_{n-2}}{t_{n-1}q_n + t_{n-2}} = \frac{s_n}{t_n},$$

using the recursive definitions of s_n and t_n. □

Finally, for this section, we explain the terminology of convergents with the following lemma and proposition.

Lemma D.2.3. *Let $q_0, q_1, q_2, \ldots$ be a sequence of integers with $q_i > 0$ if $i > 0$, and let s_i and t_i be defined for $i \geq -2$ as in equations (D.1) and (D.2). Then*

$$s_{n-1}t_{n-2} - s_{n-2}t_{n-1} = (-1)^n \tag{D.3}$$

for all $n \geq 0$.

Proof. We use induction on n. If $n = 0$, then $s_{-1}t_{-2} - s_{-2}t_{-1} = 1 \cdot 1 - 0 \cdot 0 = 1 = (-1)^0$. Suppose that we know equation (D.3) holds for some $n \geq 0$. Then, using the recursive definitions of the numerator and denominator sequences, we have

$$\begin{aligned} s_n t_{n-1} - s_{n-1} t_n &= (s_{n-2} - s_{n-1}q_n)t_{n-1} - s_{n-1}(t_{n-2} - t_{n-1}q_n) \\ &= s_{n-2}t_{n-1} - s_{n-1}t_{n-1}q_n - s_{n-1}t_{n-2} + s_{n-1}t_{n-1}q_n \\ &= -1 \cdot (s_{n-1}t_{n-2} - s_{n-2}t_{n-1}) = (-1)^{n+1}, \end{aligned}$$

by the inductive hypothesis. The claim holds for all $n \geq 0$ by induction. □

Lemma D.2.3 has the following implications for the numerator and denominator sequences of $q_0, q_1, q_2, \ldots$ for all $n \geq 0$:

$$\gcd(s_{n-1}, s_n) = 1, \quad \gcd(t_{n-1}, t_n) = 1, \quad \gcd(s_n, t_n) = 1.$$

Proposition D.2.4. *Let $q_0, q_1, q_2, \ldots$ be a sequence of integers with $q_i > 0$ if $i > 0$, and let s_i and t_i be defined for $i \geq -2$ as in equations (D.1) and (D.2). Then for all $n \geq 0$,*

$$\frac{s_{n+1}}{t_{n+1}} - \frac{s_n}{t_n} = \frac{(-1)^n}{t_{n+1}t_n}. \tag{D.4}$$

It follows that the sequence of quotients $\{s_i/t_i\}_{i=0}^{\infty}$ converges to a real number.

Proof. We have that

$$\frac{s_{n+1}}{t_{n+1}} - \frac{s_n}{t_n} = \frac{s_{n+1}t_n - t_{n+1}s_n}{t_{n+1}t_n} = \frac{(-1)^{n+2}}{t_{n+1}t_n} = \frac{(-1)^n}{t_{n+1}t_n},$$

using equation (D.3). It follows, from the increasing nature of the t_i sequence, that

$$\frac{s_0}{t_0} < \frac{s_2}{t_2} < \frac{s_4}{t_4} < \cdots < \frac{s_5}{t_5} < \frac{s_3}{t_3} < \frac{s_1}{t_1},$$

that is, the sequence of odd-indexed quotients is strictly increasing, the sequence of even-indexed quotients is strictly decreasing, and each odd-indexed quotient is strictly smaller than each even-indexed quotient. Since $t_i \geq i$ for all $i \geq 0$, it follows that $\{s_i/t_i\}_{i=0}^{\infty}$ is a *Cauchy sequence*, that is, for all positive real numbers ϵ, there is a positive integer N so that $|\frac{s_i}{t_i} - \frac{s_j}{t_j}| < \epsilon$ if i and j are both larger than N. (Given ϵ, we can take N to be any integer such that $N^2 > \frac{1}{\epsilon}$, for example.) It is a standard fact that every Cauchy sequence converges to a real number. □

D.3 Infinite continued fractions

Based on our definitions and results from Section D.2, we now define an infinite continued fraction as follows.

Definition. Let $q_0, q_1, q_2, \ldots$ be a sequence of integers such that q_i is positive if $i > 0$. Then we define the *infinite continued fraction* $\langle q_0, q_1, q_2, \ldots \rangle$ to be the limit of the sequence of rational numbers

$$\langle q_0 \rangle, \quad \langle q_0, q_1 \rangle, \quad \langle q_0, q_1, q_2 \rangle, \quad \ldots.$$

We call $\langle q_0, q_1, q_2, \ldots, q_n \rangle$ the n-th *convergent* of this continued fraction.

If the sequences s_i and t_i are defined for $q_0, q_1, q_2, \ldots$ as in equations (D.1) and (D.2), then Proposition D.2.2 shows that $\langle q_0, q_1, q_2, \ldots, q_n \rangle = s_n/t_n$ for all $n \geq 0$. Proposition D.2.4 implies that this sequence converges to a real number.

Example. Consider the infinite continued fraction $\langle 1, 2, 1, 2, 1, 2, \ldots \rangle$ in which the terms alternate between 1 and 2. In the following table, we calculate the numerator and denominator sequences for this continued fraction, for $n \leq 10$.

i	0	1	2	3	4	5	6	7	8	9	10
q	1	2	1	2	1	2	1	2	1	2	1
s	1	3	4	11	15	41	56	153	209	571	780
t	1	2	3	8	11	30	41	112	153	418	571

The sequence of decimal expansions for s/t begins as

$$1, 1.5, 1.3333, 1.375, 1.3636, 1.3667, 1.3659, 1.3661, 1.36601, 1.36603, 1.36602, \ldots.$$

Based on this limited numerical evidence, it appears that this sequence converges to a real number near 1.36602. Notice how the convergents alternate between a number smaller than and a number larger than the eventual limit, as predicted in the proof of Proposition D.2.4.

Thus, we have established that every infinite continued fraction equals a real number. It is also true that every real number v has a continued fraction expansion, which is unique if v is irrational. The terms of this continued fraction are found using the same method we illustrated for rational numbers: Let $v_0 = v$ and, for each integer $n \geq 0$, let $q_n = \lfloor v_n \rfloor$ and let $v_{n+1} = \frac{1}{v_n - q_n}$. (The difference between v_n and q_n cannot be 0 if v is irrational.)

In practice, it can be difficult to calculate more than the first few terms in the continued fraction of an irrational number. However, when v is a *quadratic number* (i. e., a root of a quadratic polynomial with integer coefficients), then we can typically describe a pattern for all terms in the continued fraction of v, using algebraic properties of such numbers.

Example. Consider the irrational number $v = \frac{-19 + \sqrt{41}}{16}$. For $i \geq 0$, and starting with $v_0 = v$, we can write each v_i as $q_i + (v_i - q_i)$ where $q_i = \lfloor v_i \rfloor$, and then let $v_{i+1} = 1/(v_i - q_i)$. Each such expression can be simplified, using properties of square roots. For instance, since $v_0 = \frac{-19 + \sqrt{41}}{16} = -1 + \frac{-3 + \sqrt{41}}{16}$, then

$$v_1 = \frac{16}{-3 + \sqrt{41}} \cdot \frac{3 + \sqrt{41}}{3 + \sqrt{41}} = \frac{16(3 + \sqrt{41})}{41 - 9} = \frac{3 + \sqrt{41}}{2}.$$

This pattern continues indefinitely, with similar simplifications, in the following table:

i	v_i	$q_i + (v_i - q_i)$
0	$\frac{-19 + \sqrt{41}}{16}$	$-1 + \frac{-3 + \sqrt{41}}{16}$
1	$\frac{16}{-3 + \sqrt{41}} = \frac{3 + \sqrt{41}}{2}$	$4 + \frac{-5 + \sqrt{41}}{2}$
2	$\frac{2}{-5 + \sqrt{41}} = \frac{5 + \sqrt{41}}{8}$	$1 + \frac{-3 + \sqrt{41}}{8}$
3	$\frac{8}{-3 + \sqrt{41}} = \frac{3 + \sqrt{41}}{4}$	$2 + \frac{-5 + \sqrt{41}}{4}$
4	$\frac{4}{-5 + \sqrt{41}} = \frac{5 + \sqrt{41}}{4}$	$2 + \frac{-3 + \sqrt{41}}{4}$
5	$\frac{4}{-3 + \sqrt{41}} = \frac{3 + \sqrt{41}}{8}$	$1 + \frac{-5 + \sqrt{41}}{8}$
6	$\frac{8}{-5 + \sqrt{41}} = \frac{5 + \sqrt{41}}{2}$	$5 + \frac{-5 + \sqrt{41}}{2}$
7	$\frac{2}{-5 + \sqrt{41}} = \frac{5 + \sqrt{41}}{8}$	$1 + \frac{-3 + \sqrt{41}}{8}$

Thus, the continued fraction of v begins as $\langle -1, 4, 1, 2, 2, 1, 5, 1, \dots \rangle$. As in the preceding example, we can calculate the convergents of this continued fraction recursively.

i	0	1	2	3	4	5	6	7
q	−1	4	1	2	2	1	5	1
s	−1	−3	−4	−11	−26	−37	−211	−248
t	1	4	5	14	33	47	268	315
s/t	−1	−.75	−.8	−.7857	−.7879	−.7872	−.78731	−.78730

It appears that the sequence $\{s_i/t_i\}$ converges, and that $v = \frac{-19+\sqrt{41}}{16} \approx -.787305$ is its limit.

Here, we can go a step further, if we note that $v_7 = v_2$. Although the process does not terminate, the first table continues the pattern of the rows $i = 2$ through $i = 6$. In particular, after the first two terms, the continued fraction of v repeats five terms indefinitely. We follow the convention of writing a bar over the repeating pattern of digits, as follows:

$$\langle -1, 4, 1, 2, 2, 1, 5, 1, 2, 2, 1, 5, \dots \rangle = \langle -1, 4, \overline{1, 2, 2, 1, 5} \rangle.$$

We could now extend the second table indefinitely, obtaining better and better rational approximations of v.

A continued fraction $\langle q_0, q_1, q_2, \dots \rangle$ is called *periodic* if there is a nonnegative integer m and a positive integer ℓ so that $q_i = q_{i+\ell}$ for all $i \geq m$. If we can select m to be 0, then we say that the continued fraction is *purely periodic*. In either case, the smallest positive integer ℓ for which this is true is called the *period* of the continued fraction. It is a fact that the continued fraction expansion of an irrational quadratic number is always periodic. The proof of this claim uses a development of what are called *reduced* quadratic numbers of a fixed positive discriminant. We do this in the context of ideal numbers in Chapter 7, so will assume this claim in the remainder of Appendix D. In Section D.4, we establish the converse. That is, every continued fraction that repeats a pattern indefinitely converges to an irrational quadratic number.

D.4 Calculating periodic continued fractions

We can calculate a periodic continued fraction in general using the following observation.

Proposition D.4.1. *If $\langle q_0, q_1, q_2, \dots \rangle$ is a continued fraction, then for every $i \geq 0$, we can write*

$$\langle q_0, q_1, q_2, \dots \rangle = \langle q_0, \dots, q_i, \langle q_{i+1}, q_{i+2}, \dots \rangle \rangle.$$

Proof. To simplify expressions in the proof, we will assume that $i = 0$. The more general argument is entirely analogous to this case. We will first show by induction on n that for all $n \geq 1$ and all real numbers $w \geq 1$,

$$\langle q_0, q_1, \ldots, q_n, w \rangle = \langle q_0, \langle q_1, \ldots, q_n, w \rangle \rangle.$$

If $n = 1$, then $\langle q_0, q_1, w \rangle = \langle q_0, q_1 + \frac{1}{w} \rangle$, while $\langle q_0, \langle q_1, w \rangle \rangle = \langle q_0, \langle q_1 + \frac{1}{w} \rangle \rangle = \langle q_0, q_1 + \frac{1}{w} \rangle$, using the definition of finite continued fractions. So, now suppose that for some $n \geq 1$, we have established that

$$\langle q_0, q_1, \ldots, q_n, w \rangle = \langle q_0, \langle q_1, \ldots, q_n, w \rangle \rangle$$

for every real number w. If $z = q_{n+1} + \frac{1}{w}$, then

$$\langle q_0, q_1, \ldots, q_n, q_{n+1}, w \rangle = \langle q_0, q_1, \ldots, q_n, z \rangle$$
$$= \langle q_0, \langle q_1, \ldots, q_n, z \rangle \rangle = \langle q_0, \langle q_1, \ldots, q_n, q_{n+1}, w \rangle \rangle,$$

using the inductive hypothesis and properties of finite continued fractions. This establishes our preliminary claim by induction. Now $\langle q_0, q_1, q_2, \ldots \rangle$ is the limit of a sequence of finite continued fractions,

$$\langle q_0 \rangle, \langle q_0, q_1 \rangle, \langle q_0, q_1, q_2 \rangle, \langle q_0, q_1, q_2, q_3 \rangle, \ldots,$$

which can be rewritten as

$$\langle q_0 \rangle, \langle q_0, \langle q_1 \rangle \rangle, \langle q_0, \langle q_1, q_2 \rangle \rangle, \langle q_0, \langle q_0, q_1, q_2 \rangle \rangle, \ldots.$$

Since $\langle q_1, q_2, q_3, \ldots \rangle$ is the limit of the sequence $\langle q_1 \rangle, \langle q_1, q_2 \rangle, \langle q_1, q_2, q_3 \rangle, \ldots$, it follows that

$$\langle q_0, q_1, q_2, \ldots \rangle = \langle q_0, \langle q_1, q_2, \ldots \rangle \rangle,$$

as we wanted to show. $\qquad\square$

Corollary D.4.2. *The continued fraction expansion of an irrational real number is unique.*

Proof. Suppose that v can be written as $\langle q_0, q_1, q_2, \ldots \rangle$ and as $\langle q_0', q_1', q_2', \ldots \rangle$ with each q_i and q_i' an integer, and q_i and q_i' both positive if $i > 0$. Then $v = \langle q_0, w \rangle = q_0 + \frac{1}{w}$ and $v = \langle q_0', w' \rangle = q_0' + \frac{1}{w'}$, where $w = \langle q_1, q_2, \ldots \rangle$ and as $w' = \langle q_1', q_2', \ldots \rangle$. Note that w and w' are both larger than 1, since each is larger than its first convergent (q_1 and q_1', respectively). Now $q_0 - q_0' = \frac{1}{w'} - \frac{1}{w}$ is strictly between -1 and 1, so that $q_0 - q_0' = 0$. Then $w = w'$ and we can repeat the argument to show that $q_1 = q_1'$, and so forth indefinitely. Thus, the two continued fraction expressions for v are identical. $\qquad\square$

We saw previously that a rational number has two distinct continued fraction expansions, of the form $\langle q_0, q_1, \ldots, q_n, 1 \rangle$ and $\langle q_0, q_1, \ldots, q_n + 1 \rangle$. An argument similar to that in the preceding proof shows that there are no other continued fraction expressions for a given rational number.

Example. Let v be the real number with the purely periodic continued fraction expression $\langle \overline{1, 2} \rangle$. Note that

$$v = \langle 1, 2, 1, 2, 1, 2, 1, 2, \ldots \rangle = \langle 1, 2, \langle 1, 2, 1, 2, 1, 2, \ldots \rangle \rangle = \langle 1, 2, v \rangle,$$

using Proposition D.4.1. By the definition of finite continued fractions, then

$$v = \langle 1, 2, v \rangle = \left\langle 1, 2 + \frac{1}{v} \right\rangle = \left\langle 1, \frac{2v + 1}{v} \right\rangle = 1 + \frac{v}{2v + 1} = \frac{3v + 1}{2v + 1}.$$

Thus, $2v^2 + v = 3v + 1$, so that v is a root of $f(x) = 2x^2 - 2x - 1$. This polynomial has two roots:

$$x = \frac{2 \pm \sqrt{(-2)^2 - 4(2)(-1)}}{4} = \frac{1 \pm \sqrt{3}}{2}.$$

Since v is larger than its first convergent, $\frac{s_0}{t_0} = 1$, then $v = \frac{1+\sqrt{3}}{2} \approx 1.366025$. This is consistent with calculations of the convergents of $\langle \overline{1, 2} \rangle$ from an example in Section D.3.

We can generalize this example, using the terminology of convergents, as follows.

Proposition D.4.3. *Let n be a nonnegative integer and let $q_0, q_1, \ldots, q_n$ be a sequence of positive integers. Let s_i and t_i be defined for $-2 \le i \le n$ as in equations (D.1) and (D.2). Then the purely periodic continued fraction $\langle \overline{q_0, q_1, \ldots, q_n} \rangle$ equals $v = \frac{-b+\sqrt{\Delta}}{2a}$, where $a = t_n$, $b = t_{n-1} - s_n$, and $\Delta = (s_n + t_{n-1})^2 + (-1)^n \cdot 4$.*

Proof. Note that

$$v = \langle \overline{q_0, q_1, \ldots, q_n} \rangle = \langle q_0, q_1, \ldots, q_n, v \rangle = \frac{s_n v + s_{n-1}}{t_n v + t_{n-1}},$$

using Proposition D.4.1 and Lemma D.2.1. Thus, v satisfies the equation

$$t_n v^2 + t_{n-1} v = s_n v + s_{n-1},$$

and is a root of $f(x) = ax^2 + bx + c$, where $a = t_n$, $b = t_{n-1} - s_n$, and $c = -s_{n-1}$. The discriminant of this polynomial is

$$\Delta = b^2 - 4ac = (t_{n-1} - s_n)^2 + 4t_n s_{n-1}$$
$$= (s_n + t_{n-1})^2 - 4s_n t_{n-1} + 4t_n s_{n-1} = (s_n + t_{n-1})^2 + (-1)^n \cdot 4,$$

since $s_n t_{n-1} - s_{n-1} t_n = (-1)^{n+1}$ by Lemma D.2.3. The roots of $f(x)$ are $\frac{-b+\sqrt{\Delta}}{2a}$ and $\frac{-b-\sqrt{\Delta}}{2a}$, the latter of which is negative. Since $\langle \overline{q_0, q_1, \ldots, q_n} \rangle$ is larger than q_0, a positive integer, we conclude that $\langle \overline{q_0, q_1, \ldots, q_n} \rangle$ equals $v = \frac{-b+\sqrt{\Delta}}{2a}$. $\qquad\square$

Example. Consider the continued fraction $\langle \overline{1, 2, 2, 1, 5} \rangle$, for which the numerator and denominator sequences begin as follows:

i	0	1	2	3	4
q	1	2	2	1	5
s	1	3	7	10	57
t	1	2	5	7	40

In the terminology of Proposition D.4.3, we have $n = 4$, with $s_4 = 57$, $t_4 = 40$, and $t_3 = 7$. Then $a = 40$, $b = 7 - 57 = -50$, and $\Delta = (57 + 7)^2 + 4 = 4100$. We conclude that $\langle \overline{1, 2, 2, 1, 5} \rangle$ is the continued fraction of $v = \frac{-b+\sqrt{\Delta}}{2a} = \frac{50+\sqrt{4100}}{80} = \frac{5+\sqrt{41}}{8}$.

Thus, we can find the quadratic number represented by a purely periodic continued fraction. Proposition D.4.1 then allows us to calculate an arbitrary periodic continued fraction.

Example. Consider the continued fraction $\langle -1, 4, \overline{1, 2, 2, 1, 5} \rangle$. From the preceding example, we know that $v = \langle \overline{1, 2, 2, 1, 5} \rangle = \frac{5+\sqrt{41}}{8}$, so we can write

$$\langle -1, 4, \overline{1, 2, 2, 1, 5} \rangle = \langle -1, 4, v \rangle = \left\langle -1, 4 + \frac{1}{v} \right\rangle = -1 + \frac{v}{4v+1} = \frac{-3v-1}{4v+1}.$$

Direct calculation shows that $-3v - 1 = \frac{1}{8}(-23 - 3\sqrt{41})$ and $4v + 1 = \frac{1}{8}(28 + 4\sqrt{41})$, so that

$$\frac{-3v-1}{4v+1} = \frac{-23 - 3\sqrt{41}}{28 + 4\sqrt{41}} \cdot \frac{28 - 4\sqrt{41}}{28 - 4\sqrt{41}} = \frac{-152 + 8\sqrt{41}}{128} = \frac{-19 + \sqrt{41}}{16}.$$

This confirms data from an example in Section D.3, in which we saw directly that the continued fraction of $\frac{-19+\sqrt{41}}{16}$ is $\langle -1, 4, \overline{1, 2, 2, 1, 5} \rangle$.

D.5 Continued fractions and rational approximations

Throughout this section, let v be an irrational real number, not necessarily a quadratic number. We have seen by example that the convergents in the continued fraction of v provide better and better rational approximations of v. In this section, we establish conversely that, under certain circumstances, rational approximations of v arise only as convergents in a continued fraction. This result has important applications to representations of integers by indefinite quadratic forms (or ideal numbers), needed in Chapter 7. We begin with the following numerical observation.

Lemma D.5.1. *Let a and b be integers with $\gcd(a, b) = 1$ and b positive. Then there are unique integers c and d with $\gcd(c, d) = 1$ and $0 < d \le b$ so that $ad - bc = 1$. In that case, $a(b - d) - b(a - c) = -1$ with $0 \le b - d < b$.*

Note that $d = b$ only when $b(a - c) = 1$, so that $b = 1$.

Proof. The equation $ax + by = 1$ has integer solutions since $\gcd(a, b) = 1$. We can let $x = d$ and $y = -c$ in any such solution. By Theorem A.5.1, then all solutions of $ax + by = 1$ have the form $(x, y) = (d + bq, -c - aq)$ with q an integer, and there is only one such solution with $0 < x \le b$. Now

$$a(b - d) - b(a - c) = ab - ad - ab + bc = -(ad - bc) = -1,$$

with $0 \le b - d < b$ since $0 < d \le b$. (We can describe $(x, y) = (b - d, a - c)$ as the unique solution of $ax + by = -1$ with $0 \le x < b$.) □

For the next claim, recall that each rational number has two continued fraction expressions, with the number of terms differing by 1. In either case, we let s_i and t_i be the numerator and denominator sequences of its convergents, as defined in equations (D.1) and (D.2).

Lemma D.5.2. *Let a and b be integers with $\gcd(a, b) = 1$ and $b > 0$, and let c and d be the unique integers with $0 < d \le b$ so that $ad - bc = 1$, as in Lemma D.5.1. Let $\langle q_0, q_1, \ldots, q_n \rangle$ be a continued fraction expansion for the rational number $\frac{a}{b}$. Then $s_n = a$ and $t_n = b$. If n is odd, then $s_{n-1} = c$ and $t_{n-1} = d$. If n is even, then $s_{n-1} = a - c$ and $t_{n-1} = b - d$.*

Proof. The first claim is immediate from our observations that $\gcd(s_n, t_n) = 1$ and that t_i is a positive integer for all $i \ge 0$. By Lemma D.2.3, $s_n t_{n-1} - s_{n-1} t_n = (-1)^{n+1}$. If n is odd, then $a t_{n-1} - b s_{n-1} = 1$. Since the t_i sequence is strictly increasing for $i \ge -1$, except that $t_1 = t_0 = 1$ if $q_1 = 1$, this implies that $s_{n-1} = c$ and $t_{n-1} = d$. On the other hand, if n is even, then $a t_{n-1} - b s_{n-1} = -1$, and we find that $s_{n-1} = a - c$ and $t_{n-1} = b - d$. □

Example. If $a = 4$ and $b = 1$, then $c = 3$ and $d = 1$ is the unique pair of integers with $0 < d \le b$ so that $4d - c = 1$. Then $a - c = 1$ and $b - d = 0$. The alternative continued fraction expansions of $\frac{a}{b} = 4$ are $\langle 4 \rangle$ and $\langle 3, 1 \rangle$. For $\langle 3, 1 \rangle$, with $n = 1$, we find that $s_0 = 3 = c$ and $t_0 = 1 = d$. For $\langle 4 \rangle$, with $n = 0$, we have $s_{-1} = 1 = a - c$ and $t_{-1} = 0 = b - d$.

Example. Let $a = 553$ and $b = 82$. In Section D.1, we saw that $\frac{a}{b} = \frac{553}{82}$ has continued fraction expansions $\langle 6, 1, 2, 1, 9, 2 \rangle$ and $\langle 6, 1, 2, 1, 9, 1, 1 \rangle$. The convergents of these continued fractions were presented in tables in Section D.2. For $\langle 6, 1, 2, 1, 9, 2 \rangle$, with $n = 5$ odd, we found that $s_4 = 263$ and $t_4 = 39$ and can verify that

$$553 \cdot 39 - 82 \cdot 263 = 1.$$

For $\langle 6, 1, 2, 1, 9, 1, 1 \rangle$, with $n = 6$ even, we found that $s_5 = 290$ and $t_5 = 43$, with

$$553 \cdot 43 - 82 \cdot 290 = -1.$$

Notice that $290 = 553 - 263$ and $43 = 82 - 39$.

Lemma D.5.3. *Let a and b be integers with $\gcd(a, b) = 1$ and $b > 0$, and let c and d be the unique integers with $0 < d \le b$ so that $ad - bc = 1$. Let v be an irrational number having continued fraction expansion $\langle q_0, q_1, q_2, \ldots \rangle$ and let s_i and t_i be the numerator and denominator sequences of the convergents of this continued fraction. Then there is an integer $n \ge 0$ for which $s_n = a$ and $t_n = b$ if and only if*

$$v = \frac{aw + c}{bw + d} \quad or \quad v = \frac{aw + (a - c)}{bw + (b - d)} \tag{D.5}$$

for some real number $w > 1$.

Proof. Suppose first that $s_n = a$ and $t_n = b$ for some $n \ge 0$. Then $\langle q_0, q_1, \ldots, q_n \rangle$ is a continued fraction expansion of $\frac{a}{b}$, and we can assume that either $s_{n-1} = c$ and $t_{n-1} = d$, or $s_{n-1} = a-c$ and $t_{n-1} = b-d$, as in Lemma D.5.2. Let w be the real number with continued fraction expansion $\langle q_{n+1}, q_{n+2}, \ldots \rangle$. Note that w is irrational, since this continued fraction does not terminate, and that $w > q_{n+1} \ge 1$. In this case,

$$v = \langle q_0, \ldots, q_n, q_{n+1}, \ldots \rangle = \langle q_0, \ldots, q_n, w \rangle = \frac{s_n w + s_{n-1}}{t_n w + t_{n-1}}, \tag{D.6}$$

using Proposition D.4.1 and Lemma D.2.1. Thus, v has one of the expressions as in equation (D.5).

Conversely, suppose that v has one of the forms in equation (D.5). We can select a continued fraction expansion $\langle q_0, q_1, \ldots, q_n \rangle$ for $\frac{a}{b}$ so that n is odd if $v = \frac{aw+c}{bw+d}$ and n is even if $v = \frac{aw+(a-c)}{bw+(b-d)}$. Let $\langle q_0', q_1', \ldots \rangle$ be the continued fraction expansion of w. Note that $q_0' \ge 1$ if $w > 1$. Thus, $\langle q_0, q_1, \ldots, q_n, q_0', q_1', \ldots \rangle$ is a well-defined continued fraction, and must converge to a real number. But we can see that this real number must be v by the same type of manipulations as in equation (D.6). $\square$

Example. If $a = 13$ and $b = 9$, then $c = 10$ and $d = 7$ are the unique integers with $0 < d \le b$ so that $ad - bc = 1$. Let $w = \sqrt{2}$ and consider the real number

$$v = \frac{aw + c}{bw + d} = \frac{13\sqrt{2} + 10}{9\sqrt{2} + 7} = \frac{13\sqrt{2} + 10}{9\sqrt{2} + 7} \cdot \frac{9\sqrt{2} - 7}{9\sqrt{2} - 7} = \frac{164 - \sqrt{2}}{113}.$$

Here, $\frac{13}{9}$ has two continued fraction expansions: $\langle 1, 2, 4 \rangle$ and $\langle 1, 2, 3, 1 \rangle$. In this example, we use the latter expression, so that $n = 3$ is odd, with

$$s_{n-1} = c = 10 \quad \text{and} \quad t_{n-1} = d = 7.$$

Direct calculation shows that $w = \sqrt{2}$ has the periodic continued fraction expansion $w = \langle 1, \overline{2} \rangle$. Following the approach of the second part of the proof of Lemma D.5.3, we conclude that $\langle 1, 2, 3, 1, 1, \overline{2} \rangle$ is the continued fraction expression for v, and has $\frac{13}{9}$ as one of its convergents. Both claims can be verified directly.

In the following proposition, we describe a case in which v has the form of Lemma D.5.3, without knowing w ahead of time.

Proposition D.5.4. *Let v be an irrational number having continued fraction expansion $\langle q_0, q_1, q_2, \ldots \rangle$ and let s_i and t_i be the numerator and denominator sequences of the convergents of this continued fraction. Suppose that there are integers a and b, with $b > 0$ and $\gcd(a, b) = 1$, so that $|\frac{a}{b} - v| < \frac{1}{2b^2}$. Then there is an integer n for which $a = s_n$ and $b = t_n$.*

Proof. Let c and d be the unique integers with $0 < d \le b$ for which $ad - bc = 1$. We consider two possibilities, based on the sign of $\frac{a}{b} - v$.

If $\frac{a}{b} - v$ is positive, consider the real number $w = \frac{-c+dv}{a-bv}$. (Note that $a - bv \ne 0$ since v is irrational.) We find that

$$aw + c = \frac{-ac + adv}{a - bv} + \frac{ac - bcv}{a - bv} = \frac{(ad - bc)v}{a - bv} = \frac{v}{a - bv}$$

and

$$bw + d = \frac{-bc + bdv}{a - bv} + \frac{ad - bdv}{a - bv} = \frac{ad - bc}{a - bv} = \frac{1}{a - bv}.$$

Thus,

$$\frac{aw + c}{bw + d} = \frac{v/(a - bv)}{1/(a - bv)} = v.$$

Now

$$0 < \frac{a}{b} - v = \frac{a}{b} - \frac{aw + c}{bw + d} = \frac{(abw + ad) - (abw + bc)}{b(bw + d)}$$

$$= \frac{ad - bc}{b(bw + d)} = \frac{1}{b(bw + d)} < \frac{1}{2b^2}$$

so that

$$0 < \frac{b}{bw + d} < \frac{1}{2}, \quad \text{and thus} \quad \frac{bw + d}{b} = w + \frac{d}{b} > 2.$$

But with $0 < d \le b$, we have that $0 < \frac{d}{b} \le 1$, and we conclude that $w > 1$. Since $v = \frac{aw+c}{bw+d}$ for some real number $w > 1$, then there is a nonnegative integer n for which $s_n = a$ and $t_n = b$ by Lemma D.5.3.

If $\frac{a}{b} - v$ is negative, the same sort of calculations as above show that

$$\text{if} \quad w = \frac{(c-a) + (b-d)v}{a - bv}, \quad \text{then} \quad v = \frac{aw + (a-c)}{bw + (b-d)}.$$

Here

$$0 < v - \frac{a}{b} = \frac{aw + (a-c)}{bw + (b-d)} - \frac{a}{b} = \frac{(abw + ab - bc) - (abw + ab - ad)}{b(bw + (b-d))}$$

$$= \frac{1}{b(bw + (b-d))} < \frac{1}{2b^2},$$

implies that

$$0 < \frac{b}{bw + (b-d)} < \frac{1}{2}, \quad \text{and thus} \quad w + \frac{b-d}{b} > 2.$$

But $0 \le b - d < b$, so we conclude again that $w > 1$. Lemma D.5.3 implies that $s_n = a$ and $t_n = b$ for some $n \ge 0$. $\qquad \square$

We conclude Appendix D with a result that, in an equivalent form, has important implications for representations of integers by indefinite ideal numbers. We will need the following fact in its proof.

Proposition D.5.5. *Let $v > 1$ be an irrational real number with continued fraction expansion $\langle q_0, q_1, q_2, \dots \rangle$, and let s_i and t_i be the numerator and denominator sequences of its convergents. Then $\frac{1}{v}$ has continued fraction expansion*

$$\langle 0, q_0, q_1, q_2, \dots \rangle.$$

If s_i' and t_i' are the numerator and denominator sequences of the continued fraction of $\frac{1}{v}$, then $s_i' = t_{i-1}$ and $t_i' = s_{i-1}$ for all $i \ge 0$.

Proof. By definition, $\langle 0, v \rangle = 0 + \frac{1}{v} = \frac{1}{v}$. Since $v > 1$, then $q_0 \ge 1$ and $\langle 0, q_0, q_1, q_2, \dots \rangle$ is a well-defined continued fraction, which must be the continued fraction of $\frac{1}{v}$. We will also write $\frac{1}{v} = \langle q_0', q_1', q_2', \dots \rangle$, so that $q_0' = 0$ and $q_i' = q_{i-1}$ for all $i \ge 1$.

Let s_i and t_i be the numerator and denominator sequences for the continued fraction of v, and let s_i' and t_i' be the numerator and denominator sequences of the continued fraction of $\frac{1}{v}$, all defined for $i \ge -2$. We find that

$$s_0' = q_0' \cdot s_{-1}' + s_{-2}' = 0 \cdot 1 + 0 = 0 = t_{-1}$$

and

$$t_0' = q_0' \cdot t_{-1}' + t_{-2}' = 0 \cdot 0 + 1 = 1 = s_{-1},$$

while

$$s_1' = q_1' \cdot s_0' + s_{-1}' = q_0 \cdot 0 + 1 = 1 = 1 = t_0$$

and

$$t_1' = q_1' \cdot t_0' + t_{-1}' = q_0 \cdot 1 + 0 = q_0 = s_0.$$

If $n \geq 2$, and we have established similar equations for all $0 \leq i < n$, then

$$s_n' = q_n' s_{n-1}' + s_{n-2}' = q_{n-1} t_{n-2} + t_{n-3} = t_{n-1}$$

and

$$t_n' = q_n' t_{n-1}' + t_{n-2}' = q_{n-1} s_{n-2} + s_{n-3} = s_{n-1}.$$

The result follows for all $n \geq 0$ by induction. □

We rephrase this result in the following form, which will be most convenient in the proof of our main result. The proof of Corollary D.5.6 is immediate and is omitted.

Corollary D.5.6. *Let v be a positive irrational number and let s and t be positive integers. Then $\frac{s}{t}$ is a convergent for the continued fraction of v if and only if $\frac{t}{s}$ is a convergent for the continued fraction of $\frac{1}{v}$.*

Theorem D.5.7. *Let $f(x) = ax^2 + bx + c$ be a polynomial with integer coefficients for which a is positive, c is negative, and $\Delta = b^2 - 4ac$ is not a square. Let $v = \frac{-b + \sqrt{\Delta}}{2a}$ and $\bar{v} = \frac{-b - \sqrt{\Delta}}{2a}$. Suppose that s and t are relatively prime positive integers and that*

$$\left| f\left(\frac{s}{t}\right) \right| < \frac{\sqrt{\Delta}}{2t^2}.$$

Then $\frac{s}{t}$ is a convergent in the continued fraction of v.

Proof. If a is positive and c is negative, then $\Delta = b^2 - 4ac > b^2$, so that $\sqrt{\Delta} > |b|$. Thus, v is positive and $\bar{v}$ is negative. Note that $f(x) = a(x - v)(x - \bar{v})$ for every real number x, since v and $\bar{v}$ are the roots of $f(x)$. For later use, we also observe that

$$v^{-1} = \frac{2a}{-b + \sqrt{\Delta}} = \frac{2a}{-b + \sqrt{\Delta}} \cdot \frac{b + \sqrt{\Delta}}{b + \sqrt{\Delta}}$$
$$= \frac{2a(b + \sqrt{\Delta})}{\Delta - b^2} = \frac{2a(b + \sqrt{\Delta})}{-4ac} = \frac{-b - \sqrt{\Delta}}{2c}$$

and similarly, $(\bar{v})^{-1} = \frac{-b + \sqrt{\Delta}}{2c}$. We now consider two possibilities for $f(\frac{s}{t})$.

Suppose that $0 < f(\frac{s}{t}) < \frac{\sqrt{\Delta}}{2t^2}$. Since a is positive, then $\frac{s}{t} - v$ and $\frac{s}{t} - \bar{v}$ have the same sign. With s and t both positive, and $-\bar{v}$ positive, we conclude that $\frac{s}{t} - v$ is positive. It follows that

$$\frac{s}{t} - \bar{v} > \left(\frac{s}{t} - \bar{v}\right) - \left(\frac{s}{t} - v\right) = v - \bar{v} = \frac{-b + \sqrt{\Delta}}{2a} - \frac{-b - \sqrt{\Delta}}{2a} = \frac{\sqrt{\Delta}}{a}.$$

Thus, we have

$$a \cdot \frac{\sqrt{\Delta}}{a} \cdot \left(\frac{s}{t} - v\right) < a\left(\frac{s}{t} - \bar{v}\right)\left(\frac{s}{t} - v\right) = f\left(\frac{s}{t}\right) < \frac{\sqrt{\Delta}}{2t^2},$$

Dividing the first and last terms of the inequality by the positive number $\sqrt{\Delta}$, we conclude that

$$0 < \frac{s}{t} - v < \frac{1}{2t^2}.$$

Therefore, $\frac{s}{t}$ is a convergent in the continued fraction of v by Proposition D.5.4.

Now suppose that $0 > f(\frac{s}{t}) > -\frac{\sqrt{\Delta}}{2t^2}$ or, equivalently,

$$0 < -f\left(\frac{s}{t}\right) < \frac{\sqrt{\Delta}}{2t^2}. \tag{D.7}$$

Here, consider the polynomial $g(x) = -cx^2 - bx - a$ and notice that

$$-\frac{t^2}{s^2} \cdot f\left(\frac{s}{t}\right) = -\frac{t^2}{s^2} \cdot \left(a\left(\frac{s}{t}\right)^2 + b\left(\frac{s}{t}\right) + c\right) = -a - b\left(\frac{t}{s}\right) - c\left(\frac{t}{s}\right)^2 = g\left(\frac{t}{s}\right).$$

Thus, multiplying by the positive number t^2/s^2 replaces (D.7) by the following equivalent inequalities:

$$0 < g\left(\frac{t}{s}\right) < \frac{\sqrt{\Delta}}{2s^2}. \tag{D.8}$$

Now we can proceed as in the first part of the proof. Note that $-c$ is positive, $-a$ is negative, and that the discriminant of $g(x)$ is $(-b)^2 - 4(-c)(-a) = b^2 - 4ac = \Delta$. The roots of $g(x)$ are $\frac{b + \sqrt{\Delta}}{2(-c)} = v^{-1}$ and $\frac{b - \sqrt{\Delta}}{2(-c)} = (\bar{v})^{-1}$, as calculated above. We can write $g(x) = -c(x - v^{-1})(x - (\bar{v})^{-1})$ for all x. Since $-c$ is positive, inequalities (D.8) show that $\frac{t}{s} - v^{-1}$ and $\frac{t}{s} - (\bar{v})^{-1}$ have the same sign. We can argue that both terms are positive, and so

$$\frac{t}{s} - (\bar{v})^{-1} > \left(\frac{t}{s} - (\bar{v})^{-1}\right) - \left(\frac{t}{s} - v^{-1}\right) = v^{-1} - (\bar{v})^{-1} = \frac{\sqrt{\Delta}}{-c}.$$

Thus, we now have

$$-c \cdot \frac{\sqrt{\Delta}}{-c} \cdot \left(\frac{t}{s} - v^{-1}\right) < -c\left(\frac{t}{s} - (\bar{v})^{-1}\right)\left(\frac{t}{s} - v^{-1}\right) = g\left(\frac{t}{s}\right) < \frac{\sqrt{\Delta}}{2s^2}$$

and, dividing by the positive number $\sqrt{\Delta}$, we conclude that

$$\frac{t}{s} - v^{-1} < \frac{1}{2s^2}.$$

Our conclusion is that $\frac{t}{s}$ is a convergent in the continued fraction of v^{-1}. But, since s and t are positive, Corollary D.5.6 implies that then $\frac{s}{t}$ is a convergent in the continued fraction of v. $\qquad\square$

References

[1] Lehman, JL. *Quadratic Number Theory: An Invitation to Algebraic Methods in the Higher Arithmetic*. Providence, RI, USA, Dolciani Mathematical Expositions, Vol. 52, AMS/MAA (MAA Press: Am Imprint of the American Mathematical Society), 2019.

[2] Ireland, K, Rosen, M. *A Classical Introduction to Modern Number Theory*. New York, NY, USA, Graduate Texts in Mathematics, 84, Springer-Verlag, 1982.

[3] Watkins, M. "Class numbers of imaginary quadratic fields," Mathematics of Computation 73, no. 246 (2004) 907–938.

[4] Dirichlet, PGL. *Lectures in Number Theory*, supplements by R. Dedekind, translated by J. Stillwell. American Mathematical Society, London Mathematical Society, 1999.

[5] Kim, SY. "An elementary proof of the quadratic reciprocity law," American Mathematical Monthly, Vol. 111, no. 1, (January 2004) 48–50.

https://doi.org/10.1515/9783111319360-012

Index

https://doi.org/10.1515/9783111319360-013